A Guide for Writing Better Technical Papers

OTHER IEEE PRESS BOOKS

Low-Noise Microwave Transistors and Amplifiers, *Edited by H. Fukui*
Digital MOS Integrated Circuits, *Edited by M. I. Elmasry*
Geometric Theory of Diffraction, *Edited by R. C. Hansen*
Modern Active Filter Design, *Edited by R. Schaumann, M. A. Soderstrand, and K. R. Laker*
Adjustable Speed AC Drive Systems, *Edited by B. K. Bose*
Optical Fiber Technology, II, *Edited by C. K. Kao*
Protective Relaying for Power Systems, *Edited by S. H. Horowitz*
Analog MOS Integrated Circuits, *Edited by P. R. Gray, D. A. Hodges, and R. W. Brodersen*
Interference Analysis of Communication Systems, *Edited by P. Stavroulakis*
Integrated Injection Logic, *Edited by J. E. Smith*
Sensory Aids for the Hearing Impaired, *Edited by H. Levitt, J. M. Pickett, and R. A. Houde*
Data Conversion Integrated Circuits, *Edited by Daniel J. Dooley*
Semiconductor Injection Lasers, *Edited by J. K. Butler*
Satellite Communications, *Edited by H. L. Van Trees*
Frequency-Response Methods in Control Systems, *Edited by A.G.J. MacFarlane*
Programs for Digital Signal Processing, *Edited by the Digital Signal Processing Committee*
Automatic Speech & Speaker Recognition, *Edited by N. R. Dixon and T. B. Martin*
Speech Analysis, *Edited by R. W. Schafer and J. D. Markel*
The Engineer in Transition to Management, *I. Gray*
Multidimensional Systems: Theory & Applications, *Edited by N. K. Bose*
Analog Integrated Circuits, *Edited by A. B. Grebene*
Integrated-Circuit Operational Amplifiers, *Edited by R. G. Meyer*
Modern Spectrum Analysis, *Edited by D. G. Childers*
Digital Image Processing for Remote Sensing, *Edited by R. Bernstein*
Reflector Antennas, *Edited by A. W. Love*
Phase-Locked Loops & Their Application, *Edited by W. C. Lindsey and M. K. Simon*
Digital Signal Computers and Processors, *Edited by A. C. Salazar*
Systems Engineering: Methodology and Applications, *Edited by A. P. Sage*
Modern Crystal and Mechanical Filters, *Edited by D. F. Sheahan and R. A. Johnson*
Electrical Noise: Fundamentals and Sources, *Edited by M. S. Gupta*
Computer Methods in Image Analysis, *Edited by J. K. Aggarwal, R. O. Duda, and A. Rosenfeld*
Microprocessors: Fundamentals and Applications, *Edited by W. C. Lin*
Machine Recognition of Patterns, *Edited by A. K. Agrawala*
Turning Points in American Electrical History, *Edited by J. E. Brittain*
Charge-Coupled Devices: Technology and Applications, *Edited by R. Melen and D. Buss*
Spread Spectrum Techniques, *Edited by R. C. Dixon*
Electronic Switching: Central Office Systems of the World, *Edited by A. E. Joel, Jr.*
Electromagnetic Horn Antennas, *Edited by A. W. Love*
Waveform Quantization and Coding, *Edited by N. S. Jayant*
Communication Satellite Systems: An Overview of the Technology, *Edited by R. G. Gould and Y. F. Lum*
Literature Survey of Communication Satellite Systems and Technology, *Edited by J. H. W. Unger*
Solar Cells, *Edited by C. E. Backus*
Computer Networking, *Edited by R. P. Blanc and I. W. Cotton*
Communications Channels: Characterization and Behavior, *Edited by B. Goldberg*
Large-Scale Networks: Theory and Design, *Edited by F. T. Boesch*
Optical Fiber Technology, *Edited by D. Gloge*
Selected Papers in Digital Signal Processing, II, *Edited by the Digital Signal Processing Committee*
A Guide for Better Technical Presentations, *Edited by R. M. Woelfle*
Career Management: A Guide to Combating Obsolescence, *Edited by H. G. Kaufman*
Energy and Man: Technical and Social Aspects of Energy, *Edited by M. G. Morgan*
Magnetic Bubble Technology: Integrated-Circuit Magnetics for Digital Storage and Processing, *Edited by H. Chang*
Frequency Synthesis: Techniques and Applications, *Edited by J. Gorski-Popiel*
Literature in Digital Processing: Author and Permuted Title Index (Revised and Expanded Edition), *Edited by H. D. Helms, J. F. Kaiser, and L. R. Rabiner*
Data Communications via Fading Channels, *Edited by K. Brayer*
Nonlinear Networks: Theory and Analysis, *Edited by A. N. Willson, Jr.*
Computer Communications, *Edited by P. E. Green, Jr. and R. W. Lucky*
Stability of Large Electric Power Systems, *Edited by R. T. Byerly and E. W. Kimbark*
Automatic Test Equipment: Hardware, Software, and Management, *Edited by F. Liguori*
Key Papers in the Development of Coding Theory, *Edited by E. R. Berkekamp*

A Guide for Writing Better Technical Papers

Edited by
Craig Harkins
Site Communication Manager
IBM San Jose

Daniel L. Plung
Technical Editor
Exxon Nuclear Idaho Co., Inc.

A volume in the IEEE PRESS Selected Reprint Series, prepared under the sponsorship of the IEEE Professional Communication Group.

The Institute of Electrical and Electronics Engineers, Inc., New York

PRINTED IN THE UNITED STATES OF AMERICA

Sole Worldwide Distributor (Exclusive of IEEE):

JOHN WILEY & SONS, INC.
605 Third Ave.
New York, NY 10158

Wiley Order Numbers: Clothbound: 0-471-86865-5
Paperbound: 0-471-86866-3

IEEE Order Numbers: Clothbound: PC01529
Paperbound: PP01537

Library of Congress Cataloging in Publication Data
Main entry under title:

A Guide for writing better technical papers.

(IEEE Press selected reprint series)
Includes bibliograhical references and indexes.
1. Technical writing. I. Harkins, Craig. II. Plung, Daniel L.
T11G84 808'.0666 81-20042
ISBN 0-87942-157-6
ISBN 0-87942-158-4 (pbk.)

Contents

Introduction

THE PROBLEM OF DOING JUSTICE to the implicit, the imponderable, and the unknown is, of course, not unique to politics. It is always with us in science, it is with us in the most trivial of personal affairs, and it is one of the great problems of writing and of all forms of art. The means by which it is solved is sometimes called style. It is style which complements affirmation with limitation and with humility; it is style which makes it possible to act effectively, but not absolutely; it is style which, in the domain of foreign policy, enables us to find a harmony between the pursuits of ends essential to us, and the regard for the views, the sensibilities, the aspirations of those to whom the problem may appear in another light; it is style which is the deference that action pays to uncertainty; it is above all style through which power defers to reason.

J. Robert Oppenheimer

As Dr. Oppenheimer, the renowned American physicist, accurately assesses, the responsibility of the scientist extends beyond the confines of the laboratory; he must, as the quotation suggests, develop and employ the ability to communicate information and concepts in a clear, concise, and effective manner. The scientist must, to use Dr. Oppenheimer's terms, develop a "style."

The development of a technical writing "style," like the development of any skill, is attained only as a result of understanding the task, studying its components, and then practicing the art. The practice of writing comes as a natural part of the engineer's or scientist's work assignment; he must prepare reports, articles, and presentations about the work he does. Unfortunately, this practice does not, by itself, necessarily lead to perfection of the art of writing. The understanding of the writing process and of the components that conjointly contribute to make that writing clear, concise, and readable does not come automatically; it must be learned.

To assist the writer in this learning process, we have designed this book to serve two principal purposes. Firstly, the book is intended to provide essential information about the writing process and its component parts; this information is the foundation upon which an appropriate "style" of technical writing can be developed. Secondly, it is intended to complement this foundation with ideas, techniques, suggestions, and guidelines culled from the volumes of literature about improving technical writing; this additional material presents a variety of perspectives and thereby affords breadth as well as quality of information.

To accomplish the dual purpose prescribed for this book, we have tried to incorporate the theoretical with the practical, the rhetorical with the empirical. Our search for articles for inclusion was not restricted to certain years or to certain publications; rather, we were guided by a single criterion in the selection of articles: how well the article contributed to completing the tasks we had assigned to the book.

Once selected, the articles were grouped into five parts. These parts allow the reader to see how pieces of information are related—to each other and to the broader concerns involved in producing quality papers and articles. Part I, "Getting Started," combines a range of considerations that are crucial in the article's initial stages. Often one's understanding of the writing process and the attitude one takes toward it determine the degree of success one will have. Consequently, the articles selected for this section are designed to promote the best attitude toward, and the fullest understanding of, the writing assignment.

Part 2 deals with the rhetoric of technical papers and articles. It is an area often ignored by technical persons—to the detriment of many communications. The articles selected for this section balance theory and application in their explanation of what rhetoric is and why its study is significant to the technical author. In particular, we have tried to focus the writer's attention on two major rhetorical concerns: the relationship between the writer and the reader, and the relationship between thinking and writing.

Part 3, "Tricks of the Trade," includes several articles of the nuts-and-bolts fundamentals of the writing assignment. Many limits for making the writing process a more pleasant, less painful, and—ultimately—more effective one are included. Perhaps, though, the title of this section is something of a misnomer. We would not want anyone to get the impression that the basic "Trick of the Trade" is anything other than good, strong, solid writing. Cumulatively, these articles should help a fledgling author achieve such an aim.

The research section of this book deals with some of the results of an extensive amount of inquiry into "what works" in areas such as readability and formatting. It is an area not usually included in books that also contain sections devoted to rhetoric. Yet, as we noted, this volume seeks to incorporate many perspectives and is designed to appeal, simultaneously, to individuals who are likely to have a bias toward either the effectiveness or accuracy of the technical message. By including both perspectives we are emphasizing that there is no reason that the conscientious technical author must compromise one for the other. Rhetorical and empirical research can be mutually supportive and can equally contribute to the preparation of "stylish" technical papers and articles.

Like a good golf swing, a good technical paper requires considerable attention to follow through. Careful attention must be given to the professional analysis of all elements of the work and to the vast array of final details that can spell the difference between success and failure in a technical communication venture. It is on such an analysis and attention that our final set of articles focuses.

Taken together, these five parts will help promote the development of a technical writing "style" capable of "doing justice to the implicit, the imponderable, and the unknown." In other words, this book is intended to help the writer learn to express himself on paper in the clear, concise, and effective manner that is the trademark of good technical writing.

We recognize that the technical paper often carries with it the responsibility to deliver it orally. However, this consideration is beyond the scope of this particular book because technical presentations have been covered so extensively in another IEEE Reprint Book, *A Guide for Better Technical Presentations,* edited by Robert M. Woelfle and published in 1975.

Finally, many people have contributed to our efforts on this book. In particular, we would like to thank Robert Woelfle, of E-Systems, Inc., Greenville, Texas, and Eileen Wilson of EG&G Idaho, Inc., Idaho Falls, Idaho, for the assistance they have given us in putting together this volume.

Part I
Getting Started

As one of the articles in this part states, "starting is usually the biggest hurdle." Accordingly, the articles in this part, which present an overview of many of the aspects of the writing process detailed in the subsequent parts of this book, are intended to help the author overcome this barrier. These articles examine the two principal steps in getting a good start and in producing a good product: 1) understanding the task, and 2) planning ahead to meet that task.

The first three articles provide an understanding of the task. Stratton develops a "needs assessment matrix" that illustrates the interrelationship among audience segments, types of information conveyed, and the purposes for communicating. This system, whose application is considerably broader than papers and articles alone, provides a useful perspective from which to examine any endeavor in technical communication. However, precisely because Stratton's system is applicable to all forms of communication, we must know when it is best to write and when it is best to use some other medium. Biscardi and Coburn analyze various conditions and situations in which "the power of a well written letter or report" is the preferred form of communication; in addition, they discuss the consequences of not using the written form when it should be used. Racker contributes to our understanding of the task by explaining about the various readers an author of technical literature may address and by demonstrating how the types of information and the purposes of communicating can be adapted to suit the requirements of each of five levels of readers.

We then turn our attention specifically to writing articles for publication. Holder explains the benefits both the author and his company may derive from publishing an article. He gives suggestions about choosing topics, gathering information, and preparing the manuscript. Olson reinforces Holder's observations on the benefits of writing and publishing. He emphasizes the importance of logically preplanning the technical article or paper and provides a checklist for evaluating it.

The final two articles address the author who wants specific guidance regarding the article's "marketability." Meyer distinguishes among the types of publications to which an author can send material; although the status of some journals he mentions may have changed, the general advice he offers remains pertinent to understanding how best to market a technical article or paper. Heald's essay ends this part as it began, on a theoretical note—albeit in this case a tongue-in-cheek one. Underlying his amusing narrative about the relationship between the value of the paper and the author's salary is a serious point: there is something to be gained by publishing a technical paper or article.

Hence, the importance of getting started.

NEEDS ASSESSMENT FOR COMMUNICATION SYSTEM DESIGN

DR. CHARLES R. STRATTON
Department of English
University of Idaho

ABSTRACT

System design for communication packages involves assessing communication needs and designing a set of communication vehicles to meet these needs. This article, which focuses on the first of these two steps, outlines three categories of data that are important to communication system design and proposes a system for gathering and correlating the data in these three categories. The tangible product of this gathering and correlating is a three-dimensional needs assessment matrix, which is used as part of the input for the actual design of the communication system. A later article will deal with the communication system design itself.

To the computer specialist, the term "system design" implies more than just the design of a system. It suggests a goal-oriented approach to a set of software that identifies a number of tasks to be performed, selects combinations of programs and subroutines to accomplish these tasks, and provides the necessary interfaces so that the goals of the system can be met in an efficient and economical fashion. The term also suggests a systematic approach to the design process—an approach that enables the design of an operating system to become as much a craft as an art form: a skill that can be described and taught and learned rather than an ability that can be acquired only through a lengthy apprenticeship.

So with system design for communications. The system design approach focuses on the goals of communication—the various segments of the audience, the various purposes for communicating—and tailors a set of vehicles for communication to achieve the goals in the most efficient and economical fashion. In this article, I will be concerned with the first of these two steps, and I will refer to it with a term borrowed from education curriculum planning: *needs assessment.*

The procedures described here were developed in conjunction with a land use planning and development study conducted by the Community Development Center at the University of Idaho. This study was conducted for the Latah County Planning Commission, Latah County being the county in which the University of Idaho is located. Additional information about the study is available from the University of Idaho's Community Development Center. Although I will make specific reference to the study and the project team in this article, the needs assessment techniques described can be generalized to any complex communication situation.

The Need for Needs Assessment

The ultimate design for a project's communication package should be based on the careful consideration of three categories of data. These are: the *types of information* being developed, the various *segments of the total audience* interested in the results of the project, and the various *purposes for communicating* information about the project. A well designed communication system offers multiple communication vehicles (publications, usually) that strive for an optimum balance among these three areas of consideration.

In all too many projects, however, the communication design is something like this: let's put all types of information together in a single report and send a copy of the entire report to whoever might be interested in it for whatever purpose we can envision. What this design does, in effect, is force each and every person in the audience to perform his own needs assignment—to wade through the entire volume of the report to identify what types of information are in it, to decide what he is interested in, and to analyze his own reasons for wanting to receive communication about the project.

In other words, communication needs assessment is something that is done in every communication situation. The only question is whether it will be done once with some degree of efficiency by the sender, or done as many times as there are receivers, with considerably less efficiency each time. On the assumption that the former is preferable, let me proceed to discuss these three categories of input data and then go on to point out how they can be identified and drawn together to form the data base for a communication system design.

Data Categories

To repeat, the three categories of data important for communication system design are:

Types of Information
Audience Segments
Purposes for Communicating

Let us take a look at each of these in turn.

TYPES OF INFORMATION

A consideration of types of information for a project or study is nothing spectacular; in the one-report design it takes the form of the table of contents. The important thing in performing a communication needs assessment is to deal with broad categories of information, rather than specific items of information. Some of the types of information identified for the Community Development Center (CDC) study mentioned above were:

Development suitability recommendations
General summary of data gathered about the study area
Detailed data about present land use
Description of data gathering methods
A brief history of the project
An evaluation of the project

If the needs assessment is conducted after the project is completed, the list of types of information is largely descriptive. If the communication needs assessment is conducted as part of the project planning, however (and I strongly recommend that it should be), the types of information list can be prescriptive, indicating the categories of information that should be developed by the project. More on this later.

AUDIENCE SEGMENTS

Communication with a homogeneous audience is rare indeed; yet most of us tend to write and speak as if we were communicating with an unstratified and undifferentiated group of people. If we give any thought at all to audience analysis, we do nothing more than trot our pet model reader out of the closet (everybody has one) and aim our communication accurately and unerringly at him. ("Of course they'll understand my terminology; they're all chemical engineers, aren't they?") It is an interesting, and enlightening, experience indeed to pause and list the various groups and categories of people that will be interested in receiving information about a particular project or study.

Again, the emphasis wants to be broad—on segments of the potential audience rather than on specific individuals. Some of the audience segments identified in the CDC study were:

The County Planning Commission members
Residents in the study area
Residents in nearby areas
County, state, and federal agencies with jurisdiction in the study area
The Community Development Center Advisory Board
Various agencies interested in the planning process
University administration
Other planning commissions, city and county, both within and outside the state
Researchers in other Community Development Centers

PURPOSES FOR COMMUNICATING

Of the three categories of data necessary for communication needs assessment, purposes for communicating is the least often considered. When purpose is considered, it is usually viewed as some vague need "to provide information about. . . ." No regard to what a reader is going to do with the information; no regard to additional purposes that the communicator himself might have. In any communication situation, there are many purposes for communicating in addition to the bare need to provide information about the subject. Some of the purposes identified in the CDC study are suggestive:

To provide data for decision-making within the study area
To provide specific examples of general procedures
To demonstrate that the Community Development Center is a capable and productive organization
To provide data of an indirect or general interest
To publicize methods and techniques developed by the study
To advocate the planning process for land use development
To help similar projects avoid certain pitfalls
To acknowledge the receipt of information and assistance
To secure future funding and support for the Community Development Center

A Procedure for Needs Assessment

Communication planning, like any other kind of planning, tends to grow better as the base of planning participation grows larger. Now, in any organization faced with a complex communication situation there will be a single person who has primary responsibility for putting the communication package together. This person may be a technical writer or editor; he may be an outside communications consultant; he may be the boss; he may be just one of the technical people to whom communication responsibilities have been assigned. For convenience I will call this person the *communications advisor*. It is quite possible, of course, for this communications advisor to perform a needs assessment and design a communications package by himself. A more productive procedure, however, is for the communications advisor to coordinate or orchestrate the activities of three key groups, as far as needs assessment goes: management, the technical experts, and the ultimate receivers of the communication.

The informal small-group discussion (i.e., bull session) can be an efficient method of identifying data in the three categories I mentioned above. The sessions should be kept short (forty-five minutes to an hour) however, and it is important that the communications advisor keep the discussion group narrowly focused on the problem. Management should be represented at least by the immediate supervisor of the project or study in question and one other management person not directly involved in the project. The number and nature of the technical experts will vary, of course, from project to project, but I would suggest at least two technical experts in the group and not more than four. Getting the ultimate receivers of the communication represented can be a bit of a problem. If you cannot get some real live readers, perhaps you can assign someone from management or from the editorial staff to sit in on the discussion for the expressed purpose of watching out for the readers' interests.

In the past, I have started with types of information and then worked on through audience segments to purposes for communicating. I do not think it makes too much difference where you start, however; the real action comes when you begin to consider all three together. New purposes for communicating will suggest additional audience segments, additional types of information will suggest new purposes for communicating, and so on. It is useful to have a large blackboard to write various ideas on, and it is further useful to have a secretary to do the writing and also to compile the final discussion notes. The communications advisor should participate actively in the discussion, throwing out ideas where he can but mostly trying to elicit ideas from the group. The more the group has a sense of participating in the identification of communication needs, the easier it will be to gain approval of a design based on these needs.

With the Community Development Center project, I spent two forty-five-minute sessions with the study project team leader, the CDC Director, two graduate students working on the project, a member of the county planning commission, and a representative of the State Planning and Community Affairs Agency. During these two rather free-wheeling discussion sessions, we generated lists of fifteen types of information, thirty-two audience segments, and eighteen purposes for communicating. I then had these lists typed up and reviewed by the original group and again by the CDC Advisory Board and a subcommittee of the County Planning Commission. These reviews netted some additions, deletions, and consolidations, but finally we had the three data category lists to serve as input to the next phase of the needs assessment.

The Needs Assessment Matrix

The lists generated at this point in the needs assessment process can be of considerable use in designing a communications system, but of even more use are the interactions among elements on each list. If we take the titles of the three lists as parameters, we can generate a three-dimensional matrix, or large cube, as suggested in Figure 1. As we move outward from the origin on any parameter (that is to say, move down the list in question) we encounter successive values (items on the list). But we can also move into the interior of the cube, looking for points where values of the three parameters intersect. In fact, the large cube, or matrix can be viewed as being made up of a great number of smaller cubes, with each small cube representing the point where values from all three parameters come together. In terms of communication needs assessment, such a small cube would represent the intersection of a certain type of information, a certain audience segment, and a certain purpose for communicating.

Now visualizing and working with a three-dimensional matrix is no problem for a mathematician, but often we mere mortals have difficulty getting the picture of what's going on. But notice, if we take any face of the big cube we have a two-dimensional matrix, or a chart—something most of us are a little more comfortable working

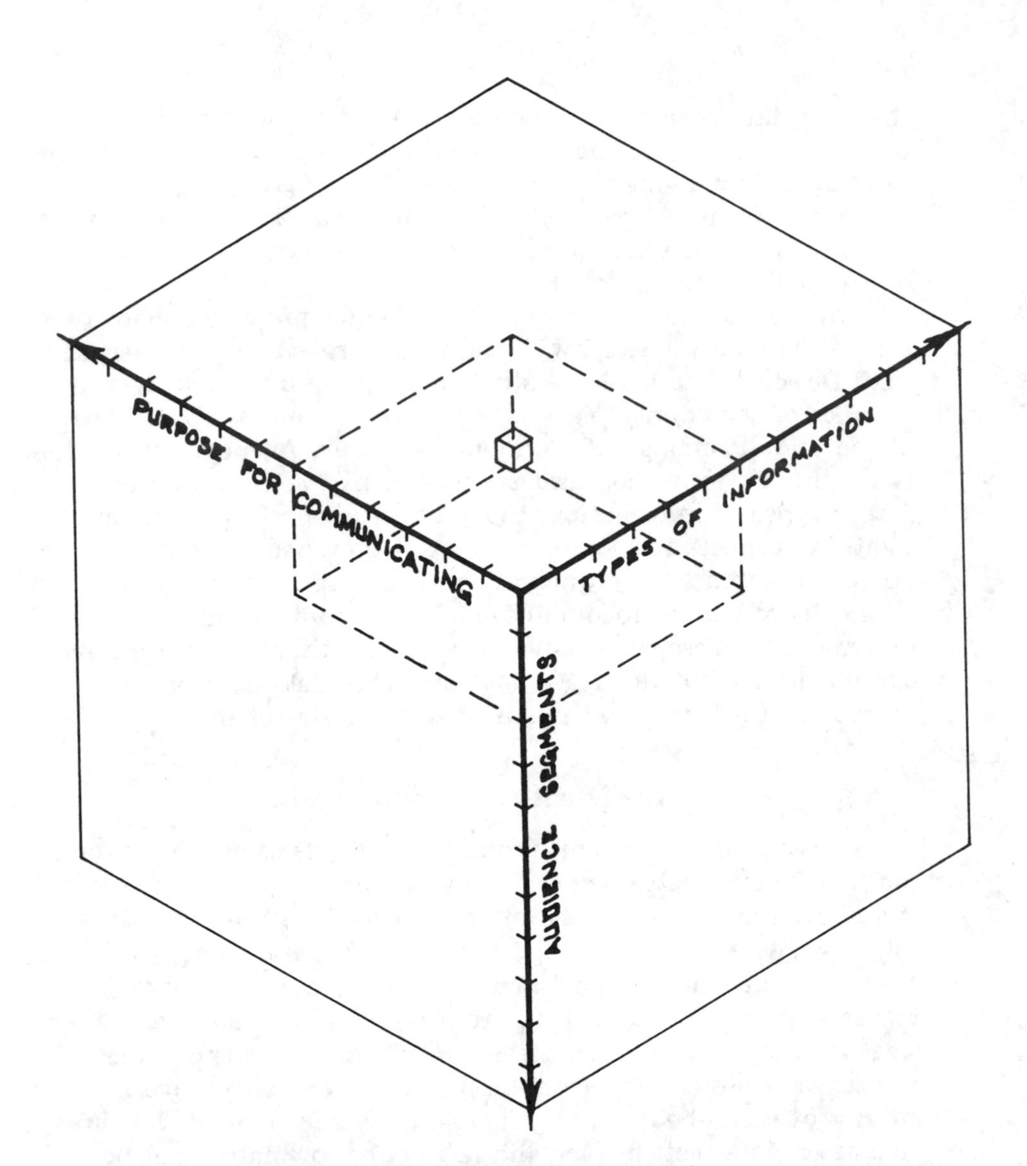

Figure 1. The needs assessment matrix.

with. In fact, we can represent all of the information in our three-dimensional matrix (our cube) with three two-dimensional matrices (charts). Again, in communication needs assessment terms, we need charts plotting the following:

Types of Information against Audience Segments
Types of Information against Purposes for Communicating
Audience Segments against Purposes for Communicating

With the charts made up, all we have done is turn a couple of the lists sideways. The payoff comes when we begin to fill x's in certain squares in the charts: when we begin to specify which audience segments are interested in which types of information for which purposes. Chances are that not all of the squares will have x's put in them; in fact, it may well be the case that relatively few squares are filled—that we have a sparse matrix, in mathematicians' terms. (It is possible of course that every square gets filled. This is no problem; it just confirms that one-report design was the best after all. But if any or many squares are left blank, this is confirmation that a more sophisticated communication system design will provide more effective communication.)

Filling in the x's is pretty much a matter of subjective judgment. The communications advisor can get a good start on this, but he should review his judgments with the original discussion group or with representatives of that group. This does two things: it broadens the base of subjectivity (which usually improves the judgment), and it enhances the sense of participation on the part of the group. Both of these are important. Whatever the method, the outcome of this phase of the needs assessment should be three charts that the group pretty well agrees represent the types of information various audience segments will be interested in, the purposes for communicating with various audience segments, and the types of information that will be important for various communication purposes. An example of such a chart is shown in Figure 2. These charts will be used by the communications advisor, who will combine them with other information to form the complete input data for a communication system design.

Concluding Comments

The procedure for conducting a communication needs assessment that I have outlined here is nothing startling. In fact, it is pretty much what any good editor does when he is asked to put together a communications package. But remember that's what system design is all about—coming at a complex decision-making process in a formal way and developing a procedure that can be followed so that persons with less than the master's experience can come up with designs that approach those of the master in efficiency and elegance. In other words, a procedure for system design is good to the extent that it doesn't surprise the master; I would be extremely leary of a procedure that led to results far afield from what a person with considerable experience and a good decision-making track record would come up with.

Throughout this article I have emphasized the importance of the sense of participation that management people, technical people, and the ultimate receivers of the communication should have in assessing communications needs and working toward a design. This is not only productive, it's good sound psychology; a person's regard for a particular program or idea is going to increase proportionally as his level of participation increases. That's the way people are.

Finally, a comment on the timing for communications planning. With all too many projects and studies, communication—or "writing the report"—is regarded solely as part of the technical staff's job and something to be worried about only after the real work of the project is completed. If a communicator is brought into the picture at all, he is brought in at the eleventh hour, given a mountain of technical data he isn't familiar with, and asked to get a report out by the end of the fiscal year—usually about three weeks away. This tends to be less than productive. A much better program, it seems to me, is to get a communicator—a writer, an editor, what have you—into the program from the very start. Let him begin right off to get familiar with the technical aspects of the project, and let him begin at day one to assess communications needs and begin work on a communication system design. The least that will happen is that you'll have better working relationships within the group and a better chance of getting the report out on time. But more likely, you'll find that the communications person can contribute to the overall project planning and execution as well, with the result that you have not only better communication but a better project to communicate about.

AUDIENCE SEGMENTS	TYPES OF INFORMATION					
	DATA	METHODS	AREAS	LAND USE	ZONING	PROJECTIONS
PLANNING COMMISSION	X	X	X	X	X	X
RESIDENTS				X	X	X
STATE AGENCIES			X	X		
CDC ADVISORS				X		
OTHER CDC'S		X				
RESEARCHERS	X	X				

Figure 2. Example of needs assessment chart.

Writing Can Still Be the Better Way

LOUIS E. BISCARDI, JR., AND WINSTON N. COBURN, MEMBER, IEEE

Abstract—**Criteria for situations in which writing is more effective as a communication tool compared to other media are discussed in this article. Emphasis is placed on "when" writing should be used instead of "how" to write. Survey-based guidelines are expressed in terms of administrative interrelation between people. Communication failure caused by implementation of the wrong medium is also discussed. An extended bibliography on oral communication, written communication, and the art of communication is included.**

THE TREND during recents years in directing technical communication has seen every medium emphasized except correspondence.

1) Face-to-face communication has been encouraged.
2) More effective use has been made of telephones.
3) Data have been regimented into printed forms to save writing.
4) Processable data have been put into forms acceptable to machines, i.e., punched cards.
5) Retrievable data are now stored in large memory banks, addressable from practically anywhere.

Such emphasis has stemmed from the need for more rapid media for today's mode of living and the widespread concern over the proliferation of paper. The advantages of the emphasized media are acknowledged, particularly when the medium is correctly applied. However, present preoccupation with the new methods and dynamic communication creates an atmosphere in which resorting to a written message is an admission of failure or a return to "horse and buggy" techniques. Overlooked in the race to automate more and more data is the power of a well-written letter or report.

Unlike this current trend of minimizing the writing effort, there are specialists in communication who see it approaching a more significant role. The noted educator in technical communication, Dr. Herman Weisman, describes the role of correspondence in these words:

> Today, despite the great capabilities and increasing use of automatic (computer) data and information-processing machinery, correspondence is the basic communication instrument in business and industry just as it was at the start of the industrial revolution 200 years ago. The post office handled almost 80 billion pieces of mail last year. The vast majority of mail is in the commercial category. The pieces of mail are increasing in excess of 2 billion a year. By 1980, mail volume is expected to exceed 100 billion annually.
>
> Increasing, automated methods necessarily will be utilized not only to handle the physical pieces of correspondence but also to originate messages of a routine and repetitive nature. However, the growing complexity and specialized nature of our technological society will demand more creative intelligence rather than machines to meet the more complicated and difficult message situations.
>
> Today (and in the foreseeable future), much of the activity of science and technology is and will be conducted through correspondence [17, p. 5].

With assurance such as Dr. Weisman's that written messages will still have a place in the future, it becomes appropriate to identify when writing is preferred; that is the purpose of this article.

ANALYSIS OF THE SITUATION

Current practices and current needs vary from company to company. The statements that follow are appropriate for a company organization that is sufficiently stratified and geographically dispersed to make communication more complex than talking to an associate at an adjoining desk. This degree of communication complexity is marked by:

1) unique communication needs for each company that develop out of the nature of its product, number of different products, number of plants, geographic separation of plants, participation in foreign markets, company organization structure, communication facilities available, and the skill of its communicators;
2) the chameleon nature of communication problems; historically, people have mistaken such problems for personality conflicts, disobedience, group dissatisfaction, interdepartmental conflicts, irresponsibility, stupidity, ineptness, lack of cooperation, parochialism, etc.

The comments in this article should be interpreted in terms of each company's situation.

In any organization the "communications climate" is a significant factor. Management's attitude can vary from the so-called "dynamic policy" to what can be called an "effectiveness policy."

"Dynamic policy" is a favorite of sales organizations and consists of beliefs such as those shown in the following list.

Manuscript received October 16, 1969.
The authors are with the Systems Development Division, IBM Corporation, Kingston, N.Y.

Reprinted from *IEEE Trans. Engrg. Writing and Speech,* vol. EWS-13, pp. 13-17, May 1970.

1) Do not telephone if it is possible to visit the party.
2) Do not write if it is possible to telephone.
3) When writing be brief.

The "effectiveness policy" stresses that a message must be created to motivate a desired action.

Top management can be counted on to have definite ideas about communications, but in a dispersed or multilevel organization much can become lost or distorted in reflecting these ideas to the working force. In looking up from the bottom of a large organization, it is not unusual for writers to indiscriminately apply a few "do's" and "don't's," because they believe it is what their management wants.

A survey of management in almost any organization would reveal that it is a believer in the value of communications. With so much stress placed on the importance of communication, it would seem that the climate in any selected company could never become wintery, but it can.

Unfavorable communication climate in an organization reflects one of two basic conditions: the organization does not recognize all of its communication problems or does not know how to solve them; or the majority of the communicators in the organization are misinformed or unaware of management's communication policies.

A striking example of the first condition is the staffing of a position dedicated to a volume of vital communications with an employee who has difficulty communicating. If the appointee is provided with skilled assistance, then obviously he was selected on the excellence of his other qualities. However, when the situation demands personal communication skill, then the appointee is expected to acquire it as readily as he acquired the responsibility.

To illustrate the second condition, good examples are found in two basic communication principles which an organization can distort into rigid rules.

First, consider the principle of "brevity" which everyone endorses and attempts to apply. What happens when this principle is distorted into an arbitrary rule of "no internal memoranda over one page in length?" One thing that can happen is that writers will refer to previous documents and will eliminate summarizing, stating the problems, reviewing previous communications, etc. The effect upon the reader is valuable time lost in looking up earlier correspondence and messages when a slightly longer memorandum could have been quite comprehensive.

Next, consider the principle of "positive thinking" and how this can become distorted by an organization into arbitrary rules against all unpleasant words, every unfavorable admission, or any poor negative. We suspect that the enthusiast who converts "failure" to "unrealized objectives" would probably describe a cloudburst as "intensive nonsunshine". The circumlocution forced upon the writers by the overapplication of these fine principles can be momentarily amusing, but amusement fades when annoyance is experienced by the reader. In a favorable communication climate the communicator is valued for his particular contribution to the team effort and is guided by published policy rather than being hamstrung by rigid rules.

When the communication climate of an organization is favorable, there may be something in the way of a policy, either official or simply an observed practice, to guide people as to when correspondence is preferred. Where policy or practices exist, the reader can compare the criteria given here with those of his own organization. More likely, any published communicating guide created by an organization will emphasize "how" to express the message and the reader must search for an implied "when".

Since much published material is available on the "how" of writing, this paper considers only the "when". Published material also presents a wealth of references on oral communication and the basic art of communicating. The Bibliography lists some well-known references under the headings of Oral, Written, and Communication Art.

Since the available reference books fail to discuss when correspondence should be used, established practices and common sense remain as the only guides except in those organizations which have published their own rules. An informal survey of a few associates produced a list of criteria which they have applied in their various work assignments. From this survey to which associates responded in terms of their specific work, all criteria were classified in the following general situations.

1) *Confirmation:* You confirm pertinent details of a conversation, a telephone call, or a group meeting, especially when action or decision will be based on such details. The purpose is to supply copies to all participants with a deadline stated for additions and/or corrections.

2) *Proposal and Acceptance:* You make a proposal that commits your department's manpower, production facilities, or funds—or you accept a proposal made to you, whereby you enter into a similar commitment.

3) *Request or Solicitation:* You request action from another department or function. In another variation, you request data or a decision which can only be supplied after significant effort. Your solicitation may be addressed to several people when your objective is to find someone who can supply data, services, or resources.

4) *Data Transfer—Selective or Mass:* You supply data to another, for example, a list of parts on hand by numbers and quantity. Such data may require mass distribution so that the correspondence must be reproduced by an appropriate method. The needed number of copies versus the reproduction systems available influences your choice of method.

5) *Command or Decision Communications:* You direct a command to one or more persons or you convey to the necessary people the management (or business) decision which you must implement.

6) *Situation Report:* You wish to define a problem, document an agreement, report the complexities of a negotiation standoff, etc. When used in an informal state, the

situation report can be stripped to the essential subdivisions, even down to one division, for example, "recommendations." When used in the formal state, the complete report outline normally would be followed, namely:

a) analysis of situation (sometimes as definition of problem)
b) possible solutions (described with advantages and disadvantages)
c) your recommendations
d) further investigation or action needed
e) references and list of attachments.

7) *Situation Estimate:* You need to predict the future situation at some specified date, as a basis for long range planning. Your method needs to be explained to permit checking and to ensure consistency in later revisions or extensions.

8) *Professional Courtesy:* You have been assigned to answer a piece of correspondence from an individual who has professional or business stature and you want to show him proper courtesy. One old rule of thumb was to "respond-in-kind." You would answer a letter with a letter; you would return roughly the same level of formality or informality. Finally, you would arrange for the reply to be signed by a person of appropriate rank in your organization. The "respond-in-kind" rule should not arbitrarily be applied when you can see how to better serve the original correspondent by other means.

9) *Sensitivity and Security:* You supply data that will be reflected in a public announcement or data that reveal the capability of a product. Both are sensitive areas, in that misunderstanding can be disastrous. To protect your personal interests, you use writing with reference to your formal suggestions, your inventions, or your other creative effort. You can also protect the interests of a fellow employee whose creative efforts are under discussion by making the matter a written record.

10) *Activity or Project Report:* You wish to advise others of the present state of a planned and scheduled enterprise. You may need to report status at various phases of a long project, so that titles, such as "Planning Report," "Interim Report," and "Final Report" are frequently used. When the planned effort is to continue indefinitely, the terms "Activity Report" or "Status Report" are often used.

11) *Upon Request:* You receive a management instruction or an associate's request to answer a specific communication by written correspondence. You may be advised to write by your legal department or the company's legal counsel.

12) *Controversial or Political Issues:* You must first consider carefully whether you need to express the thoughts in question. If you must say it, can you avoid a public declaration? If you put it in writing, it can come back to haunt you forever. However, putting it in writing will be preferable to being accused of verbal statements that you cannot disprove. In a tense situation you will not be heard correctly. In a purely political situation your opposition will listen mainly for the purpose of discrediting you.

A variation of this situation is a letter of protest written by an employee who has been ordered to proceed with an action which he is convinced will be harmful to his company. It is a rare company that has a proper channel for a written protest of this nature. Generally, the employee should take advantage of the counseling, hearings, and appeals available to him in his company rather than attempting a written protest.

13) *The Meeting Agenda:* You find that an involved situation necessitates a meeting of responsible and busy people. After the initial flurry of arrangements and approvals, you are obligated to conserve the time of the attendees and to make the meeting effective. You prepare and distribute an "Agenda" for the meeting which can include items such as the following:

a) a summary of the problem or current developments
b) list of prior documents and/or decisions
c) the approximate level of complexity when technical discussion is involved
d) the management functions to be represented and the level of management, when appropriate
e) an opportunity for the addressees to make suggestions prior to the meeting—on a tight schedule, telephone response can be encouraged
f) some indication that the scope and nature of the problem can be resolved only by a meeting
g) further explanation for the meeting, which has already been briefly stated in the meeting notice.

Through the preceding 13 criteria which have been expressed in terms of the administrative interrelation between people, some basic similarities can be sensed. One excellent reason for finding the common denominators of these criteria is that they could be a guide to identifying new criteria not listed here. One set of such common denominators can be found if the criteria are analyzed in terms of theoretical communication problems like the following.

1) Communication in breadth—or conveying the same information to a number of people. Examples: criteria 1) and 13).
2) Communication in depth—or giving information a relative permanency as in the case of vital records, etc. Examples: criteria 2), 6), and 7).
3) Communication analysis—best illustrated by a complex written message which can be studied in whole or in part as much as needed to gain understanding. Examples: criteria 6) and 7).
4) Communication with authentication—or indicating the authority. This can be illustrated by the power of a signature to represent a function or department. Examples: criteria 5) and 2).
5) Communication at length—or conveying long and complex messages. Example: criterion 4).

In Dr. Weisman's quotation, he predicts application in the "more complicated and difficult message situations." Both the 13 basic criteria and the 5 common denominators derived from them seem consistent with his prediction.

Consequences

What are the consequences of failing to use written communications when they should have been used?

First, a communication failure is much more apt to occur when the appropriate medium is not selected for the communication. Failure to use written media when they should have been used means that the best medium was not selected.

Second, the form of the communication failure will differ according to the situation. For example, in criterion 13), a meeting agenda could be telephoned to the people on the list, but then some of the recipients could misunderstand; some would not take sufficient notes thus forgetting part of the message; and some would not give the matter sufficient thought before the meeting since they have no written notice as a reminder.

Third, the misunderstandings, forgotten details, and inaction described above are typical primary results of a communication failure. In turn, they may cause failure, delay, or damage to the work project underway; it is at this secondary level that dollar values can be best assigned. When communication failure does not cause loss or damage to the work project, then people have made some additional effort which served to offset it. This offsetting effort can represent the time of valuable technical or administrative personnel whose services were needed elsewhere; therefore it should not be assumed to be without cost.

Consequences cannot be discussed without a brief explanation of how closely the human factor is related.

1) Theoretically, the choice of a less desirable communication medium can be offset by human effort. For example, a long message can be delivered by telephone if both parties are patient, deliberate, and the recipient does not mind taking the message down in longhand or by typewriter. Practically speaking, people soon tire of doing a task the hard way, so that their error rate can increase to the point of breakdown. Time is not available for inefficient operations.

2) Practically and theoretically, even the best choice of medium will be somewhat offset by human failure. When written communications have been properly selected, the writer can fail to express a necessary thought, or the recipient can read it hurriedly to arrive at an incorrect interpretation.

3) Analysis of past operations can be made with the assumption that every error found was caused partially or fully by communication failure. Such an assumption is least critical of the employees involved. In some instances, when an investigation is unusually fruitful, sufficient evidence may be found to classify a given error as caused by human failure.

Summary

In summary, this article is based upon a conviction that most organizations presently have particular communication situations which are best served by the written media. For the future, Dr. Weisman's quotation indicates that repetitive and routine messages will be largely taken over by machines, but human creativity will be needed for the remaining ones. An organization can therefore assume that some future application of written messages can be planned. The 13 criteria and the 5 common denominators presented here are guides to the situations when written communication is preferred. To best apply these guides to the problems of your organization, the following factors should be considered.

1) The general criteria expressed here need to be restated in terms of your work and organization structure. Any conflict with existing rules must be resolved.

2) Your organization guides need to be published and thoroughly distributed. Provision must be made for updating guides as conditions change.

3) Within the organization the guides need to be applied uniformly. Implementation can follow a planned schedule.

4) Your operations need to be surveyed for the "complicated and difficult message situations" mentioned by Dr. Weisman, which have not been classified by this article.

5) For most organizations, there will be direct savings resulting from improved efficiencies of communication sufficient to greatly offset the cost of the effort outlined here.

6) Even when direct savings are insignificant, the effort may obviously have contributed to the smooth momentum of a going operation—a factor to be considered even though it is difficult to evaluate in dollars.

Despite the conviction that most organizations could benefit from developing organization guides, it is hoped that some few organizations have so excelled in communications, that they will find nothing of interest here. If there is any organization that feels it can ignore communications problems, then it may need to consider the words of Rosenstein *et al.*:[1] "Any corporation, business, or for that matter any human organization, is in reality a communication system. Its effectiveness at all times is dependent upon the efficiency with which it communicates."

Bibliography

Oral Communication

[1] H. Bonner, *Group Dynamics: Principles and Applications.* New York: Ronald, 1959.

[2] D. Carnegie, *Public Speaking and Influencing Men in Business.* New York: Association Press, 1955.

[3] J. E. Dietrich and K. Brooks, *Practical Speaking for the Technical Man.* Englewood Cliffs, N.J.: Prentice-Hall, 1958.

[4] E. J. Hegarty, *How to Run Better Meetings.* New York: McGraw-Hill, 1957.

[1] A. B. Rosenstein, R. R. Rathbone, and W. F. Schneerer, *Engineering Communications.* Englewood Cliffs, N.J.: Prentice-Hall, 1964, p. 7.

[5] R. L. Kahn and C. F. Connell, *The Dynamics of Interviewing.* New York: Wiley, 1957.
[6] G. M. Loney, *Briefing and Conference Techniques.* New York: McGraw-Hill, 1959.
[7] G. S. Nutley, *How to Carry on a Conversation.* New York: Sterling, 1953.
[8] L. Surles and W. A. Stanbury, Jr., *The Art of Persuasive Talking.* New York: McGraw-Hill, 1960.
[9] Sutter *et al., Discussion and Conference.* Englewood Cliffs, N.J.: Prentice-Hall, 1954.

Written Communicaton

[10] W. P. Boyd and R. V. Lesiker, *Productive Business Writing.* Englewood Cliffs, N.J.: Prentice-Hall, 1959.
[11] T. G. Hicks, *Writing for Engineering and Science.* New York: McGraw-Hill, 1961.
[12] S. Mandel and D. L. Caldwell, *Proposal and Inquiry Writing.* New York: Macmillan, 1962.
[13] J. H. Manning and C. W. Wilkinson, *Writing Business Letters,* rev. ed. Homewood, Ill.: Irwin, 1959.
[14] R. L. Shurtur, *Written Communication in Business.* New York: McGraw-Hill, 1957.
[15] H. J. Tichy, *Effective Writing for Engineers, Managers, Scientists.* New York: Wiley, 1966.
[16] J. N. Ulman, Jr., and J. R. Gould, "Informal reports" in *Technical Reporting.* New York: Henry Holt, 1959, ch. 6.
[17] H. M. Weisman, *Technical Correspondence.* New York: Wiley, 1968.

Communication Art

[18] C. Cherry, *On Human Communication.* New York: Wiley, 1957.
[19] R. Flesch, *How to Write, Speak, and Think More Effectively.* New York: New American Library, 1963.
[20] R. Gunning, *The Technique of Clear Writing.* New York: McGraw-Hill, 1952.
[21] S. I. Hayakawa, *Language in Thought and Action.* New York: Harcourt, 1949.
[22] R. O. Kapp, *Presentation of Technical Information.* New York: Macmillan, 1957. A United Kingdom viewpoint.
[23] R. R. Rathbone, *Communicating Technical Information.* Reading, Mass.: Addison-Wesley, 1966.
[24] C. W. Redding and G. A. Sanborn, *Business and Industrial Communication—A Source Book.* New York: Harper and Row, 1964.
[25] T. E. R. Singer, Ed., *Information and Communication Practice in Industry.* New York: Reinhold, 1958. Treats communicating through processable data.

Of prime importance to anyone engaged in written communication is the ability to determine the proper level at which the material should be written—this article tells how this may be accomplished.

Selecting And Writing To The Proper Level†

By JOSEPH RACKER*

GENERALLY IN ANY form of objective writing, and particularly in technical writing, the writer tries to convey information in the most direct and concise manner. To do this, the writer assumes that the reader has a certain amount of basic information and that this information does not have to be repeated. As a result the writer may omit a great deal of pertinent data, relying upon the reader to supply this data. This point can best be understood by the two simple examples shown in accompanying Tables I and II. Summarizing: the *writing level* actually represents the level, or general background, that the *reader* must have to understand the writer.

Importance of Selecting The Proper Level

Technical writing is a form of instruction and many of the techniques that represent good teaching practice also apply to good writing practice. It must be remembered that readers are human and subject to human limitations. Do not assume that if the reader is sufficiently interested in your paper he will "puzzle out" all of the information you have failed to supply, or conversely that he will read through a great deal of extraneous material to find an occasional bit of useful information.

MEASUREMENT OF A D-C VOLTAGE

Consider the instruction: "Check that the voltage at pin 5, V101 is +160 volts, ± 10%. If it is not, adjust R101 until this reading is obtained."

In order to perform the instructions indicated in this statement, the reader must know and fill in the following bits of information:

1. **The locations of pin 5, V101 and R101.**
2. **That voltage must be measured between pin 5, V101 and ground.**
3. **The location of ground.**
4. **That a d-c voltmeter must be used and the type that should be used.**
5. **The procedure for using the d-c meter.**
6. **That a reading of between +144 and +176 volts, d-c, is acceptable.**
7. **That R101 must be adjusted if the reading is less than 144 or more than 176.**
8. **How to adjust R101.**
9. **That if it is necessary, R101 should be adjusted for a reading of +160 volts, dc.**

Readers that can fill in all of this information can perform the instructions given without any difficulty. Readers that do not have all of this information may find it difficult, if not impossible, to perform this instruction correctly.

Table I

Generally a reader will not read material that is above his level. Obviously if he is required to supply information that he does not have, he cannot follow the text. Few readers will read several other background books or articles in order to understand the one you have written. Most readers will dismiss the text that is above their level as being incomprehensible.

Most readers will absorb material best that is written

†Presented at IRE-PGEWS Second National Symposium.

*Vice President and Chief Engineer, Caldwell-Clements Manuals Corp., New York 17, New York.

Reprinted from *IRE Trans. Engrg. Writing and Speech,* vol. EWS-2, pp. 16–21, Jan. 1959.

slightly below their level. However, if it is considerably below the reader's level, he will tend to skim over large portions of it and generally will miss some vital information.

The over-all result is that he will often end up with the wrong results or misleading information and attribute it to the poor technical background of the writer. Human nature is such that a reader will often assume that material written to a much lower level is also inaccurate or inadequate.

The ideal level is the one that allows the reader to supply all information readily available to him and provides only information that is new or not fresh in his mind. Facts that are generally available but not used frequently (and likely to be forgotten) should be included in this ideal level of writing.

As in every ideal situation, it may be approached but can never be reached. The writer could not know all of the information that is readily available to the reader. Even if he had this information, many readers must be considered and each has his own level. The best that the writer could hope to do is to write at a level that is satisfactory to the average reader.

As an aid in writing to the average reader, five levels of technical writing, and the assumed background of the average reader for each level, will be defined. Then I will describe some general rules that should be followed for each level. Before defining these levels, I would like to emphasize a general rule for writing at any level.

LOGARITHMIC AMPLIFIER RESPONSE CURVE

Consider the instruction "Check that the amplifier response is linearly logarithmic (within 1 db) over an 80 db range".

In order to perform the instructions indicated in this statement, the reader must know and fill in the following bits of information:

1. **What the expression "linearly logarithmic (within 1 db) of an 80 db range" means.**
2. **What test equipment is required to measure such a response.**
3. **How to connect and operate such test equipment.**
4. **The starting point of the 80 db range.**

Readers that can fill in this information can perform the instructions given without any difficulty. Readers that do not have all of this information will find it IMPOSSIBLE to perform this instruction correctly.

Table II

General Considerations in Writing at any Level

The important consideration in writing at any level is to be fully cognizant of all the information necessary to understand or carry out the written material. When the material involves instructions, this can readily be accomplished by actually performing the instructions and carefully noting every step that is taken. If this cannot be done because the equipment is not available, then the writer must rely upon his experience and visualize every step that is taken.

For example, the statement "Tune Coil L8 for maximum output" may be as simple as it sounds or may be a very critical adjustment requiring special instructions. The writer should know all of the steps necessary to tune L8 properly.

It is equally important, though not as easy to visualize, to determine all the information necessary to follow papers on equipment design and theory. A common occurrence in this category is when an author gives one equation and then states that another equation can be derived from this first one. In many cases the derivation is arrived at by complex mathematical operations and possibly by elimination of some terms deemed negligible for the problem being considered. Certainly the author should be fully cognizant of all the steps and assumptions necessary to derive one equation from another.

I am not trying to state at this time that all information must be used. At some levels of writing omissions of information are justified. However, before an author can determine what information can be omitted he should be *fully cognizant of all the information necessary to understand the subject matter.* This is the point being emphasized at this time. The selection of what information is or is not required for a particular level is often far easier than determining all the information required.

Now we come to the definitions of five levels of technical writing. First I will define them, then I will describe how to write to each level. The five levels are as follows (See Table III):

FIVE LEVELS OF TECHNICAL WRITING

1. **Operator's or Non-Technician level.**
2. **Field or Technician's level.**
3. **Depot, advanced Technician's or Junior Engineer's level.**
4. **Engineer's level.**
5. **Advanced Engineer's or Scientific level.**

Table III

Definition of Operator's or Non-Technical Level

In an operator's or non-technical level of writing, the reader is assumed to have no specialized training in the subject matter of the text. Note that this does not necessarily mean that the reader does not have any technical background at all. Examples of writing that fall into this category are as follows:

1. Describing the operation or maintenance of an equipment to personnel who have no training on this or similar equipment (for example, a military trainee with no technical training).

2. Preparation of a handbook for instructing pilots on the operation of an electronic auto-pilot. While the pilot has an extensive technical background, he may have little or no knowledge of the theory and operation of electronic automatic landing systems. Consequently, from the viewpoint of his knowledge of electronic equipment, the pilot is considered a non-technical person.

3. Preparing a brochure or proposal directed at high-level corporation or military personnel who are either not interested in or not qualified to understand the technical details of a project. However, they are interested in the important over-all results such as reduced costs, weight, or other advantages that they can readily

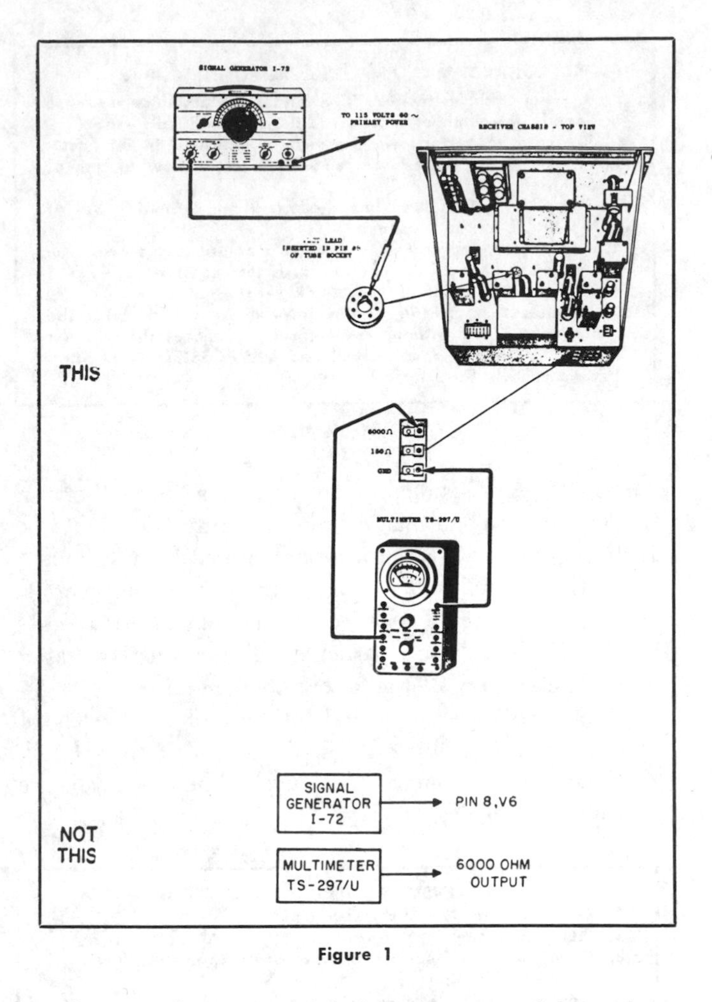

Figure 1

evaluate and in a simplified explanation of how these advantages are obtained.

Definition of Field or Technician's Level

In a field or technician's level of writing, the reader is assumed to have special training on a specific equipment or class of equipment with a specified number of tools or test equipment. Generally the reader has been trained to maintain the equipment while it is in the field, home or airport. Examples of personnel that fall into this category are:

1. A television repairman who services calls in the home. Generally he is capable of repairing simple, common troubles, making some adjustments, and removing major assemblies.

2. An aircraft mechanic who checks the operation of aircraft on a routine maintenance basis at the airport and makes periodic adjustments and minor repairs.

3. A technician who services electronic or mechanical equipment in the field, by following routine preventive and corrective maintenance instructions such as lubrication, adjustments, and replacement of readily accessible and removable parts.

Definition of Depot, Advanced Technician's or Junior Engineer's Level

In a depot, advanced technician's or junior engineer's level the reader is assumed to have extensive background information, either in experience or schooling, on the subject matter. Generally the reader is capable of thoroughly understanding the theory of operation of the equipment and can select the proper tools or test equipment necessary to maintain the equipment when it is sent to the depot, shop or factory or aircraft hangers. Example of personnel that fall into this category are:

1. The factory or shop television serviceman who repairs and aligns sets that cannot be handled by the field man.

2. Technicians who overhaul electronic or aircraft equipment.

3. Technicians and junior engineers who build, test, and "debug" developmental models of equipment. These personnel generally can design test setups and make minor design modifications.

Definition of Engineer's Level

In the engineer's level, the reader is assumed to have an engineering degree, or equivalent, and at last several years experience as a practising engineer. Generally at this level the primary emphasis is in the development of new equipment or techniques rather than in the maintenance of existing equipment.

Definition of Advanced Engineer's or Scientific Level

In the advanced engineer's or scientific level, the reader is assumed to have an advanced engineering degree or extensive experience in a particular field. Generally at this level, the primary emphasis is in theoretical calculations, new concepts, or basic research.

Writing to Operator's or Non-Technical Level

I will describe some general rules to be observed in writing to an operator's or non-technical level. (See Table IV)

GENERAL RULES FOR WRITING TO OPERATOR'S OR NON-TECHNICAL LEVEL
1. Include all information.
2. Back text with simple, pictorial illustrations.
3. Use step-by-step procedures.
4. Keep reference to other procedures or books to a minimum.

Table IV

1. *Include all information.* Do not expect the reader to supply any technical information in order to understand the text. Thus, in example given in Table 1 covering taking of a voltage reading, all 9 items of information would be supplied.

2. *Back text with simple, pictorial illustrations.* Extensive use of illustrations to emphasis and clarify text is highly recommended. Illustrations should be simple and duplicate the physical appearance of the equipment. Schematic or symbolic representations should not be used. This is graphically illustrated in Figure 1 on this page which shows a pictorial and a symbolic diagram for making test connections for i-f alignment of a receiver. Note that the pictorial diagram shows the physical location of every connection point, while the

symbolic diagram does not give any physical information at all. Another effective method of emphasizing an important instruction to the reader is through cartoons. Cartoons which properly supplement the text add interest and effectiveness to material written at this level.

Figure 2 shows a pictorial representation of a complex pattern that is used to enable a non-technical person to visualize a technical phenomena.

3. *Use step-by-step procedures.* Whenever, possible, instructions should be presented in short, distinct steps. When a large number of instructions are involved, the operational steps should be subdivided into procedural steps. For example:

A. Turning meter on
   1. Place meter power cord into a-c outlet.
   2. Place meter POWER switch in the ON positon.
   3. Allow meter to warmup for 30 seconds.

B. Setting of Meter Controls Prior to Operation
   1. Place AC-DC switch in AC position.
   2. Place RANGE SELECT switch in 100 VOLTS position.
   3. Adjust ZERO knob for a zero reading on meter.

C. Connection of Meter to Set
   1. Locate terminal marked AUDIO OUTPUT at back of set.
   2. Connect red lead of meter to AUDIO OUTPUT terminal.
   3. Connect black lead to chassis.

4. *Reference to other procedures or books should be kept to a minimum.* As much as practical, all steps necessary to perform an operation should be listed in text. References to other procedures or manuals such as "turn equipment on as described in manual covering meter" or "repeat steps a to c, paragraph 1-15" or "turn power off" should be used sparingly. Unless it becomes unduly repetitious, include all steps given in the manual covering the meter, literally repeat steps a to c, and provide the procedure for turning power off.

Writing to a Field or Technician's Level

I will now describe some general rules to be observed in writing to a field or technicians level. (See Table V)

1. *Specify test equipment normally available in the field, when possible; provide detailed procedures on special test equipment.* A widely used method of determining where field level maintenance ends and depot level maintenance begins is to consult the list of test equipment available in the field. In military applications a list of field test equipment is usually made up for each equipment; in commercial applications this list is determined by common practice. If maintenance can be performed with test equipment available in the field, it is considered field maintenance. All maintenance requiring test equipment not available in the field is generally considered depot maintenance. It is not necessary to provide detailed operating and test setup instructions on field test

Figure 2

equipment. However, for the exceptional case where it is necessary to use special test equipment in the field, detailed information should be supplied on how to setup and operate this equipment.

2. *In theory of operation, provide sufficient information to permit proper performance of maintenance but do not repeat basic theory or include detailed theory not required for maintenance.* In preparing maintenance information, such as trouble shooting charts, it is impossible to list every possible trouble that may occur in the field. Only general procedures can be given together with some common causes of failure. To enable the technician to deal with many of the specific problems he will face, a theory of operation section is generally included. This section should not repeat basic theory, or describe details of theory that will not aid in the maintenance of the equipment at the field level. For example, in radio equipment, it would not be necessary to describe the operation of a conventional amplifier or rectifier, nor would it be

GENERAL RULES FOR WRITING TO A FIELD OR TECHNICIAN'S LEVEL

1. **Specify test equipment normally available in the field, when possible; provide detailed procedures on special test equipment.**
2. **In theory of operation, provide sufficient information to permit proper performance of maintenance but do not repeat basic theory or include detailed theory not required for maintenance.**
3. **Use graphs or nomographs instead of equations when possible.**
4. **Repetition of instructions is not necessary, reference can be made to other paragraphs or manuals.**

Table V

necessary to describe the detailed operation of a sealed motor which, if inoperative, would have to be replaced entirely. The theory section should include a description of any special circuits and indicate the key components of all circuits. For example, stage V101 is an amplifier which utilizes a feedback resistor, R101, to reduce distortion and improve the stability of the amplifier. Or, V102

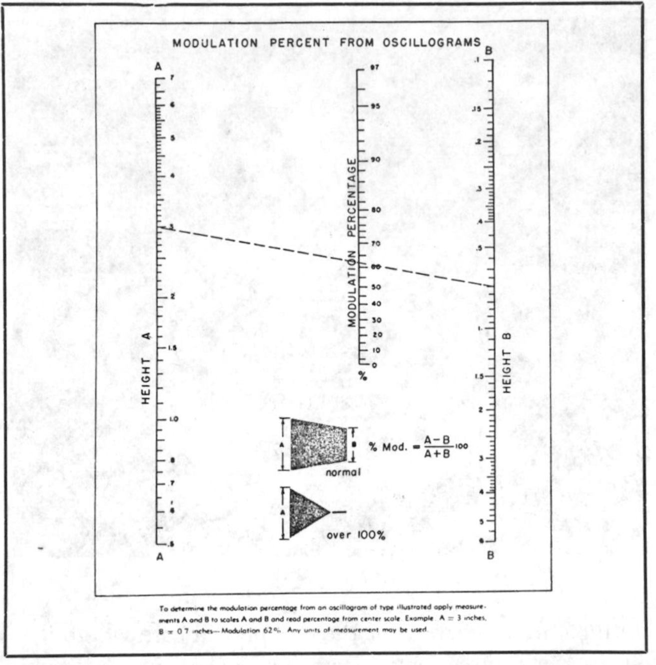

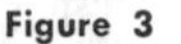

Figure 3

is a multivibrator whose repetition rate depends primarily on the r-c network of R103, C101, and R102. The effect of any variable components, which the technician must adjust, should be particularly emphasized. For example, the audio output is determined by the setting of potentiometer R105 (AUDIO OUTPUT) which varies the amplitude of the signal applied to the grid of audio amplifier V103. Liberal use of simplified schematics should also be employed.

3. *Use graphs or nomographs instead of equations when possible.* In some instances, it is necessary to use mathematics to explain theory or perform maintenance. Generally the field technician can be expected to understand some simple mathematics, such as sine and cosine functions, but even simple mathematical formulas and certainly those involving higher mathematics should be supplemented with illustrations that enable solution of the equations by means of a straight edge. For example, Figure 3 on this page is a nomograph which permits calculation of modulation percentages by means of a straight line.

4. *Repetition of instructions is not necessary, reference can be made to other paragraphs or manuals.* At this level it is not necessary to repeat information included in other procedures in the book or those contained in other manuals normally available in the field.

Writing to a Depot, Advanced Technician's or Junior Engineer's Level

I will now describe some of the general rules that should be observed in writing to a depot, advanced technician's or junior engineer's level. (See Table VI)

1. *The operation of any available military (for military applications) or commercial test equipment need not be covered.* At this level it can be assumed that any available piece of test equipment can be procured and that the reader can become, or is, familiar with the operation of this equipment. Furthermore, it may be assumed that the reader can select equivalent test equipment that he may have available (modifying it when necessary to make it equivalent to the equipment specified).

2. *Special or fabricated test equipment can be specified and should be described.* Any special test equipment that is available can be specified and if such equipment is not available, the fabrication of such equipment should be covered. Any operating instructions that are unique to such special test equipment should be described.

3. *Describe all special disassembly and assembly techniques, supplement such procedures with diagrams.* Complete disassembly and reassembly procedures are often included in books written at this level. While it is not necessary to describe obvious disassembly or reassembly steps, any special techniques should be covered. That is, if it is just a matter of loosening some nuts or unscrewing some screws to disassemble equipment, a detailed procedure is not required. However, if it is necessary to set a gear in a specific position before a holding screw is loosened in order to slip the gear out, this instruction should be described. Diagrams, preferably exploded views, which illustrate the disassembled equipment, identify each part, and indicate by index number the order of disassembly, should be used to supplement the text. Such a diagram should be provided for even simple disassembly procedures because after an equipment is disassembled the reassembly may not be apparent, at least to the extent of replacing all the lockwashers, using the proper sized nuts and bolts in the right places, etc.

4. *The complete theory of operation of the equipment must be included; conventional circuitry may be covered by a block diagram description.* The theory of operation should cover every circuit and include a complete set of diagrams to permit servicing every item in the equipment. Text description of all conventional circuits can be on a block diagram basis such as "the receiver consists of an r-f stage V101, mixer-oscillator V102, three i-f stages V103 to V105, detector V106, and audio amplifier V107". However, a detailed schematic of all of these circuits should be given as well as a description of any circuits that are unique to this equipment.

GENERAL RULES FOR WRITING TO A DEPOT, ADVANCED TECHNICIAN'S OR JUNIOR ENGINEER'S LEVEL

1. **The operation of any available military (for military applications) or commercial test equipment need not be covered.**
2. **Special or fabricated test equipment can be specified and should be described.**
3. **Describe all special disassembly and assembly techniques, supplement such procedures with diagrams.**
4. **The complete theory of operation of the equipment must be included; conventional circuitry may be covered by a block diagram description.**
5. **Use graphs and nomographs instead of equations when possible.**
6. **Repetition of instructions is not necessary, reference can be made to other paragraphs or manuals.**

Table VI

5. *Use graphs and nomographs instead of equations when possible.* The comments made on this subject for field level also apply at this level.

6. *Repetition of instructions is not necessary, reference can be made to other paragraphs or manuals.* The comments made on this subject for field level also apply at this level. (See Table X)

Writing to an Engineer's Level

I will now describe some of the general rules to be observed in writing to an engineer's level. (See Table VII)

1. *Start text with basic information that reader must know.* A brief review of the basic information that the reader should know should be included in one of the early paragraphs. When limitation of space does not permit repetition of basic information, then an indication of what basic information is required and where it may be obtained should be included.

GENERAL RULES FOR WRITING TO AN ENGINEER'S LEVEL
1. Start text with basic information that reader must know.
2. All information should be objective.
3. Information should be complete.
4. Simple calculus can be used freely; advanced calculus should be avoided, if possible, or referenced in an appendix if it must be used.
5. Define all terms in equations; use standard symbols when possible.
6. Use same units throughout text, when possible, or clearly indicate when new units are being used.

Table VII

2. *All information should be objective.* Material written at this level should be objective and should not include digressive and unnecessary information. For example a description whose purpose it is to gain acceptance of a new equipment by potential users should emphasize performance characteristics and need not detail design problems unless they are important to the evaluation of the equipment performance.

3. *Information should be complete.* All information necessary for the understanding of the material should be included in the text either directly or by reference to other source material. The terms "it can be shown that" should be used only in conjunction with a reference that does show it. All assumptions, approximations, or short cuts used to arrive at conclusions should be stated. In other words, it should be possible for any engineer reading this material with all referenced source material to fully understand and evaluate the material presented. Many an author has been embarrassed by the question, "How did you arrive at this conclusion?" coming from a reader several years after a paper was written. The conventional reply is "I knew it when I wrote the article, but now I must look at my notes to refresh my memory." If it is necessary to consult the author's notes in order to understand the text material, then the material has not been properly written.

4. *Simple calculus can be used freely; advanced calculus should be avoided, if possible, or referenced or placed in an appendix if it must be used.* Equations involving mathematics no higher than simple integration or differentiation can be used freely. However, if advanced differential equations or integrations are involved, it is advisable merely to state the over-all conclusion and refer to another text for the derivation of this conclusion (when it is derived in another text) or include the derivation as part of an appendix. Of course, such mathematics should only be included when it is necessary to achieve the objective of the article. Some authors believe that it adds to their prestige to include complex mathematical formulas in their papers. This is *not* true if these formulas are not necessary to accomplish the objective of the article. In this latter case, it merely provides a major obstacle to the understanding of the material and often completely discourages readers from reading the paper.

5. *Define all terms in equations; use standard symbols when possible.* When using equations clearly define each term in the equation including its units: for example, v is the velocity in feet per second. When possible use the standard symbol, as defined by the IRE or equivalent engineering authority, for each term.

6. *Use same units throughout text, when possible, or clearly indicate when new units are being used.* A great deal of confusion can result when the author goes from one set of units to another set; for example, inches to centimeters, grams to ounces, degrees to mils. If possible all of the information should be presented in one set of units. If this is not possible, then the fact that a conversion is being made and the conversion factor used (if it is not well known) should be clearly specified.

Writing to an Advanced Engineer's or Scientific Level

1. *Introduction should reference sources where basic information can be obtained.* At this level of writing the author generally cannot start with basic information and keep text within reasonable limits. He should reference source material where this basic information can be obtained.

2. *Higher mathematics can be used freely; however, it should be placed in footnotes or the appendix when possible.* Any mathematics necessary to develop the subject matter of the paper can be used. When possible, detailed derivation of an equation involving higher mathematics can be placed in a footnote or appendix.

3. *Information should be complete.* The comments made on this subject for an engineering level also apply at this level.

4. *Define all terms in equations; use standard symbols when possible.* The comments made on this subject for an engineering level also apply at this level. However, since a great number of equations may be used and some terms repeated several times, it is convenient to define all terms at the beginning of the article.

5. *Use same units throughout text, when possible, or clearly indicate when new units are being used.* The comments made on this subject for an engineering level also apply at this level.

Try Writing a Technical Article*

FRED W. HOLDER

Abstract: It can be rewarding to have your own article published—recognition for both you and your company, improvement in your writing skill through practice, better understanding of your subject, and even—in some cases—financial award. Mr. Holder tells you how to go about researching and writing your article, finding a publisher, and preparing your manuscript for submittal.

"Why should I spend my own time writing technical articles for publication?" I could say, because it will help to improve your company's technical image. I would be right, but you wouldn't be greatly impressed. You're more interested in, "What does it do for me?" I believe you, the employee, will gain much more from writing technical articles than your company.

As an author, you will gain recognition from your company's management, from your associates, and possibly even from your customers. Also, publication credits look good on your resume if you are looking for a better position. Of course, there is an educational factor to be considered. I can't think of a better way to get a full understanding of a subject than to write about it. Let's face it, you have to understand the subject fully before you can write logically and clearly about it. A sideline effect derived from the effort you put forth in writing a technical article is improvement of your writing ability.

You may receive a final benefit from writing technical, or for that matter non-technical, articles for publication—money! Most commercial magazines pay for the articles they publish. For example, Electronics World pays about 5¢ a word. If you can make the pages of Playboy, you can earn up to $3000 per article. Also, for the person who has had an unusual experience, Reader's Digest publishes first person articles of about 2500 words; for this they pay almost $2800.

SELECTING A SUBJECT

"All right," you say, "I've seen the light, but where do I start? How do I decide what to write about?" Well, starting is usually the biggest hurdle, especially for the beginning writer. Before you can write, you must have an idea; you must know what to write about. Newspapers and technical magazines can supply many ideas. New subjects, new concepts, and new products are covered in the newspapers. You might also try listening to conversations of friends and coworkers.

If you are a beginning writer, it is best to stick with a familiar subject. For instance, if you are working in the electronics field, it is a good idea to write about electronics subjects, at least in your first or second article. The subject you select should have a fairly wide interest, especially if you are aiming it toward one of the more widely distributed publications. Finally, your subject should have current appeal. It should cover state-of-the-art items, new developments, new techniques, etc.

CHOOSING A MAGAZINE

Once you've decided to write an article and have selected your subject, find out if the article has a chance of being published. To do this, write to an editor; just any editor won't do. The magazine you contact must publish the kind of material you are proposing to write. You can choose a magazine with which you are familiar or leaf through the magazine racks at your company and public libraries. If you do a great deal of writing, you may want to look for a wider range of markets, or possibly, to determine how much the magazine will pay for your efforts. A document published by Writer's Digest called "Writer's Market" is available in most libraries.

> In August 1961, I was a $100 a week electronic technician with General Dynamics Electronics in San Diego, California. Layoff was eminent. So, I left General Dynamics to take a job as a trainee technical writer in Los Angeles, California. My knowledge as a technician really paid off. In four years I progressed from trainee to supervisor of technical writing.
>
> Many of the successful technical writers I know were one time electronic technicians. They wanted something better and tried technical writing because, as a profession, it offers many of the challenges of the engineer's job, without requiring an engineering degree; the professional benefits enjoyed by the engineer; and a salary that is comparable when one considers the difference in educational requirements.
>
> This is how I begin my 2000-word article, "Technician to Technical Writer—A Change Worth Considering," which tells what the technical writing profession has to offer the sharp technician who is willing to learn new ways of applying his technical knowledge. It also tells about the educational requirements, salaries, and promotional opportunities. Information presented in the article is based on both research and personal experience. Interested?

Fig. 1. Sample text from a query letter.

*Reprinted by permission of Society for Technical Communication from TECHNICAL COMMUNICATIONS, First Quarter 1970.

Reprinted from *IEEE Trans. Prof. Commun.*, vol. PC-19, pp. 38-41, Dec. 1976. (Reprinted by permission of Society for Technical Communication from *Technical Communications*, First Quarter 1970.)

Yes, we are interested in an article titled "Technician to Technical Writer—A Change Worth Considering." Might I suggest, however, that instead of writing it from one man's viewpoint, you include the ideas of others who have done the same.

Illustrations are important to articles in RADIO-ELECTRONICS, and you should give some consideration to some means of illustrating your article.

Enclosed are tearsheets of recent articles in our "Career" series. They might give you some idea as to how we have treated the topic in the past, although they shouldn't limit you, in case you have an idea for original treatment. The idea is to present a practical overview of "What's in it" for the technician who aspires to reach further.

Let me know if you have any questions. If you'd like to send an outline first, I'll be glad to comment on it.

Fig. 2. Sample text from an editor's reply.

CONTACTING AN EDITOR

After you've selected a magazine, draft a query letter to the editor. This letter helps determine whether the editor is interested in an article on your subject. If the first editor rejects your idea, don't despair. It is quite possible that he has either recently published an article on the subject or has one somewhere in the mill, or even possibly he may have already assigned one to another author. To be of greatest value, the query letter must be written somewhat like a proposal: you are proposing to the editor an article that you feel will be of interest to him and to his readers and will merit publication in his magazine.

A good way to start the query letter is to use your proposed lead paragraph (see Figure 1). If it is interesting enough to hold the editor's attention, he will read on through your letter and perhaps ask to see the article. The second paragraph of your query should give the title of your proposed article, the approximate number of words you expect it to contain, the approach you plan to use, how and whether it will be illustrated, and any other pertinent facts that you feel would be of interest to him. If you are an authority on your subject, be sure to tell him. If the editor is interested in your article, he'll write you, perhaps making suggestions on how the article should be handled. I received such a letter on my first article (see Figure 2). On occasion, he may request an outline. If the editor responds to your query, write him another letter thanking him for his interest and giving him an approximate date on which you expect to complete the article. At this time, it would be well to advise your company's public information office of your intent to write an article, its subject, and whether or not you have an editor interested. If you just want to write an article, and don't want to try to market it yourself, you might contact your company's public information office and tell them what you would like to do (the same type of general information that you send to the editor). They may be able to place the article for you.

GATHERING MATERIAL

You've gotten past the major hurdle. You've decided to write an article and you've decided what to write about. The next step, research, is the most frustrating and, at the same time, the most rewarding phase of the writing game. If you are an authority in your field, you may merely write, shall we say, from the top of your head on a subject with which you are thoroughly familiar. For most of us, however, writing requires a great deal of usually extensive research.

Once I've decided to write an article on a particular subject, I usually start with a public library or the company library. Most libraries contain telephone directories from almost every major city in the United States and many minor cities. The yellow pages can direct you to manufacturers or dealers who handle a particular type of product. For example, when I wrote my article "Electronic Fire and Smoke Detectors" (Electronics World, February 1968), I began by having my wife call companies in the Los Angeles area which specialize in installing fire and smoke detectors. From these companies, we obtained names of manufacturers and, where possible, their addresses.

The library also has a fine reference document titled The Reader's Guide to Periodical Literature. This is an excellent source index for articles on any subject you may be planning. The guide refers to the back issues of magazines which the library may have available. The subject index for books is a very good reference. Finally, for information on industrial firms, Thomas' Register is hard to beat.

Some unusual subjects are covered in the newspapers. Sometimes I get information, or leads to information, from the advertising in newspapers or magazines.

May I ask your help? I am a freelance writer in the Los Angeles, California area. I am working on an article that surveys the "state of the art" in electronic smoke and fire detectors. My deadline is 12 July 1967, so I will need information by 15 June 1967.

I will need the following types of information on your system:

1) A brief theory of operation with block diagram or schematic.
2) A brief discussion of any elements of the system which advances the state of the art.
3) Functional specifications of the device or system.
4) A press release photo (or photos) of the device or system.

I hope the information requested is not proprietary and, in addition, has already been prepared for some other purpose. Any assistance you can give me will be greatly appreciated. Thank you.

Fig. 3. Sample text from a letter requesting information.

One of the best sources of information for an article is an interview with an authority on the subject. If you should be lucky enough to know an authority or have access to such a person, prepare a list of specific questions, use a tape recorder, and take up as little of his time as possible. If you don't have a tape recorder or if the individual prefers that you not use one, take extensive notes. If the person is an authority, he is not going to be pleased that you don't think enough of what he has to say to write it down.

Finally, one of the data-gathering methods I prefer is writing letters to request information from companies or authorities who deal with the subject in question (see Figure 3). I usually write from two to twenty letters to gather information for one article. It is important to ask for specific information, otherwise you may not get the information you want.

WRITING THE FIRST DRAFT

When you finally gather all the information for your article, you will have a mass of newspaper clippings, notes, information supplied from manufacturers, books that you've checked out of the library, etc. The next logical step is to write the article. Of course, you don't just sit down and start writing. You have to read and digest all the material you've gathered, analyze it carefully, and make rough notes for use in the actual writing of the article.

Now that you know all about your subject, determine a logical order in which to present the material, and write this down in outline form. If you've done all of your homework—read the material, analyzed it, decided how to present it, and drafted yourself an outline—you are ready to write the first draft. Keep in mind where to place the beginning, middle, and end. The beginning is your lead. If you initially prepared a good lead paragraph for your query letter, then this can be the lead for the actual article. The best articles make use liberally of anecdotes, quotes of authority, facts and figures, and above all, they are technically accurate and specific.

While writing your first draft don't worry about grammar, punctuation, and all of the niceties of writing. Write as fast as possible; get your thoughts on paper. Don't spend too much time deciding how to construct a sentence . . . you will never get past that sentence. As you write, you'll find the writing become easier. About half way through the article, it flows smoother and reads much better than in the beginning. Be conscious, as you write, of the need to illustrate certain points of your discussion. If you feel an illustration would best emphasize or clarify a point, reference the illustration and worry about it at a later time. Get those words down on paper.

POLISHING THE MANUSCRIPT

Set the first draft aside for several days before trying to revise it or doing anything more with it. This allows you to look at the manuscript objectively. Now, review your first draft very carefully: Does it carry you along? Does it have continuity? Is there a thematic thread running throughout to lead the reader from beginning to end? Is there transition from paragraph to paragraph and part to part? Is the general writing quality good? If you've followed my advice and written the first draft as quickly as possible, you'll likely find the general writing quality to be rather poor.

You've reviewed your manuscript and decided what's wrong with it and where it needs to be improved. Now, rewrite it to put in anything you left out or to take out anything that you shouldn't have put in. This is also the time to smooth and polish your grammar, sentence structure, etc.

If you have time, set the article aside for a few more days to give yourself a new point of view before polishing if further. After the third or fourth rewrite, the article should sould like you just thought of it. Polish your lead very carefully, making sure that it's interesting enough to grab the reader, get his attention, and hold his attention.

Don't treat your title lightly. It's just as important, if not more important, than your lead. It should tell the reader what the article is about or sound interesting enough to stop him and make him at least read your lead paragraph. When writing your title, review the titles of articles published in a recent issue of your target magazine. This technique also applies for lead paragraphs.

Evaluate the length of your final manuscript. If it runs between 2000 and 4000 words, you're in pretty good shape. The The average technical article runs somewhere between 2500 and 3000 words. An article too long or too short may require rewriting.

ILLUSTRATING THE ARTICLE

Good illustrations are the backbone of the article, and they play a heavy role in getting an article accepted. Most magazines like to have several illustrations accompanying an article. For example, in my article "Technician to Technical Writer" (Radio-Electronics, February 1967), I included a photograph of the technical writer I interviewed and a flow diagram illustrating the steps involved in writing a manual. If you furnish photographs with your article, they should be 8- by 10-inch glossy, black and white prints; they must be clear, have a sharp focus, and be well lighted.

If you're researching an article on equipment or new developments, you may be able to get photographs from the companies furnishing the information. You just have to ask for them. For example, in my article on technical writing, I called the company where the technical writer worked, explained to them that I was interviewing him for an article, and asked if there was a possibility of getting a photograph of him at work. They were glad to send one. He went to their photolab, they made a dozen black and white shots, sent me a proof sheet to select the view I thought would be most suitable, and then furnished a glossy print of the selected view. In other published articles, companies furnished photographs at no cost. After all, it's good publicity.

Line drawings such as block diagrams and schematics can also enhance a technical article. If you're discussing theory or how a piece of equipment works, they become essential. When you're requesting information from a company, you can also request a schematic and permission to use it. In some instances, where you have researched from other publications such as magazines, you can often obtain permission to use illustrations from that magazine. You can also draw your own, since the magazine probably will redraw line drawings to meet their own specifications before publication. A clean pencil or pen-and-ink drawing suffices.

One important point to remember about illustrations: if they don't illustrate an important point or feature of your article, don't include them. As the old saying goes. "A picture is worth 1000 words"—but not the wrong picture.

An illustration has little value if it doesn't have a good caption. Some magazines use brief captions containing only a few words, while others use extensive captions describing the illustration. Follow their example and you won't be far off.

OBTAINING OUTSIDE HELP

When you've prepared your illustrations and polished your manuscript to the point where you can no longer do anything more to improve it, ask someone else (perhaps your wife or husband, a friend, or a coworker) to read your manuscript. The value of having someone else read your material, providing they will give you constructive criticism, cannot be overemphasized. They can help you pinpoint unclear areas and help to improve your manuscript. After you have considered and possibly incorporated any comments made by the person who reads it, prepare the manuscript for mailing.

PREPARING THE MANUSCRIPT

Manuscripts must be typed, double spaced, on heavy white bond. Type on only one side of the paper generally with 1 inch margins. Always indent paragraphs. Make it as easy on the editor as possible. He has to read your manuscript along with many hundreds of other manuscripts. It's a good idea to use a new ribbon, or at least one that will print good and black on the page. For your own protection because editors are human and sometimes lose manuscripts (and so does the U.S. Mail), make at least one copy of the manuscript for your files.

Make sure that you proofread your manuscript carefully for both grammatical and typographical errors, misspellings, etc. There is nothing that can turn an editor off more quickly than bad grammar and poor spelling, or masses of typographical errors.

CONCLUSION

At last the job is done. It is ready to mail, but don't forget to get your company's clearance. Chances are that you signed a paper that gives them title to any writing you do as well as patentable items. You need their clearance. Now, package your article in a cardboard box or between heavy pieces of cardboard and put it into a heavy envelope. Be sure to send a stamped, self-addressed return envelope just in case the editor rejects your masterpiece.

Finally, writing is hard work, but it can be quite rewarding to see your name in print. Rewarding not only in status, but perhaps rewarding in your pocketbook too.

Plan Ahead for Publication

Having mastered calculus, thermodynamics, mechanics, electronics, and other complicated subjects, an engineer certainly has the potential for expressing himself clearly. He has been trained to think logically, and he should be able to put his thoughts and ideas on paper logically. If you are not satisfied with your present writing skill, or if you are reluctant to write at all, consider the method suggested here.

DEWEY E. OLSON
Manager, Engineering Publications
IBM Systems Development Division Laboratory
Rochester, Minn.

THROUGHOUT INDUSTRY there is a growing movement to improve the communications skills of engineers. Foremost among those interested in this endeavor are engineers themselves—and for a good reason. Sooner or later, almost every engineer has some knowledge which should be of wide interest to others in his profession. The best way to share this knowledge is by publishing a paper or article.

Reasons for Writing and Reading

According to Dwight E. Gray,[1] most engineers and scientists write for one of four reasons:

1. To disseminate useful information; to share specialized knowledge with others. The Professional Engineer's Code of Ethics encourages this.
2. For professional recognition and advancement. The prestige and professional recognition gained by both the author and his company or institution are immeasurable. Publication is one of the fastest ways of becoming recognized as a knowledgeable leader in your field.
3. For money. Small honorariums are often paid by technical publications. This is motivation enough for some.
4. Because it is required. Corporations and institutions realize that good papers, which reflect advanced technologies and concepts, lead to industry-wide acceptance of new products and services. They often require their engineers to publish.

1. References are tabulated at end of article.

The last of these reasons is weak, because it represents the least interest on the part of the author. A primary requirement for a good technical paper—whatever the motivation—is interest.

Why do other engineers read what you have written? There are three reasons, according to one analysis.[2]

1. They want to keep abreast of technological advances.
2. They need specific information for a research or development project.
3. They want to see what the competition is up to.

The Proper Publication

More than 3 million articles will be published this year in an estimated 52,000 technical publications.[2] It is extremely important to find the proper place for your message. You must consider first whether your writing effort should be a *paper*, destined for publication in the journal of a professional society, or an *article* for a technical magazine such as MACHINE DESIGN.

Both of these publishing mediums offer much in prestige and professional recognition—for the author and for his company or institution. The professional journal is the better choice for a paper that actually advances or significantly adds to the state of the art in a fairly narrow segment of a given discipline.

The technical magazine is for the article of more general interest. You reach a wide audience here, but it is also a less specialized audience. Readers of technical *articles* usually want to know what is going on in the technological world; they are primarily interested in applications rather than theories. They will want to know how they can apply your information and will expect you to stress the practical aspects and the usefulness of whatever you are writing about.

To be accepted for publication in a technical magazine, an article should present new, useful information on a timely subject. It must be:

- **Appropriate.** You must know the publication's circulation and slant your article toward its readers.
- **Accurate.** Your statements must be supported by fact.
- **Adequate.** The article must contain all necessary detail, but only what is necessary.
- **Clear.** It must be well organized and well written. It must flow logically and be written in terms the reader can understand.
- **Concise.** Today's busy reader wants facts quickly and completely. He will not be impressed nearly as much by a scholarly approach as with a direct one.

Preparation for Writing

A potential author must understand the communication process, Fig. 1. The information source is your warehouse of knowledge on the subject. It might also be references that you will use. The transmitter is you, the author. The channel is the published paper. The receiver is your reader. The destination is the use he makes of your data.

An important element in the diagram is the "noise," which can be defined as anything that intereferes with the reader's understanding of what you intended. It may be a missing signal. If you don't supply details for the reader, he will be forced to fill them in himself—and they may be wrong. Noise may be a term that the reader doesn't understand, or a formula or drawing that isn't clear. It may be just involved writing. Your job as author is to design into your communication system the lowest possible noise level.

There are several procedures that will produce good communication with a low noise level. First, relax. Converse with your reader when you write. Don't let style, format, or fear of criticism bewilder you or form a mental block. Keep it simple. Many engineers are able to discuss complicated subjects with great ease, but when they try to put the same ideas on paper, the result is confusion.

Before starting your manuscript, examine previous writings on the subject. A complete literature search will help determine the scope of your paper and the significance of your work. It will also show you how others have approached the same problem. Discuss your proposed paper with your supervisor. He can help determine the scope

Checklist for Technical Article or Paper Preparation

1. Formulate idea for paper or article. Discuss with your supervisor and colleagues to determine if paper should be written.
2. Search the literature to determine what has been written on the subject.
3. Write a comprehensive outline. A good outline reads like a table of contents.
4. *Think* the article through. Ask yourself if your outline will allow you to present the right amount of data in the best manner.
5. Gradually expand outline headings into sentences and paragraphs. Keep one idea to a paragraph.
6. Smooth transitions and expand on key words and ideas.
7. Rough out illustrations.
8. Write the rough draft, then see if you have answered these questions:

Introduction
Did you properly orient the reader?
Did you tell why the study (device, etc.) was needed? Why it is significant or unique? What problem did you solve?
Are the scope, limitations, and problems of the study well defined?
Does the introduction generate enough interest in the reader for him to read the entire paper?

Body of Paper
Have you given necessary background material? Too much?
Is the problem, concept, or system adequately and accurately covered theory, test results, applications, methods of implementation?
Did you make a point?

Conclusion
What was the original problem?
How was it solved?
Has a conclusion really been made?

9. Revise the draft as required.
10. Have it typed double-spaced with at least one copy.
11. Proofread manuscript carefully.
12. Review with your supervisor.
13. Submit.

of your writing and will be in a position to know the proprietary nature of the work. He can also help select those phases of your project which are really worth publishing.

Think of the paper as another design task. Before beginning a complicated system design, you'd probably sketch a block diagram. Do the same with your paper. Begin with an outline—ideally, begin the outline while you are still involved in the project. Make notes of points you particularly want to cover.

The outline, which will serve as the framework of your article, should be built carefully.

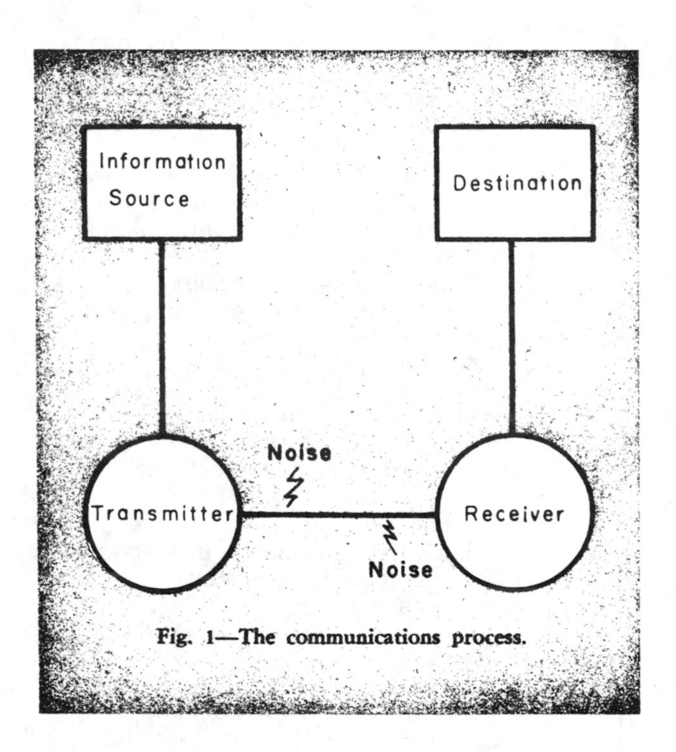

Fig. 1—The communications process.

- Jot down on paper each idea as it comes to mind without regard to its importance or place in the outline.

- Classify these ideas into major groups and subgroups.

- Arrange the main groups into logical order suitable for use in the final paper.

- Arrange sub-subjects in logical order under the main groups.

- Check the completed outline on the following points: Sequence, importance, duplications, omissions, clarity. Rearrange and make changes as needed.

Look over the data you have been gathering as your project progressed and ask yourself what is relevant. Now, just stop and *think* the whole paper through. Think about how you are going to get your message across. A little time spent in the thinking stage will eliminate a lot of time and effort in corrections later.

Fig. 2—Simple words can be misunderstood.

Again, consider your reader and what he will be interested in learning about your subject. Ask yourself what points you want to stress, and how you can best get your message across.

Developing the Article

Now sit down and start writing. You might think of this stage as the breadboard step in building a circuit. Try to write without interruptions, if possible. Prepare a rough draft first, writing as rapidly as possible without any attempt to edit while writing. And don't worry about mechanics such as sentence structure, punctuation, and wording at this stage.

The rough draft is simply an expansion of your outline. Major headings of the outline are broken down into subheadings, and subheadings are broken down into key words. Now write sentences about the key words in the outline, and you have a rough draft. If you can't follow the outline, it's a poor one. Make another and try again.

Start your writing with a solid introduction, which should orient the reader to the subject. Sell your idea in the introduction. Tell why your approach offers advantages. Explain briefly the needed improvements and how you made them. Briefly describe the old method (if necessary) and highlight your improvements. State your conclusion broadly in the opening paragraphs. Prove it later.

The main body of your paper tells your technical story—how the device works, the experimental setup, the method, the analysis. Use it to provide informative copy and to anticipate and answer the reader's questions about your project. The text should describe the operating theory of the device or mechanism covered in the paper, state precisely the object of *your* study and what prompted the study, and describe the method *you* chose to study the subject and explain why this method was chosen over others.

Often it will be necessary to describe the experimental or analytical setup used to obtain the data, because your reader will accept or reject the validity of your results largely on his opinion of how you attained them. For the same reason, it may be necessary to explain how you actually *performed* the experiment or *made* the analysis.

Of course, you must present the details. The scientific approach says you first simply present the data accumulated in your experiment or project—through use of tables, illustrations, words, and statistics—before you analyze the data for your reader and explain how you reached your conclusions.

The conclusion is one of the most important parts of your article. Don't present page after page of technical data and then ask the reader to interpret all of it. You're the author. You must have reached some conclusions. State them as simply and as forcefully as you can, but in more detail than the broad statement given in the introduction.

Remember to acknowledge professional guidance or advice that you received from colleagues on the project. In the bibliography section, list papers, documents, books, and other printed technical data that had significant value in your investigtaion or development. You may also list other important or pertinent publications that provide substantial supporting data for your report. If you quote an author or refer directly to a particular publication, be sure to give proper credit by citing a reference number in the text. After the manuscript is completed, you should write a title and, if needed, an abstract. Abstracts are becoming important in reliable information retrieval. They should emphasize the important *findings* of the report rather than the *techniques* used, and should be informative rather than descriptive.

Smoothing Out the Rough Draft

After all of the ideas have been committed to paper in the form of a first draft, the detail work begins. You must put the manuscript through a number of processing and finishing operations:

Rewrite: Revise the first draft of your manuscript by adding explanatory material and smoothing the transitions between paragraphs and sentences. Rewrite as many times as you can—you'll make improvements every time. If possible, let a couple of days elapse between rewrites so that you can look at the manuscript "cold."

Clarify: Read the draft to see if the meaning of any sentences or paragraphs seems vague. Try to state each new idea as simply as you can. Rewrite long, complicated sentences into two or more easy-to-read sentences. Generally, short sentences contain less noise. Longer sentences are often required when your conecpts are too complex to be expressed effectively in a short sentence, or when short, declarative sentences become monotonous.

Check Punctuation: Proper punctuation is extremely important in technical writing. A misplaced comma or hyphen can often completely

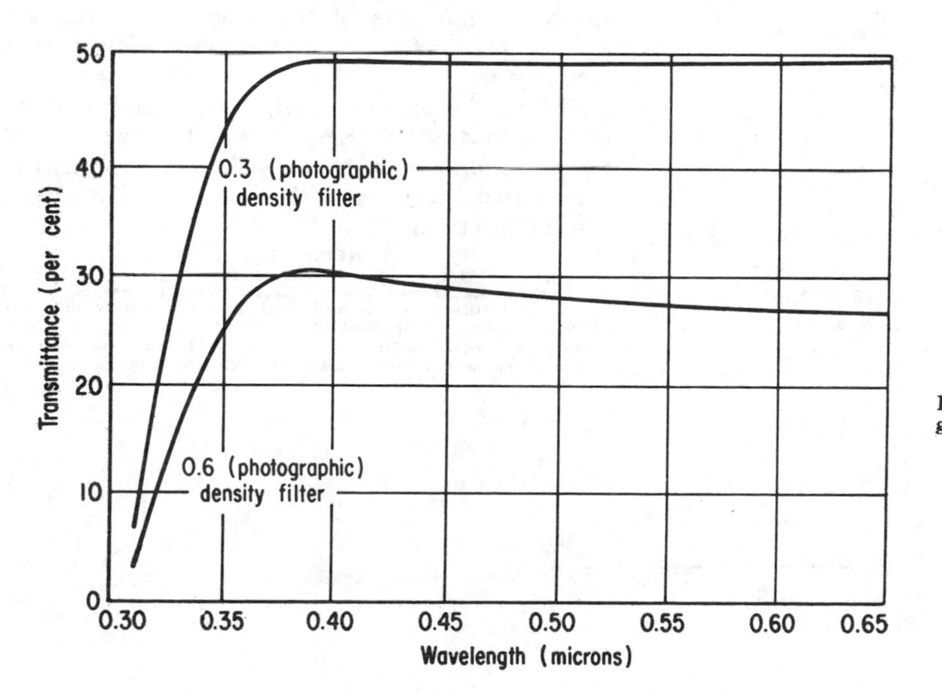

Fig. 3—Example of a good, simple illustration.

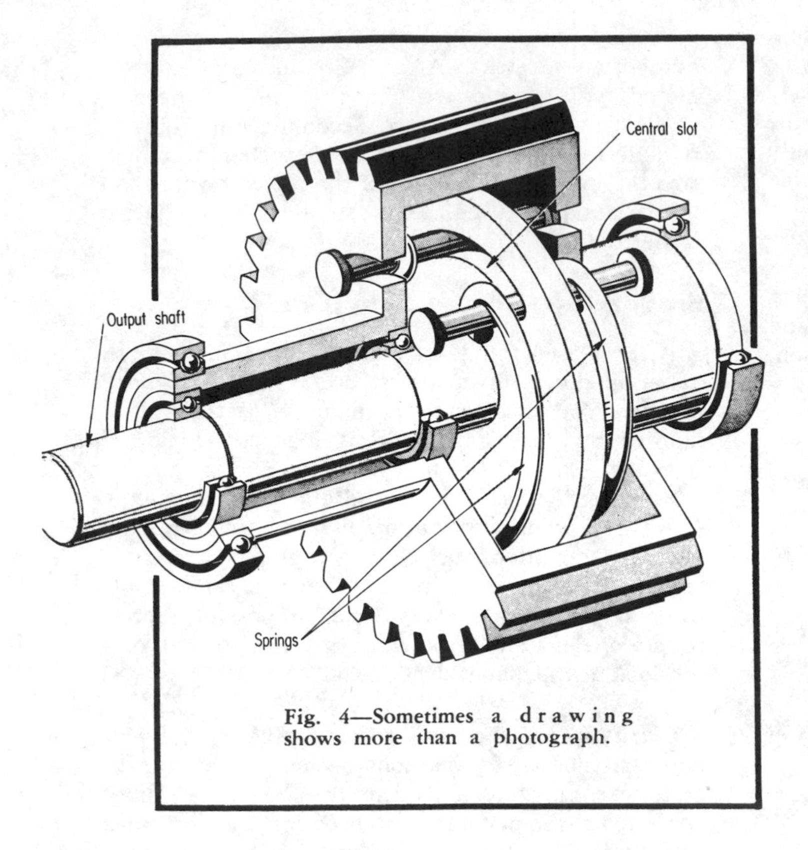

Fig. 4—Sometimes a drawing shows more than a photograph.

change the meaning of a sentence. You can reduce punctuation problems by constructing your sentences so that little punctuation is needed.

Enliven the Writing: Use an active voice rather than a passive voice. For example, instead of saying "x is exceeded by y," say "y exceeds x." Try to avoid weak verb forms such as are, am, was, were, is, and been.

Avoid phraseology. Favor the short word and the specific expression. Instead of saying "at the present time," say "now;" instead of "it is interesting to note," simply say "note."

Watch Your Language: Avoid unnecessary jargon which often impairs meaning. Don't try to avoid technical terms—you're writing a technical paper—but use the most familiar word that is exact. Be sure that your readers will interpret the term the way you mean it. If you have any doubt, however, define the term, even if it adds to the length of your paper.

Even simple words contribute to the noise of our communication, Fig. 2. Combinations of simple words in simple sentences can cause as much trouble. For example, if you say, "Engineers can't publish too many papers," what did you mean? On one hand you might mean that "Papers are so good you just can't publish enough of them." However, you could mean, "Papers are dangerous—don't let engineers publish too many."

Often you can improve your paper simply by eliminating unnecessary words. The redundancy of our language alows constant restatement of an idea, but it is a main cause of noise and often fogs meanings. Keep your writing simple.

Illustrations

Good, clear illustrations should not only support your text, but should eliminate the need for much of it. A good illustration is usually a simple one, Fig. 3. Eliminate unnecessary grid lines on a graph. Label both ordinate and abscissa and, where necessary, the curves themselves. State units of measurement. If a photograph doesn't show what you want, consider a drawing, Fig. 4, because it often does a better job.

The Last Steps

After you've completed your manuscript, ask your supervisor and colleagues to review it. This may help save you later embarrassment over minor details after the article appears in print. Don't let fear of criticism prevent you from preparing an article or paper that you believe is needed, however. Invariably, criticism will help you do a better job next time.

Have your manuscript typed in the form required by the publication to which you are submitting it. If no form is specified, type it double-spaced on one side of the paper. Proofread the manuscript carefully. You are responsible for its accuracy.

Number the pages and clip them together. Don't fold the manuscript. Keep a copy to answer questions the editor may have. Submit the manuscript with a transmittal letter that briefly explains your reason for writing it.

REFERENCES

1. Dwight E. Gray—"Science Writers and Editors vs. Readers: What Do the Former Owe the Latter?" *Proceedings of the 6th Annual Institute in Technical and Industrial Communications*, 1964, Colorado State University, Fort Collins, Col., pp. 51-58.
2. Evan Herbert—"Finding What's Known." *International Science and Technology*, January, 1962, Conover Mast Publications, New York, N. Y., pp. 14-23.

PLACEMENT OF TECHNICAL PAPERS FOR MAXIMUM EFFECTIVENESS

Charles A. Meyer
Electron Tube Division
Radio Corporation of America
Harrison, N. J.

SUMMARY

This paper supplies answers to the many questions engineer-authors have raised with regard to selecting a suitable publication for their technical papers. It includes a description of the three major categories of technical magazines related to the electronics field: (1) the professional society publications; (2) company publications of a professional caliber, and (3) commercial technical magazines. Differences among the publications in these three categories are discussed. Subjects covered include the editorial review and technical evaluation procedures, submission requirements as to illustrations and number of copies, handling of publications, payment policies, availability of reprints, and the like.

The engineer who spends many hours writing a technical paper wants to and deserves to reach a large group of readers. The readers, however, must be a "knowledgeable" group so that the information will not only be of benefit to them but will also bring credit to the author and his sponsoring company.

With the hundreds of technical publications and dozens of professional society meetings each year, the engineer-author often has a problem in selecting the right publication or platform for his material. This paper describes some of the technical publications and their methods of operation so that the choice will be an easier and more profitable one.

Types of Publications

On a broad basis, the technical publications that electronics engineers read and write for can be divided into three categories: professional, company, and commercial. Professional publications are those issued by the professional societies and include the PROCEEDINGS OF THE IRE, the CONVENTION RECORDS, the AIEE ELECTRICAL ENGINEERING, the IRE Professional Group TRANSACTIONS, and the like. The second group includes publications of a professional nature issued by industrial companies. Examples of publications in this group are the RCA REVIEW, BELL SYSTEM TECHNICAL JOURNAL, and IBM JOURNAL. The third group, publications issued by commercial publishing organizations, is the largest and includes publications like ELECTRONICS, ELECTRONIC DESIGN, ELECTRONIC EQUIPMENT ENGINEERING, ELECTRONICS WORLD, AUDIO, and many others.

There are several major differences among these groups of magazines. They differ in readership, emphasis, editorial review procedure, submission requirements, and payment practices. In addition, there are substantial differences among the magazines within the groups. With a knowledge of these differences, the engineer-author will be better able to select the right publication for his writing efforts.

Professional Society Publications

Among the aristocrats of the technical publications field are the major publications of the professional societies, such as the IRE PROCEEDINGS or AIEE ELECTRICAL ENGINEERING. These publications are considered by many to be essentially publications of record. Papers are accepted and published with the expectation that they will be of permanent value as reference material. Papers submitted to a publication in this group are usually reviewed by volunteers from the society membership who are experts in the particular field of the paper under consideration. The review cycle may be fairly long (several months and occasionally much longer), and may include several reviewers. If changes or additions are suggested, the author is asked to provide them. To facilitate the review procedure, the author is asked to submit three to six copies of his paper. Upon its acceptance, he is also asked to submit reproducible copies (including inked master drawings) of his illustrations. The professional societies do not remunerate the author; in fact some of them, such as the Physics societies, make a publication charge of up to forty dollars a page. The publication of a manuscript by a professional society journal provides the author and his sponsoring organization, company, or university with prestige, an intelligent readership, and, of course, a substantial amount of professional satisfaction.

A number of other professional societies besides the IRE and the AIEE publish papers related to the electronics field. Publications of some of these societies are shown in Chart I. This chart shows the publication, sponsoring society, frequency of issue, and approximate circulation figures. The circulation figures are based on average circulation in the first half of 1961 as obtained from STANDARD RATE AND DATA.

Apart from the PROCEEDINGS, the IRE is a very prolific publisher of a wide variety of technical material. It sponsors the TRANSACTIONS of 28

Reprinted from *IRE Int. Conv. Rec.*, vol. 10 (pt. 10), pp. 44-48, 1962.

different professional groups, the RECORDS of its major international meetings, and the publications resulting from a host of regional or specialized meetings. In addition, it is responsible for the Professional Group newsletters, the STUDENT QUARTERLY, and the section publications ranging from newsletters to glossy magazines.

NAME	SOCIETY	ISSUED	CIRCULATION*
PROCEEDINGS OF IRE	IRE	Monthly	76,000
PG TRANSACTIONS	IRE	Varies	Varies
ELECTRICAL ENGINEERING	AIEE	Monthly	53,000
TRANSACTIONS	AIEE	Varies	Varies
PHYSICS TODAY	AIP	Monthly	32,000
JOURNAL OF APPLIED PHYSICS	AIP	Monthly	9,000
REVIEW OF SCIENTIFIC INSTRUMENTS	AIP	Monthly	9,000
ISA JOURNAL	ISA	Monthly	21,000
SCIENCE	AAAS	Weekly	68,000
ASTRONAUTICS	ARS	Monthly	22,000
JOURNAL OF SMPTE	SMPTE	Monthly	
JOURNAL OF AES	AES	Quarterly	3,000
AEROSPACE ENGINEERING	IAS	Monthly	20,000

* Circulation figures are based on average circulation in first half of 1961 as obtained from STANDARD RATE & DATA.

Chart I - Some Professional Society Publications Covering the Electronics Field.

The TRANSACTIONS of the Professional Groups do not necessarily have the same standards and/or practices as the PROCEEDINGS. The papers accepted may be highly specialized and, perhaps, even less comprehensible to the nonspecialists than some of the PROCEEDINGS papers. The individual PG TRANSACTIONS vary in their acceptance and review practices, although some operate on a formal basis very much like that of the PROCEEDINGS. The circulation of most of the Professional Group TRANSACTIONS is under 10,000 copies. The TRANSACTIONS of the Professional Group on Electronic Computers (PGEC) has a circulation of 11,500 at this writing, which is the largest figure for all the Group publications. Most Professional Groups publish their TRANSACTIONS 4 to 6 times a year.

The IRE Section publications usually seek semi-technical or non-technical material. Because of the scarcity of this type of material and a plethora of advertisements, these publications are often obliged to reprint a considerable amount of material from other publications.

Company Publications

A second group of publications of interest more as a reference source than as a publication medium to most engineer-authors is the company or house-organ type publications. Authors who work for companies or organizations having such publications should study them carefully, for they offer a distinct opportunity. Most of these publications are issued to serve a number of important purposes. One, of course, is to present a worthy company image. Another is to provide an outlet for material, that cannot readily obtain publication elsewhere because it is either highly specialized or extremely long. The BELL SYSTEM TECHNICAL JOURNAL, for example, has published many papers on telephone-communications subjects that no other periodical could afford to consider. On the other hand, it has also published pioneering articles in fields that other publications were not ready for at the time.

Further attractions of a house organ are that it can publish promptly, is flexible as to number of pages, and, if necessary, can hold up an issue for an important article. Review and payment practices of these publications vary. The RCA REVIEW uses RCA scientists as reviewers and makes a small token payment to each author on publication. The IBM JOURNAL, on the other hand, pays its reviewers, who are recruited from outside the company, but does not pay its contributing authors. The BELL SYSTEM TECHNICAL JOURNAL and the SPERRY ENGINEERING REVIEW do not pay either the authors or the reviewers a special remuneration.

Commercial Publications

The most varied and the largest group of publications is the commercial group, the magazines issued by commercial publishing organizations. This group is often the most attractive to the engineer-author. The editors vigorously solicit articles. When published the articles are often glamorized by eye-attracting editorial or pictorial treatment, and there is some financial remuneration for the author. Quality standards for this group of publications differ from those of professional society publications, but are not necessarily lower.

The commercial publications have members of their staff review submissions. Although these editors have varying degrees of expertness in technical areas, they usually have a good idea of what their readers want. When submitted articles are not up to the quality standards of the publication or do not fit the space requirements, the editors will, if the subject matter warrants, work with the authors (or in some cases without them) to upgrade the material. In some cases the magazine staff will rewrite the article submitted to a considerable extent. The author may be consulted on the changes, but he is often offered a "fait accompli" in the form of galley or even page proofs and a very close deadline. Although this procedure may disturb the feelings of some authors, the resultant articles are usually a considerable improvement over what was submitted. Occasionally, however, some embarrassing technical errors result from cavalier editing; therefore, the author should review proofs very carefully.

The best and safest procedure is to submit the article or paper in good shape and with the right slant for the particular publication so that no editing will be required and acceptance will be practically on sight. The review cycle can be fairly short. Some of the commercial publications accept or reject articles in less than two weeks. Others take even longer than the professional societies.

A further mark of distinction for the commercial technical publications is that almost all of them pay the author for his material. Payment averages about $25 a printed page, including illustrations. Payment alone is usually not a sufficient incentive for the busy engineer to write an article. Satisfaction, prestige, and commercial position are more significant rewards.

New technical fields have spurred new magazines at a rapid rate even surpassing the growth of the IRE Professional Groups. Chart II lists a number of commercial magazines grouped according to major field. Circulation figures are based on the same source as Chart I. Some of these magazines have specialized in narrow technical areas which have considerable growth potential. In the microwave field, for example, there are the MICROWAVE JOURNAL and the new one MICROWAVES. In the semiconductor field, there are SEMICONDUCTOR PRODUCTS and SOLID STATE DESIGN. In the space-electronics field, there are MILITARY SYSTEMS DESIGN, SPACE AERONAUTICS, and MISSILES AND SPACE. In the data-processing or control field, there are AUTOMATIC CONTROL, AUTOMATION, CONTROL ENGINEERING, and several others.

There is another group of commercial technical publications, not listed in Chart II because they are of practically no interest for technical articles, which are a good source of information on new products and technical literature issued by manufacturers. These publications include INDUSTRIAL EQUIPMENT NEWS, ELECTRICAL DESIGN NEWS, and INSTRUMENT AND APPARATUS NEWS. ELECTRONIC PRODUCTS started out in this category, but has enlarged its scope and is now publishing several technical articles in each issue.

The commercial publications which do not specialize in a limited technical area include ELECTRONIC DESIGN, ELECTRONIC EQUIPMENT ENGINEERING, ELECTRONIC INDUSTRIES (temporarily called TELE-TECH), ELECTRONIC PRODUCTS, ELECTRONICS, and WESTERN ELECTRONIC NEWS. These publications operate in a business-like manner; they usually acknowledge receipt of manuscripts, review them promptly, and publish them promptly. They are not all equally efficient, and even the best have occasional lapses, but on the whole authors find them cooperative and competent.

Another part of the technical publications spectrum includes the journals that are read by service technicians, experimenters, radio amateurs, audiophiles, as well as by many engineers who may double in the above categories. Publications in these groups include ELECTRONIC TECHNICIAN, ELECTRONICS ILLUSTRATED, ELECTRONICS WORLD (one-time RADIO AND TV NEWS), POPULAR ELECTRONICS, RADIO ELECTRONICS, PF REPORTER, AUDIO, and the radio amateur publications QST, CQ, and "73". (It could be argued that QST belongs in the category of professional society publications because it is the organ of the American Radio Relay League, a membership group. It has high standards of technical quality, good editing, and many distinguished contributors, and, like the publications of the Professional Societies, it does not pay its contributors.)

Each of the magazines mentioned above has a distinct personality. Some of them, like ELECTRONICS WORLD and RADIO-ELECTRONICS, have circulations running into the hundreds of thousands. Most pay quite well for articles and may even sustain a "bull pen" of part-time free-lance writers who are regular contributors. The serious engineer with a talent for designing electronic novelities ranging from transistorized metronomes to audio equipment or communications receivers can work up a profitable hobby writing for these publications.

Selection of Publication

Even though the engineer-author is familiar with the publications and literature of the field in which he is working, he does not necessarily know how to reach the readers who will be interested in using or applying what he has developed. For example, the engineer who contributes to the design of a new electron device wants recognition from his scientific peers, the people working in the field of designing such devices. He probably knows how to reach these people because he reads the same technical publications that they do. But more important than recognition from his associates and competitors, he and his employer want and need to reach the people who may buy his device for use

NAME	ISSUED	CIRCULATION*
GENERAL		
ELECTRONIC DESIGN	Biweekly	38,000[c]
ELECTRONIC EQUIPMENT ENGINEERING	Monthly	41,000[c]
ELECTRONIC INDUSTRIES	Monthly	58,000[c]
ELECTRONIC PRODUCTS	Monthly	57,000[c]
ELECTRONICS	Weekly	54,000[a]
WESTERN ELECTRONIC NEWS	Monthly	19,000[c]
MILITARY-SPACE		
MILITARY SYSTEMS DESIGN	Bimonthly	38,000[c]
MISSILES & SPACE	Monthly	24,000[c]
SPACE/AERONAUTICS	Monthly	58,000[c]
INDUSTRIAL		
ELECTRONIC PACKAGING & PRODUCTION	Bimonthly	15,000[c]
ELECTRO-TECHNOLOGY	Monthly	32,000[c]
INDUSTRIAL RESEARCH	Monthly	41,000[c]
PRODUCT ENGINEERING	Biweekly	51,000[a]
RESEARCH DEVELOPMENT	Monthly	34,000[c]
CONTROLS, AUTOMATION, DATA PROCESSING		
AUTOMATIC CONTROL	Monthly	37,000[c]
AUTOMATION	Monthly	33,000[c]
COMPUTERS & AUTOMATION	Monthly	4,000[c]
CONTROL ENGINEERING	Monthly	32,000[a]
DATAMATION	Monthly	31,000[c]
INSTRUMENTS & CONTROL SYSTEMS	Monthly	32,000[c]
SERVICE TECHNICIAN		
ELECTRONIC TECHNICIAN	Monthly	82,000[a]
ELECTRONICS WORLD	Monthly	240,000[a]
PF REPORTER	Monthly	74,000[a]
RADIO-ELECTRONICS	Monthly	156,000[a]
HOBBY, EXPERIMENTER, POPULAR		
CQ	Monthly	92,000[a]
ELECTRONICS ILLUSTRATED	Bimonthly	187,000[a]
ELECTRONICS WORLD	Monthly	240,000[a]
POPULAR ELECTRONICS	Monthly	388,000[a]
POPULAR SCIENCE	Monthly	1,260,000[a]
QST	Monthly	105,000[a]
RADIO-ELECTRONICS	Monthly	156,000[a]
SCIENTIFIC AMERICAN	Monthly	300,000[a]
73	Monthly	34,000[a]
OTHER		
AUDIO	Monthly	25,000[a]
BROADCAST ENGINEERING	Monthly	8,000[c]
MICROWAVE JOURNAL	Monthly	22,000[c]
NUCLEONICS	Monthly	20,000[a]
SEMICONDUCTOR PRODUCTS	Monthly	13,000[b]
SOLID STATE DESIGN	Monthly	20,000[c]

* Circulation figures are based on average circulation in first half of 1961 as obtained from STANDARD RATE & DATA.

[a] Paid circulation.

[b] Paid circulation about 75%.

[c] Controlled circulation.

Chart II - Commercial Technical Publications Listed by Major Field.

in the design and development of new equipment or systems.

At this point the writer crosses paths with the advertising man (the space buyer or media specialist) who faces the same problem in placing ads to reach the maximum suitable readership. In evaluating a publication, the media specialist considers size of readership, caliber of reader, prestige of publication, effectiveness or reputation of editorial content, and of course cost. The engineer-author would do well to consider most of these same factors.

The hypothetical designer of a new electron device could write a paper at several different levels. If the design is based on a significant new concept, perhaps the first paper should be a theoretical one outlining the development of the new concept, describing the reasons for its significance, and perhaps showing mathematically how he would have developed the new concept if he knew exactly where he was heading when he started. Such a paper would be best suited for a professional publication: the IRE PROCEEDINGS, the PHYSICAL REVIEW, or the like.

The next level considered could be a paper describing the possible applications of the new device and giving some performance data based on its use in a typical circuit, piece of equipment, or system. For such a paper, the commercial-type technical publications will show a very strong interest. If the title or abstract of the paper were to be submitted for presentation at a national meeting of one of the engineering societies and it then appeared in the meeting program, many magazine editors would write or phone the author or his company asking for an opportunity to consider the article for publication. When such a request comes to the engineer-author, he should proceed with deliberate care. An impulsive commitment to the first editor who reaches him may not result in the placement of his material where it will have the maximum effectiveness.

The experienced engineer-author decides the publication for which he is aiming before his paper reaches final form. Because most engineers don't write papers often enough to become very experienced in paper placement, however, they must rely on fortune or the assistance of a more experienced supervisor or a publications man.

For the author-engineer who wishes to determine the most effective publication for his article, the Charts accompanying this paper should be helpful as a first step. After the choice has been narrowed down to a few publications having the desired field of coverage, the author can list the possibilities in order of choice based on relative circulation, timing, prestige, or any other factors considered important. One such factor could be, for example, whether or not a periodical has recently published anything on the author's specific subject.

After a publication is selected, engineer-authors sometimes query the editor by phone or mail to determine his interest. This practice may be helpful, but the editor will generally reserve his opinion and suggest that the completed paper be submitted for consideration. Like a good doctor, the editor is reluctant to diagnose over the telephone without seeing the patient. He will, however, give instructions on the magazine's submission requirements as well as other helpful information.

How to Submit

The mechanics of submitting an article to a commercial publication are relatively simple. Required are one clean original double-spaced copy of the paper, a complete set of illustrations - not necessarily reproducible masters - and a letter of transmittal stating that the article has not been published before and has not been accepted by another publication. If a security or proprietary clearance is involved, it is the author's responsibility to obtain it prior to submission.

Submission to a professional society for publication in its major periodical usually involves multiple copies and sets of illustrations. The IRE, for example, requires three sets of each to facilitate its review procedure. In addition, master copies of illustrations suitable for reproduction are also required. These master copies, which should be inked, drawn, and lettered professionally, may be submitted after the paper has been reviewed and accepted.

The receipt of the article or paper will be acknowledged by the publication, often with a form postal card, and then the wait begins. The professional societies may take several months. The commercial magazines will vary considerably. Among the prompt reviewers are ELECTRONICS, ELECTRONIC DESIGN, and ELECTRONICS WORLD. Such publications will accept or reject a paper in a matter of a few weeks and, if the editors have changes or additions to suggest, they will communicate with the author promptly. After an article is accepted, publication may take several months or longer depending on the frequency of the publication. Weeklies, of course, can publish much more rapidly than monthlies.

Once the article is accepted, the author may not see it until it is in print. Usually, however, he will receive a galley proof and an opportunity to evaluate any editorial changes. He will probably not see page proofs or proofs of the illustrations.

The professional society publications move more deliberately. After the review, the interchange of comments, and eventual acceptance, the Society will submit galley proofs, proofs of the illustrations, and often page proofs. The TRANSACTIONS of the Professional Groups of the IRE, however, do not all follow the same procedure as the PROCEEDINGS OF THE IRE. The TRANSACTIONS are edited by volunteers, and their practices vary considerably. Some of the Professional Groups will, like the CONVENTION RECORD and other meeting records, require the authors to submit their manuscripts typed on prepared master sheets which are then photographed and reproduced directly by a photo-offset process. In such cases no editing can be done and final results are entirely the author's responsibility.

Reprints

Once an article or paper is in print, the author may receive many requests for reprints. The professional societies may solicit a reprint order when they send the author proofs to review. The commercial publications usually do not solicit reprints, but will arrange to supply them if requested. Reprints are not expensive, varying in price from about $5.00 a hundred copies and up, depending upon the number of pages. Reprints ordered from the printer of a professional society journal prior to publication are usually the lowest priced.

The average engineer-author is stimulated and encouraged by requests for his article and will want reprints to fill these requests. Often, the company or institution he works for will pay for a quantity of reprints, supply a number of them to the author to take care of requests, and distribute the bulk of them where they can contribute to the company's prestige or commercial interests.

Most business organizations recognize the benefits to them, as well as to the authors, of getting good articles published in the technical periodicals. The appearance of a well-written article or paper in an appropriate publication is by far the most effective way of reaching a large group of readers eager for knowledge of technical advances in their special fields of interest.

An Equilibrium Theory for the Meeting Paper

J. H. HEALD, JR.

Abstract—**Some theoretical guidelines are given for answering such questions as: How long should the paper be? Before what audience should I present it? And, how far should I go to give it?**

THERE IS A well-known dictum facing today's aerospace engineer which is often popularly phrased as "publish or perish." Accordingly, the publication and presentation of our accomplishments does occupy a substantial portion of our time and effort.

Unfortunately, there exists a very real technology gap between the techniques we use in engineering our work and in presenting the results. For example, the modern aerospace engineer can now handle such heretofore unmanageable problems as radiative heat transfer through nonequilibrium boundary layers, while the modern-day flow-field analyst loves nothing better than to sink his teeth into a juicy nonlinear second-order partial differential equation.

However, nowhere in the literature of our field can one find answers or theoretical guidelines to such professionally critical problems as: How long a paper should I write to justify a conference trip to San Francisco? Is my salary in keeping with the length of my paper? And what determines the proper conference at which to present my paper?

As a first step toward reducing this technology gap, I propose a simple equilibrium theory to answer these and many other conference paper questions.

Consider the equilibrium condition when the value of a technical paper being presented at a conference is equated to the value of the author presenting the paper. In symbolic notation, the equilibrium condition may be represented by

$$V = CT \tag{1}$$

where V is the value of the paper in dollars, C is the hourly cost rate of the author, and T is the total time involved.

The proper expressions for V, C, and T must now be developed. The total time involved can be simply stated as

$$T = t + l \tag{2}$$

This paper is reprinted from *Astronautics and Aeronautics*, December 1968.

The author is with Arnold Research Organization, Inc. (ARO), Arnold Air Force Station, Tullahoma, Tenn. 37389.

where t is the travel time to and from the conference in hours, and l is the length of the conference in hours. The travel time t can also be expressed as d/S where d is the total distance traveled round trip to and from the conference, and S is the average travel speed in miles per hour.

The total time can then be written as

$$T = d/S + l, \tag{3}$$

while the hourly cost of the author is simply

$$C = P_a + D \tag{4}$$

where P_a is the author's pay rate in dollars per hour and D is the per diem rate in dollars per hour.

The effective value of the paper being presented can then be expressed as

$$V = L/R \times N \times P_l \tag{5}$$

where L is the length of the paper in pages, R is the average speaking rate in pages per hour, N is the number of people listening to the paper, and P_l is the average salary of the listeners.

Substituting (3), (4), and (5) into (1) give the equilibrium condition as

$$L/R \times N \times P_l = (P_a + D)\ (d/S + l). \tag{6}$$

This relation can then be solved to evaluate any one of the parameters for its equilibrium value. As an example, solving for d, the distance traveled gives

$$d = S[L \times N \times P_l/R(P_a + D) - l]. \tag{7}$$

As a numerical example, assume you write a 10-page paper to be presented at a 3-day conference (24 working hours) attended by 200 people whose average salary is \$10 000 (\$5 per hour). If your average reading rate is 40 pages per hour, your salary \$10 000 (\$5 per hour), your per-diem allowance \$2 per hour (based on an 8-hour day), and you travel at an average speed of 200 mph (a combination of airline and taxi), then (7) gives a travel distance d of 2200 miles, or 1100 miles one way.

If you wanted to go to a conference a little farther away, (7) makes it quite clear what you have to do to increase your travel distance. First, you can write a longer paper; increasing L increases d. You can pick a conference attended by a larger, higher paid audience, which increases N and P_l. Another possibility is to pick a shorter conference, since reducing l increases d. If you can establish a reputation for speaking more slowly, then reducing R will

Reprinted from *IEEE Trans. Engrg. Writing and Speech,* vol. EWS-12, pp. 18-19, May 1969. (Reprinted by permission of the American Institute of Aeronautics and Astronautics from *Astronautics and Aeronautics,* Dec. 1968.)

also increase d. Or, in final desperation, you can take a cut in pay.

Now the applicability of this theory is further verified by its ability to predict certain naturally occurring phenomena. First, it points out the advantage of speaking slowly, hence increasing the travel distance. Likewise, it is advantageous to speak before a large, high-paid audience. This not only increases the travel distance but certainly encourages the presentation of higher quality material. Also, it is noted that the theory is a strong argument for an increase in per-diem allowance. This would require writing longer papers for a given travel distance, hence increasing the technical reporting output of the author's company.

Equation (7) also provides a useful means of determining the state-of-the-art efficiency of technical reporting. For the example given above, the travel efficiency is about 11 miles per word. (Considering 200 words per page.) As an historical comparison, by best estimates Paul Revere did no better than 0.01 miles per word during his famous 20-mile trip. By the age of Lincoln, the efficiency had significantly increased to about 1 mile per word as evidenced by Lincoln's own 200-mile trip to present a 200-word report at Gettysburg. Efficiency appears to be improving at an ever-increasing rate, and the present record was established in December 1968, when three men travelled 240 000 miles into space for the terse report "A-OK."

As another example of the power of this equilibrium theory, (6) is now solved for the author's own salary P_a, which gives

$$P_a = L \times N \times P_l / R\,(d/S + l) - D. \qquad (8)$$

This relation allows one to examine his equilibrium salary status. For example, if your boss routinely asks you to travel to a conference 1000 miles away to deliver a 10-page paper at a 3-day conference attended by 200 engineers whose average salary is $10 000, then your salary should be $11 000 per year. If your salary is less than this, you are obviously underpaid for the work you're doing and are justified in asking for a raise. It is expected that much

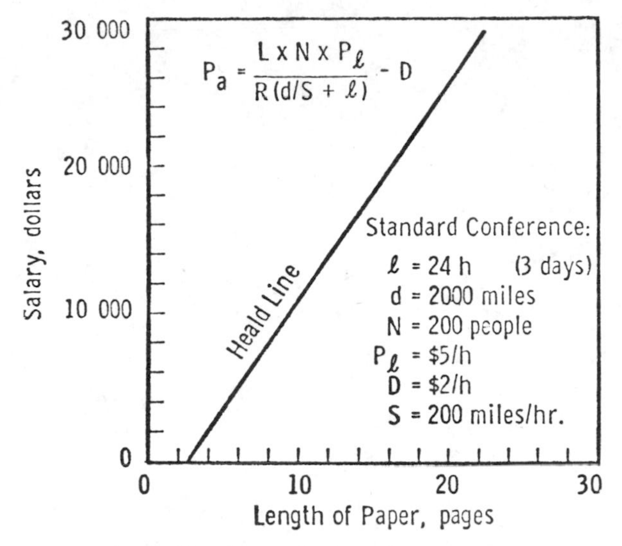

Fig. 1. Heald's standard conference curve.

reference will be made to this theory on forthcoming employe performance evaluation sheets. As a result, reprints of this article will be made available on request.

To help demonstrate the usefulness of this theory (8) is graphically represented in Fig. 1 for the conference mentioned above. Plotted along the vertical axis is the author's salary and along the horizontal axis is the length of the paper required to attend this standard conference. For a given salary, the appropriate length paper can be immediately determined. Very modestly, the line representing this equation is called the Heald line; and it can be seen that, for a paper of given length, if your salary falls above this line you are obviously well-Heald.

The power and validity of this theory is further substantiated in one last observation. For the first time we have theoretical verification of a technical reporting phenomenon which has long been observed. This is the observation that the higher paid personnel in any organization are the long-winded talkers.

And so, in keeping with my salary, I must now conclude.

Part II
The Rhetoric of Papers and Articles

THE ROLE OF RHETORIC, which for our purposes may be defined as reasoning, in modern communication has its antecedents in the teachings of Aristotle. Reasoning, in turn, has two principal components that require the technical communicator's careful attention: the relationship between the writer and the reader, and the relationship between thinking and writing. Both of these issues are very complex and often very theoretical. Yet the articles in this section afford a balance of theoretical and practical information while clearly explaining the significant role rhetoric plays in producing first-rate documents.

Sawyer begins this section with an analysis of the unique "rhetoric of the scientist." The rhetoric of science must deal with the "explosion of information" and must determine how most effectively to communicate scientific and technical concepts. In so doing, there are problems that must be resolved. Davison provides information on three categories of problems that can develop between the writer and the reader and offers advice on how to eliminate these problems. Lufkin focuses on the need to maintain a high quality in our reports; he discusses how quality requires clarity as well as correctness and that to achieve this the writer must always consider the reader's point of view.

The next two articles examine the relationship between thinking and writing. Bernheisel suggests that writing should not be done only after a project is completed, but should proceed in step with the design process. The attention to detail, logic, and precision required in writing will benefit the project by causing the writer to see the project in a new and clearer light. Similarly, just as Bernheisel sees writing as an aid to clearer thinking, so Lang—who also emphasizes logic and precision—argues that clear thinking is an aid to clearer writing.

In the following three articles, this relationship between thinking and writing is explored in terms of the report's organization and style. Perkins, in a five-step sequence, discusses the logic of each of the report's sections and shows the sections' relationship to the steps involved in problem solving. Marder adds to this discussion of the report's internal logic; he states that the "methods of science and exposition are parallel" and therefore the principles of rhetoric offer a natural means of organizing. Miller then discusses how context (the situation surrounding a statement) operates in communication; she provides valuable insights into the dynamism of communication and adds to our understanding of how reports can achieve the "efficiency that technical communication seeks."

We end this section with Michaelson's article, which, in effect, synthesizes the main points made by the other articles in this section. He proposes that structure, content, and meaning are interdependent and that these, not just literary style or technical merit, are the criteria for gauging the success of our reports. His conclusion is one this section urges us to remember: "the paper of quality...will have the earmarks of fresh insights and of the inspired thinking that characterizes good work."

All told, this collection of articles is a strong indication that the ancient art of Aristotle still plays a vital part in today's world.

Rhetoric in an Age of Science and Technology

THOMAS M. SAWYER

I SUPPOSE THAT I NEED present no argument that we are indeed living in an age of science and technology. The very fact that we are the first people in history to have viewed, via electromagnetic signals, men driving a vehicle on the surface of the moon over 200,000 miles away in space would seem to be proof enough.

Despite this evidence, I believe that many people teaching English at the college level today consider this and similar accomplishments to be the results of the labors of a relatively small number of specialists trained in esoteric skills while the majority of college students are headed for the more comprehensible humanistic disciplines of Law, Art, Philosophy, History, Foreign Languages, Education, and Literature. Such an impression is understandable when we remember that most English teachers come in contact with a representative sample of the student body only in the freshman composition course, before these students have decided upon a field of specialization. The upper-level courses in English departments are usually devoted to the study of literature, are required of majors in English, and are often elected as cognate courses by students planning to go into some other branch of the humanities. The number of students specializing in science and technology that the English teacher meets in his classes tends to be rather small and it is natural that he should assume that training in the analysis of literature and of essays devoted to social criticism, which appears to be the focus of most freshman composition courses, will be useful and pertinent to the majority of the students since they seem to be headed for the humanities.

However, it is a fact that the majority of college students will specialize in disciplines like Mathematics, the Social and Biological Sciences, Business, and Engineering. Moreover, the professional rhetoric that graduates of these disciplines will be required to employ is in many ways quite different from the rhetoric taught to them when they were freshmen. For them the freshman composition course has only marginal utility.

Each year the U. S. Office of Education reports on the number of degrees awarded by institutions of higher education. In 1968-69 about 729,000 Bachelor's degrees were awarded. Of these, a little over 300,000 were in the humanistic disciplines like those I have listed above, while nearly 400,000 were in the scientific and technological disciplines. If we consider all degrees—Bachelor's, Professional degrees, Master's, and Ph.D.'s—the total in 1968-69 approached 1,000,000 and the ratio of humanities graduates to science and technology graduates remained constant at all levels. Fifty-six percent of all degrees granted in 1968-69 were in science and technology and 44% were in the various humanities.

I was interested to note that about 1,200 of the Ph.D.'s or 7%, were in English and Literature. In view of the present job market for English teachers that now appears to be about 1,000 too many. But most of these Ph.D.'s were trained in literature. The demand for people trained in rhetoric, especially in the sort of rhetoric employed by the 56% who received degrees in science and technology, remains high and is likely to increase.

Why do I forecast an increasing market for people in rhetoric? First of all, because these scientists and technologists do a lot of writing; in fact, they may write more on the average than do graduates of the humanities, and secondly they are much concerned with improving their writing. Just out of curiosity I examined the *University of Michigan Bibliography,* which lists the publications of the faculty. I should point out immediately that this is not a very accurate measure of faculty publication because the editor has some difficulty in getting everyone to reply to his questionnaire. Moreover, I checked the listings for only three departments and counted only the entries, paying no attention to the number of pages. Even so it appears that in 1968 a little over 50% of the professorial staff of the English Department published something, while 64% of the professors of Psychology and 73% of the professors of Physics listed publications. The English Department produced 88 articles, poems, reviews, and 12 books, for a total of 100 entries; while the small Physics Department produced 122 articles and abstracts, and the Psychology Department produced 185 articles and 11 books for a total of 196 entries—nearly twice as many as the English Department. You will note that the English Department produced more *books* than either Psychology or Physics, an interesting feature of scientific and technological rhetoric to which I will refer a little later.

Now because we are producing so many graduates in science and technology, and because they write so much, we are caught in what many scientists have called an "explosion of information." Please note that word carefully—an explosion of *information,* of new facts, of new knowledge. I think few people in the humanities have really given this explosion very much consideration, but the scientists have. They have devoted at least two international conferences to it, and interestingly enough, the proceedings of the first conference sponsored by the Royal Society in 1948 could be bound up in one relatively slim volume, while the proceedings of the second conference in 1958 sponsored by the National Science Foundation had to be bound in two thick volumes. Not only had the amount of information increased in those ten years, the number of people concerned about it and the number of papers dealing with it had doubled. The reason for their concern is that these new facts will affect *our* lives, not just the lives of our children and grandchildren, but *our* lives, for the simple reason that the time lag between theoretical formulation and practical application has been decreasing steadily. It was forty years between Einstein's Special Theory of Relativity and the atomic bomb; it was twelve years between Sputnik and Neil Armstrong's footprint on the moon; and it was only seven years between John Bardeen's discovery of the principle of the transistor to the solid-state radio and TV.

Some idea of the tremendous size of this information explosion may be gathered from the special issue of the *Times Literary Supplement* of 7 May 1970. In that issue, D. J. Urqhart, of the British National Lending Library for Science and Technology, wrote:

> Last week the N.L.L. (National Lending Library) received about 8,000 new, separate items containing probably more than one million pages in perhaps 50 languages. This week the N.L.L. will almost certainly receive another million new pages. No private mansion in the kingdom would be able to accommodate all this material. No one man could possibly read everything and he would have the gravest difficulty in selecting the small fraction of this material which was of special interest to him. . . .

> Today there are more than 2,000 periodical publications devoted to abstracts and indices of scientific literature. More periodicals are now concerned with this activity than the total number of scientific periodicals which existed 100 years ago. . . . A simple measure of the rate of increase of the output can be obtained by looking at the Royal Society Catalogue of Scientific Literature. This eventually covered all the scientific literature in all subject fields in the 19th century. It occupies about the same shelf space as last year's output of Chemical Abstracts which relates to one year's output in chemistry alone.

To bring this point home, I wonder how many of you know what *holography* means? How many of you have seen a *hologram?* When I completed the draft of this paper, I had no idea that on that very day the Swedish Academy of Science would award the Nobel Prize in Physics to Dr. Dennis Gabor for his theoretical work leading to holography. Despite the fact that the resulting publicity should have made the word *holography* better known, I repeat my question: how many of you know what *holography* means? We now have in Ann Arbor at least three different companies actively engaged in producing and selling holograms and holographic techniques, both in the United States and overseas. Ann Arbor has become the holography center of the United States since Professor Emmett Leith of the University of Michigan perfected the technique less than ten years ago, and it may not be long before further improvements will enable you to have three-dimensional, as well as color, television in your homes. But I have long since ceased to be surprised that practically no visitor to Ann Arbor has even heard of the word, let alone seen a hologram.[1]

This explosion of information has led to another sort of problem that is beginning to be of serious concern to the scientist and technologist. Because new facts and new devices are appearing in such profusion, the ordinary laymen simply cannot digest them all. They begin to feel bewildered and confused. Where are science and technology taking us? And because they are bewildered and confused, some of these people are resentful, suspicious, and antagonistic. Like the Luddites of the Industrial Revolution, they seek to break up this new machinery, to bring science and technology to a halt.

In a recent article published by the United Nations Educational, Scientific, and Cultural Organization (UNESCO), Dr. Paul Couderc, formerly head astronomer of the Paris Observatory, points out that this is a dangerous situation—not because science will be hurt, although this antagonism will indeed hurt science—but because it leaves us laymen at the mercy of unscrupulous charlatans who are always ready to prey upon human credulity and ignorance. Some way must be found, he argues, to make the findings of science known to everyone. And in the *Times Literary Supplement* to which I have previously referred, Sir Rudolph Peierls writes:

> In this situation it is more important than ever to have access to good clear expositions of knowledge on many levels,

[1] Earl Ubell has an article on the subject of holography in the *New York Times* Sunday edition of November 7th in which he reports Dr. Gabor as saying that holography is best left a mystery, simply because the average person needs a great deal of arcane information about light and its properties to understand the holographic techniques.

I am not in the habit of picking quarrels with Nobel Laureates, but it seems to me Dr. Gabor is adopting a defeatist attitude. I was hired to write an explanation of the high resolution radar which Emmett Leith and his colleagues developed. It wasn't easy, but the resulting paper seemed to satisfy both the radar men who developed it and the laymen who read about it. Holography is only a few steps further along the same track, and I think it can be explained in terms that anyone can understand.

> ranging from reviews for experts in one branch of a subject to broad and readable literature for non-experts.

And on the same page, C. P. Snow, who first called our attention to the gap between the two cultures, adds:

> . . . the old academic resistance to popular exposition is now not only outdated but anti-intellectual; and it is the duty of any decent scientist to take an interest, and if possible to play a part in serious popular exposition . . . it begins to sound as though this is the present climate of the subject.

It is in this serious, popular exposition of science and technology that I believe the teacher of rhetoric and composition can play a useful and important part. I have been asked on several occasions to assist my colleagues in the College of Engineering and in the science departments of the literary college to explain for the benefit of the laymen who have contracted for their expert services precisely what it is they have accomplished. I have invariably found these engineers and scientists receptive to my suggestions and grateful for my help. Thus I am persuaded that the demand for people who can teach students of science and technology how to explain what it is they are doing to the educated layman, while at the same time satisfying their professional peers, is bound to increase. My own department has recently suggested to the graduate program advisors of the College of Engineering that we might be useful to them in teaching their Ph.D. students how to write more readable dissertations, but we were embarrassed to discover that the response was so enthusiastic that we could probably not supply sufficient staff to satisfy the demand and a shortage of funds might prevent us from acquiring additional staff.

I am convinced that it is impractical to offer during the freshman year the sort of rhetorical training required to produce readable scientific exposition. Freshmen simply do not have sufficient background in science or in technology. For this reason, I have argued that training in composition should be moved to the senior year. Those of you who did not hear my argument at the NCTE meeting in Washington in 1969 may read it in the April 1971 issue of the *Journal of Technical Writing and Communication.*

Perhaps a more difficult objection to overcome is that an English teacher without training in science and mathematics cannot be expected to deal with the rhetorical problems of physicists, mechanical engineers, or medicinal chemists. I confess that I was much worried about this when shortly after the beginning of my first attempt to teach our senior-level course in scientific and technical communication a student in electrical engineering loftily rejected my comment that I couldn't understand what he had written with the rejoinder: "Well, I wrote this for other engineers, not for an English teacher. They will understand it even if you don't." What could I say? Out of desperation I reproduced his paper and distributed it to the rest of the class, and fortunately the class was a mixed bag of civil engineers, naval architects, mechanical engineers, and so forth. To my great relief, none of them understood what he had written either.

This experience taught me an important lesson. A layman is simply someone from a different discipline. The corollary is this: if a non-scientific English teacher can't understand the explanation, the chances are excellent that a scientist or engineer from a different department will be just as baffled.

Now the important question is: how does one approach the problem of clarifying scientific and technological concepts? It seems to me that the rhetorical principles involved are considerably different from those ordinarily taught in most composition courses.

As Tom Wilcox and Bonnie Nelson have pointed out, it is extraordinarily difficult to pin down what is being done in freshman composition, but I will assume that most English teachers, trained as they are in literature, and faced as they are by freshmen who have only limited experience to draw upon for their writing assignments, will naturally incline toward the critical analysis of literary works as theme subjects. I don't object to this at all; in fact I would encourage it. I believe that no college student should be permitted to graduate without a firm background in the literature which makes up a large part of our culture, though certainly not the whole of it.

But it is a serious mistake to dilute such important subject matter with considerations of the comma splice, the run-on sentence, and the dangling modifier. If we think literature is important for freshmen, why not come right out and admit it? Why not simply state: "We assume that anyone with an educational background sufficient for admission to this institution can write a grammatical sentence and a well-organized paragraph. If he can't, he doesn't belong here. We are going on to more important things. No college graduate should be ignorant of the drama of Oedipus upon which Freud draws so heavily, or of Aristotle's analysis of the weaknesses of a democracy as opposed to a republic, or of Descartes' argument leading to his statement 'Cogito, ergo sum'. We have our hands full just trying to introduce him to the literature of our culture. Our colleagues teaching in the high schools and elementary schools are fully competent. We trust them to send on to us only those students who can express themselves competently in the English language."

Much as I enjoy reading and analyzing literature, and much as I would urge you to make it the whole subject of the freshman course, it will not help the scientist and technologist explain his subject matter. The rhetorical principles are different. It seems to me that most of our memorable literature is based upon a form of dramatic confrontation. We pit one man, or group of men, against another man, or group of men. We pit Achilles against Hector, the Greeks against the Trojans, Saint Joan against the clergy and the feudal barons, Robert Jordan against the Spanish fascists. The antagonist and protagonist seek to outwit, thwart, and defeat each other.

But science and technology are simply not dramatic. Nature does not confront us. Nature is neutral. The universe is benign, Augustinian; not evil or Manichean. For this reason we have very few literary examples of man pitted against Nature. Jack London's *To Build a Fire,* Stephen Crane's *The Open Boat,* and Hemingway's *The Old Man and the Sea* come to mind, but I am hard pressed to think of other examples.

A scientist is not a protagonist; he is more like the pilot of a small ship at sea. The sea has no designs to thwart him; it doesn't even notice him. But let our pilot make an error in his navigation, and the sea will do nothing to help. It will remorselessly drive him further and further off course, but if he never makes port, it is because of his own ignorance or negligence, not because of the machinations of the sea.

So scientists and engineers are concerned with the intensely practical and mundane problems of how to stay afloat and how to reach port safely, rather than with intelligence estimates of an enemy's capabilities and intentions. This is not nearly so interesting perhaps as plot and counterplot, but it is important to us because we are passengers on that ship and if our pilot misses his star sight, we are likely to end up where we didn't want to be, wishing we had studied some navigation too.

I am convinced that my colleague, Professor W. Earl Britton, put his finger on the really important distinguishing characteristic of scientific writing when he noted in his chapter in the book *On Writing By Writers* that technical or scientific rhetoric is functional. It is meant to guide action, to be acted upon. For this reason it must convey one meaning and one meaning only. It cannot be ambiguous; it cannot be open to more than one interpretation. There must be no room for doubt. In Britton's words, it might be a bugle call, not a symphony.

It would appear at first glance that it should not be hard to devise a terminology which is non-ambiguous, is open to only one interpretation, but the fact is that it is very difficult and the way in which it is achieved is an important characteristic of the rhetoric of the scientist. As Hannah Arendt has pointed out, scientists *know* through *process*. A term for them has meaning only if it is expressed as a unique set of operations which will physically produce the referent, the object being defined. How does the physicist know that the atom, once considered the smallest unit of matter, consists of a number of even smaller particles? He knows this because he can perform a set of operations which will actually produce them or their effects. Why does the physicist believe that the ether, which Huyghens postulated as necessary for the propagation of waves of light, does *not* exist? Because no operations can be performed which will produce ether, and no operations will produce its presumed effects. Michelson's experiments designed to produce the effects of ether on the speed of light have sometimes been called the most famous negative results in the whole history of science. The existence of the neutrino, which Fermi postulated in order to balance his equations, is still in doubt, although Frederick Reines and his colleagues believe that they have found a set of operations which will demonstrate that the neutrino is real. For several years I have been pleading with Macmillan to re-issue Percy Bridgman's *The Logic of Modern Physics,* which discusses operational definitions in detail and in terms that I think a reasonably curious layman can understand. Perhaps you will help me in my efforts to get that book back in print.[2]

Now if a unique sequence of operations will permit us to produce a phenomenon invariably, we have a natural tendency to give the phenomenon a name, a symbolic label. It simply makes it easier to remember and easier to talk about. We can describe all the steps of the operations George Simon Ohm went through with his compass, wire, and battery, but it is much simpler to describe the phenomenon he invariably produced with the phrase: Ohm's Law.

Once we have assigned a label to a phenomenon, it begins to take on the characteristic of the middle term of the major premise of a syllogism. For example, if we go through a series of operations immersing electrodes in water, we invariably produce two gases. We choose to label one oxygen, the other hydrogen. Our labels now naturally lead to the premise: all water is composed of two parts of hydrogen and one part of

[2] Professor Zoellner's rejection of the behavioral objectives offered by the NCTE strikes me as well founded because they cannot be operationally defined. For this reason I cannot resist offering my own simple, reliable, operational definition of a competent teacher of composition. Let the composition teacher instruct his students in the writing of instruction manuals. As a final examination, ask the students to write a manual explaining how to wire up an electric chair without injuring the strapped-in occupant. Seat the composition teacher in the electric chair and strap him in. Hand the manual written by one of his students to a reasonably stupid freshman. If the composition teacher avoids electrocution, he is defined as competent and rewarded with tenure. If there is a shower of sparks and the smell of burning, he is defined as incompetent and his department may use the salary vacated by his demise to employ someone else on a trial basis.

oxygen. Notice how easy it now becomes to deduce a host of conclusions by combining this premise with others similar to it. All water will decompose under electrolysis into hydrogen and oxygen. All salt will decompose under electrolysis into sodium and chlorine. All four of these elements combine with other elements. Therefore, it should be possible to rearrange these natural combinations into new ones, to combine hydrogen and chlorine and to combine sodium and oxygen. It was by deductive reasoning something like this that Mendeleev was able to devise the periodic table of elements and predict the existence of elements which we have only recently been able to actually produce.

It is important to notice that you cannot construct such a deductive network unless you can begin with a sharply defined phenomenon, something that happens invariably under the same conditions. Such phenomena are much more common in the physical sciences than in the biological. One has only to recall the Harvard Law of animal behavior: "Under the most rigorously controlled experimental conditions that can be devised by the human experimenter, the animal will do just as he damn well pleases." And, since human beings are just as ornery as animals, invariant human phenomena in the social sciences are also relatively rare. Almost all philosophers of science seem to agree that the physical sciences are the most developed; that is, they can predict with great accuracy what will happen as the result of certain operations. The biological sciences are less well developed, they can predict less accurately; and the social sciences are the least developed and least capable of accurate prediction; but both are striving to achieve the deductive character and the predictive accuracy of the physical sciences.

Once the scientist has established a deductive system, two other rhetorical characteristics begin to emerge. First, he often takes it for granted that his reader understands the major premises with which he began. It seems perfectly natural to speak of the Pythagorean theorem without going back to explain what it means, just as it is natural to refer to Ohm's Law without going back to describe the operations which Ohm went through to arrive at this law. As Thomas Kuhn has pointed out in his *The Structure of Scientific Revolutions*, the book which explains the paradigm, the basic premises, has tended to disappear in the developed sciences. Those working in the field don't need it; they already take the premises for granted. It is the report of the experiment which tests the deduced conclusion which matters to them. It is only the layman who needs to have the chain of deductions leading up to the experiment mapped out for him. This is why the Physics Department, working within a deductive system, produced no books, while the Psychology Department, still in the inductive, observational, and natural history stage of their science, produced eleven books.

And once the scientist has established a deductive system he naturally begins to use symbols to keep track of his chain of propositions and deductions. It is time-saving. He need not write out whole statements in words each time. Thus he can say $H_2O = 2H + O$, and Ohm's Law becomes $E = I \times R$. This is the reason the most developed sciences are so mathematical.

I don't speak mathematics, but I often wish I did. It seems to me that it is the poetry of science. It is in mathematics that a scientist dreams and engages in flights of fancy. I can only get a glimmer of the delight they must feel through such books as Tobias Dantzig's *Number: The Language of Science*, which Albert Einstein called one of the best books on mathematics he had ever read. Dantzig illustrates, for example, the new horizons of imagination opened up by Vieta and Descartes, who chose to use the letter X to designate the unknown in algebra. It sounds like a simple thing, but here is how Dantzig describes it:

> In vain, after this will one stipulate that the expression a-b has a meaning only if a is greater than b, that a/b is meaningless when a is not a multiple of b . . . The very act of writing down the *meaningless* has given it a meaning; and it is not easy to deny the existence of something that has received a name. . . .
>
> What distinguishes modern arithmetic from that of the pre-Vieta period is the changed attitude towards the "impossible." Up to the seventeenth century the algebraists invested this term with an absolute sense. Committed to natural numbers as the exclusive field for all arithmetic operations, they regarded possibility, or restricted possibility, as an intrinsic property of these operations. . . .
>
> Today we know that possibility and impossibility have each only a relative meaning; that neither is an intrinsic property of the operation but merely a restriction which human tradition has imposed on the field of the operand. Remove the barrier, extend the field, and the impossible becomes possible.

It is not surprising that most scientists and engineers talk to each other, and talk to their students, in mathematics. And very naturally their students start to use the same mathematical language when they try to explain things to a layman. To a layman like me who still counts on his fingers and has trouble balancing his checkbook, this is terribly discouraging. I am easily baffled by mathematics. But then I recall Einstein's warning that in order for a mathematical statement to have meaning in the real world, it must refer to phenomena that can be sensed; it must eventually refer to something we can see, touch, taste, smell, or hear. And I also recall that the physicist, Otto Frisch, wrote in his book, *Modern Physics Today,* that anything you can say in mathematics you can also say in English, although it may take you several pages to explain what is contained in one mathematical formula.

With this in mind, I simply ask my engineering student what his mathematical symbols mean—in English. His first answer is likely to be of no help at all. He will probably reply, "Why *V* stands for velocity, *T* for time, and *D* for distance." But now, remembering that these concepts *must* be based on operations, I simply ask him *how* he measures each one. It will soon become clear that each of these terms will be defined differently in different contexts. One can use a tape measure to define (D), the distance of a broad jump, but how does one measure (D), the distance to the moon? One can use a stop-watch to measure (T), the time it takes to run a mile, but Galileo, who had no stop-watch, measured (T), the time it took cannon balls to roll down an inclined plane, in terms of the weight of water pouring out the spigot of a bucket. His thumb turned off the flow of water just as one would stop a stop-watch, and he was accurate to within one-tenth of a second.

Two final comments. There is one rhetorical device commonly used in the humanities that I would urge scientists and engineers to avoid; that is analogy, or simile and metaphor. There is a great temptation to try to demonstrate that a new and unfamiliar phenomenon is analogous to something with which we are already familiar. The electrical engineer feels it may be helpful if he says that electricity flows like water in a pipe. But in the long run such an analogical explanation gets him into deep trouble. He is brought up short when he comes to induced currents and to vacuum tubes and transistors. Now he has to confess that electricity is nothing at all like water in a pipe, and he has to go back to the very beginning and start all over again. He would do much better to stick to operational definitions.

And it may come as something of a

surprise to people who are not familiar with technical writing to learn that scientists and engineers are multi-media conscious. There are some things that are exceptionally difficult to describe in words but which can be clearly illustrated in the graphic arts. The editorial office of nearly every research and development concern employs a group of illustrators and photographers. But alas, we find that the tendency to use abstract symbols gradually invades even this art. As Richard L. Gregory points out in his book *The Intelligent Eye,* over time scientific illustrations become less and less literal representations of the equipment involved and, like the evolution of Egyptian pictograms, become more and more abstract and symbolic until today the radio engineer's circuit diagram is probably incomprehensible to most of us.

In conclusion, this functional, non-ambiguous, operational, bugle call rhetoric of science and technology means that scientists the world over can communicate accurately and precisely with one another without having their meanings distorted. Operations can be performed anywhere—given adequate equipment. A scientist in Japan or India can communicate precisely and unambiguously with his colleagues in Belgium or Brazil. He need only say, "Perform the following operations and you will see exactly what I mean." And the tests of the truth of a scientific deduction are the same in every country. In a very real sense the scientific and technological community is probably the only true international community in existence today. Those Luddites who would seek to hamper and destroy it are making a serious mistake. People sharing common standards and speaking a common language are our best hope for international peace and tranquility. These people now need our help to bring others into an understanding of their community. It is time we came to their aid by helping them and their students to communicate their knowledge to the public at large.

Humanities Department
College of Engineering
University of Michigan

Anatomy of Uncommunicativeness

Roy Davison

Executive Assistant
Technical Services, System Development Corp.

Very much in evidence in meetings of this kind is the assumption that people wish to communicate better. Of those persons who clearly do not wish to communicate better, we say, "Well, they ought to! " and sail on as though we had in fact shot every cloud out of our blue and euphoric sky. True, we talk about barriers to communication. But the barriers that we talk about are those that operate below the level of awareness (like prejudice, or a failure to tie to the listener's experience) or else are barriers that we are struggling to surmount (like shortages of time or money with which to produce our communication). We have no truck with the chap who stands legs braced, chest bared to the wind, saying, "I won't!" My theme here is that we cannot turn our backs on those persons who do not wish to communicate better and who take this stance deliberately. For they are everywhere. And to turn away from them is to refuse to grapple with some of the fundamental problems that beset the technical and industrial community. I've come here to talk about three or four of these problems and what we might do about them.

This paper is about communication, or the lack of thereof, between persons who are in some necessary relationship to one another. We have this condition wherever one man is required to act on information produced by another, and the other is required to produce information for the one to act on. This case prevails in organizations. It exists to some extent between scientists who are not in any organizational relationship, but who are professionally obliged to keep in touch with one another. At least we say this is a professional obligation. Let me note, too, that our attention will be centered on documented communication. So much for the more distant boundaries to this paper.

Any discussion about the deliberate and blatant impairment of communications ought to touch on the subject of classified defense information — that is information marked confidential, secret, or top secret. And so I will touch on it, to the extent of saying that I am not going to treat of it further. I think the topic is precisely relevant to meetings like this. But the proper form of discussion would be a symposium or round table presenting speakers who can illuminate the conflict between (as President Kennedy put it) " the need for far greater public information" on one hand, and "the need for far greater official secrecy" on the other. I am not qualified for this.

My subject is different. I wish to talk about people. I think, for example, about the scientist whom Dr. Light spoke about here last summer -- the one who refuses to read. Dr. Light said,

> The possibility must be faced that the horse can be led to the trough but cannot be made to drink therein. If scientists and practitioners are surrounded by every conceivable kind of goodie — in the form of attractively packaged scientific data such as indexes, abstracts, critical reviews, programmed teaching — learning devices, closed circuit television and communications techniques yet undreamed of -- a substantial number may still refuse to sit down and eat. They're just not hungry for this diet! Make no mistake -- any communications packaging is worth the effort. Still it must be clearly recognized — and I repeat -- that the millenium of perfect rapport and understanding among everyone concerned cannot be won solely by perfecting the communications process.
>
> Stated differently, we should expect not to get through to significant numbers of "receivers" of messages, no matter how competent the senders, no matter how important the information, and no matter how ingenious the transmission, because of limitations peculiar to these receivers or consumers or our data. (4)

Before we reach too many conclusions about this scientist, we must ask some questions:

1. Does his refusing to read really hurt anything? Or is he, in spite of his disability, a competent and productive worker?

2. If his work is impaired, what is the cost?

3. If the cost is excessive, how can we improve the situation?

4. More to the point, what can we communications specialists do to improve things?

You will find, as we proceed here, that we become increasingly occupied with the role of the communication specialist in these problems. He is almost always on the scene. Ostensibly, he is there in the role of physician -- of one who has come to work a cure. But when we burrow into these problems, we more and more seem to encounter him not in the physician's chair at the bedside, but rather in bed, burning with the same fever as the patient, and we begin to wonder in which of the two the ailment is really embodied.

With that image established, we can sort the elements of the Anatomy of Uncommunicativeness, at least those elements to be discussed here, into three groups: transmission elements, reception elements, and Hippocratic elements. The first two groups are concerned with such problems as that of the scientist who won't read. The third represents the area in which these problems must be solved —if they are to be — and in it we encounter still other problems. That is the outline of the material to follow.

TRANSMISSION ELEMENTS

The mirror image of the scientist who won't read is the scientist — or engineer or technician — who won't write. He is party to a paradox. We hear much, these days, about information glut. The report of the President's science advisory committee entitled Science, Government, and Information (5) devotes a section to the proposition that "Unnecessary publication should be eliminated." Another section says, "What is useless must be kept out." In these lights, our man who won't write appears as a public benefactor, who should be rewarded. But the paradox that I spoke of a moment ago is the familiar one of starvation in the midst of plenty. Witness an Air Force publication entitled A Summary of Lessons Learned from Air Force Management Surveys (1) This document lists many symptoms of basic deficiencies in contractor management on a variety of projects. Notice the word "symptoms." Laced throughout, are symptoms like these:

> "Support data not developed."
>
> "No written procedures to cover standard practices..."
>
> "Basic management decisions requiring action by several departments not documented or publicized."
>
> "Development of technical data late and of poor quality."
>
> "Lack of communication between quality, reliability, engineering, manufacturing, and test personnel."
>
> "Late delivery of and inaccurate technical order publications."
>
> "Inability of the contractor to produce technical publications on time."
>
> "Incompatibility of technical data (publications) and the related hardware."
>
> "Long delay in delivery of manuals as compared to hardware (not concurrent or even closely concurrent.)"

These symptoms all come back to one thing: Information glut not withstanding, someone, somewhere, has elected not to communicate something that he should have communicated. I said "elected." No one is ignorant of the need for the communications that are somehow not produced. This Air Force document, as its title indicates, relates these symptoms to inadequate management practices. Within the abstraction "management practices" are people, three in particular, who must not leave the room until we have answered some questions:

One is the manager who finds reasons -- good reasons, I suspect -- for not giving documentation a higher priority on the scale of things that he manages. Another is the scientist or engineer or technician who possesses the information we want, but whose tail the manager has not twisted. And making a trio of it is the communication specialist — the technical writer or editor, the information scientist, the technical publications engineer, or whatever. In whom does the negligence lie?

We are comforted to point to the manager and say, "He must do something." If the problem is to be solved, not just palliated, the manager must indeed play an active role. But I suspect that this will be some time coming. Let me venture that most, if not all, managers in the technical and industrial community will freely state that documentation is important, and those that I've encountered in this act have been patently sincere. But they do exactly as you and I would do. They place this very important matter of documentation on their yardstick of other important matters, and it finds itself nowhere near the top. Important, yes! Meanwhile — first things first.

The military have inagurated data management programs to alleviate inadequate documentation. Surely, improvement will result. But so far, we have not witnessed the pivotal event that will bring about substantial and enduring change. Some major value shifts must occur first. A customer must someday refuse to accept, and pay for, a much desired piece of hardware because the supporting documents are not on time or up to snuff. Or, a bidder must lose a contract or suffer dollar penalities because of history of poor documentation. I cannot imagine any smaller event conveying the message. Let me emphasize, I do not think this will happen tomorrow, despite the increasing enthusiasm we discover all around us for doing something about documentation.

We are all, I suspect, aware of exceptions to this. I know of people whose salary growth has slowed because they are negligent about documenting their work. Surely this speaks of managers taking action for better documentation. But does this reflect wisdom or does it reflect stability — quiet times (relatively speaking) in which we find opportunity to worry about documentation? When our projects are young and unstructured, when our people are untrained, when our production schedules live on the edge of disintegration, we are not this diligent about documentation, even though the need for written communication is, if anything, greater than now. In reduced salary growth here and there, I do not see a general management concern about documentation of the kind I am talking about.

If my assumption is sound that we cannot today expect real help from the manager, then we must look elsewhere. Let me suggest, for a starter, that we go and take a closer look at the scientist or engineer or technician from whom we must somehow squeeze our needed drop of information.

He is apt, we discover, to draw a sharp line between doing and writing. He asks, "Do you want me to finish this equipment design? Or do you want me to stop work and write about it ?" His position is clear. And here we are, standing at this desk, insisting it's important that he produce a report on his work in, say, ten days.

If we are to cope with him, we must recognize that report writers divide into three classes:

1. Those who are anxious to improve their skill and who are willing to work hard doing it.

2. Those who have a report to write in ten days and whose sole wish is to survive this one job. They couldn't care less about writing improvement in general.

3. Those who are anxious to improve their skill (like class 1) but who hanker for the

Golden Key. They do not intend to work at it.

I haven't the slightest idea how we can help the latter. Always ahead of them is a new book or article promising "Writing Can be Easy!" and away they go, out of earshot of those of us who say that it isn't really quite like that.

And those of the first class are not our concern in this conversation. If they predominated in our constituencies, our task would be altogether different. It would be what we commonly think it is — the betterment of persons whose motivations square with our ambitions for them. They are the persons for whom we conduct our writing seminars and workshops. They are the ones to whom we address our writing handbooks and about whom we talk in gatherings like this. But we concern ourselves with them not because they represent the larger problem or are most in need of attention, but because they are the easiest to attend to. As is so often the case, we prefer those problems for which we have answers.

That leaves us in the company that we have preferred to ignore — our unmotivated class two. The chap who has a report to write in ten days and who doesn't like it at all.

Look at him there. Holding his head in his hands. Perhaps sobbing quietly. And then think of the writing handbooks that you know, the whole, long shelf of them, and the catalog of available writing courses, workshops, and seminars, and ask, is there anything in the lot of them that will meet him where he is — in terms of his language background, his motivations, his workload, his present schedules? The answer is no.

What can we do? If this man is to be reached it will be by someone who lightens his work, who somehow reduces the total task that exists here — that is, the composite task of finishing the equipment design and writing the report. It will not be someone who invites him to attend a writing seminar two hours a day. It will not be someone who directs him to a hand book on report writing or who lays before him the corporation's sixty-page style manual.

Whatever help is offered must appear to him as help and not as an added burden. I think this means that the guidance or instruction we give him ought not to take more than fifteen minutes in the giving. I am thus stringent about it because our choice, it seems to me, is between giving him something in fifteen minutes or giving him nothing at all.

We can start by recognizing that what is required here is not a good report but a passable one. Perhaps that statement is gratuitous. But I suspect that we have all known the communications specialist who, in the face of this problem, would fall to worrying about the editorial quality of the forthcoming report and about the kind of covers it should wear. For shame!

Well, there is the problem. I'm not suggesting that we toss him crumbs; I happen to think that in fifteen minutes we can give him real help. This requires that we sift through our stores of information and folklore about writing, discarding many ideas that we had thought important and retaining only those that count in this situation. This bare handful will very likely contain some thoughts that we had not bothered to think before. Possibly the most salutary concomitant of this exercise is the demonstration to someone else that we are able to see his problems as he sees them.

You will emerge from this exercise with fifteen minutes worth of what? I don't know. I know what I emerged with, but I'm not sure about much of it — except that the process gave me a clearer appreciation of the problem here. And so it is the process that I recommend, and not the particular result that I happened to arrive at.

Let's enlarge on this case. Suppose we think not in terms of a specific report to be finished in ten days, but in terms of generally bettering our engineer's writing ability. I have postulated that he is not at all interested. That suggests where we must begin. We must find a way to play on his motivations, and we will not do this in the way that we conventionally assay it — that is by proclaiming that learning to write well will contribute to recognition of his work, or to the successful completion of his work, or to the advancement of his work by facilitating his thought processes — for these protestations, however true they may seem in our experience, will not register as truth in the lights that he steers by — his salary, his promotions, the esteem of his boss, the satisfaction that comes from a well-designed electronic circuit, or machine, or computer program, or whatever. These are the things that count: money, position, esteem, satisfaction (in some other order?). Very likely, he will not have suffered a setback in any of these particulars. In fact, his public abhorrance of "paperwork" (the gutter term for documentation) may have speeded his advancement on the ground that he is devoted to Getting the Job Done, and not to piddling around with diverting matters.

When we look at this chap — at the way he writes now, at his working environment, at this schedules and commitments — and when we mix what we see there with what we know, or think we know, about his language background, about his motivations, about the craft of writing, about learning and teaching processes, we find that we face a disheartening tangle of cross-purposes. And yet the criteria for our writing program must engage with the whole of this tangle, if we are to do anything at all with the man whom we have established as our problem case. Let me describe some of the criteria that I think I see:

The first concerns content. What is the writing problem? Faulty grammar? We encounter problems behind problems. There emerges on paper a pervasively muddy prose in which ideas are hidden, so that the reader must grope, or are distorted, so that the reader goes forth innocently with the wrong information. Faulty grammar may contribute. So may a parade of other faults. The question for us is this: Are there ways for a writer to spot and remedy a cloudy passage without bringing to the task the professional skills of a grammarian, a lexicographer, a journalist? I think there are ways to do this — to provide tools by which the layman can recognize and improve the unclear passages in his writing — and so let that be our first criterion for a successful writing improvement program:

Provide tools especially suited to the layman.

Were I privileged to view writing improvement apart from life's other aims and its commitments and pressures, I should prescribe a three-part program:

1. A thorough grounding in the elements of composition.

2. Frequent review of one's own writing by an expert — a teacher or a professional editor.

3. Extensive reading among authors noted for clear expression.

A program of that kind requires time. Too much time, in that we're talking about internal trai ning and therefore about time carved from one' job. The costs of a writing-improvement program ought not to outweigh its rewards. And thus we arrive at our second criterion:

Require minimum time from the job.

Compressed writing courses are a common response to that criterion. A manager arranges a writing class or workshop for his people — a class that requires, say two hours a week for six weeks. But this compression becomes itself a problem. The student receives too much information. He touches on many writing aids, but emerges with no key by which to know where in the bundle to search for particular help with a particular problem. He possesses a book without an index. Thus our third criterion:

Avoid compression.

That the program should fit local needs is axiomatic. Sucha fit means, first, that practical betterment, however, small is realized right from the first, It also means that the participant can relate the program to himself — to his needs. Our fourth criterion, then, is this:

Fit local needs.

As we look at the man in his job, the thought grows that perhaps the job itself -- that is, the pattern of work and of human associations -- can help bring about improved writing. Before a document is finished, many persons see it, handle it, work on it. These persons include the writer's professional associates, his typist, his supervisors, perhaps his department or corporate editors. Maybe all of these people could, in addition to their other reasons for handling the text, be encouraged to look at it editorially, and with an eye not wholly uninformed about editing. Again we should expect two gains: Betterment in fact, plus a reinforced urge for betterment in the writer, resulting from group interest and group effort. Here is our fifth criterion:

Pave the way for participation by everyone in a given working unit.

Our sixth criterion is implicit in the fifth. But it is of such importance, and so easily by-passed, that I wish to set it out separately:

Pave the way for participation by the unit's head.

About that sixth criterion: The impelling force behind a writing-improvement program is ordinarily a manager who has concluded that he can no longer afford bad writing. He roars, "Writing is important!" and pushes buttons to provide a writing workshop or seminar for his people.

Assume that the workshop or seminar is a good one. Assume further that you are among those attending and that you do, in fact, learn to write well. You return to you office and write a report in your new and fluent manner. A good report. You deliver it to your boss for his review and approval.

But did he attend the writing workshop or seminar? No, he didn't. He said that other demands on his time were such that, regrettably, he could not participate.

The consequences that I see are two. First, he edits to rules different from those that governed your writing. He does violence to your text. Second he does violence to your opinion about the importance that he attaches to good writing. This comes not only from your mauled text, it comes also from the circumstance that he spent none of his own important time on this allegedly important matter. By not personally joining the writing program, whether by choice or by press of circumstances, the manager immeasurably weakens the program's bite. That is why it seems to me, a writing program must pave the way for managerial involvement.

The heat, pressure, and light generated by these means will go for nothing if our writing program flourishes for a week or two and then burns out, like a Fourth-of-July rocket. Let us, then, design a program that will --

Run continuously.

Finally, the participant in this program must know how he is doing. On this depend his satisfaction, his interest in the task, and his ability to improve in deficient areas. In other words, the program must --

Provide feedback.

Of these criteria, only three, it seems to me, are consistently honored by the writing programs with which we are all familiar. That is, these programs seek to require minimum time from the job, by running only one or two weeks; they endeavor to fit local needs, by using materials prepared on the job; and they arrange to provide feedback, through group discussions and instructors' evaluation of student work. But the other criteria do not receive sufficient attention. Let me list the others again.

Provide tools designed for the layman

Avoid compression

Pave the way for group participation — not omitting the boss

Run continuously.

I list these criteria, all of them, as a first step toward enlarging what appears to be our conventional notions about writing improvement programs. Until we meet these criteria, all of them, we are not going to make much of an impression on the scientist or engineer who simply is not interested in writing.

Whatever the foundation problem here in the case of the man who won't write, one of the middle problems in the structure is that he has come to view documentation as a by-product of his main work. Very likely, this view is going to cause trouble in other of his relationships with the information system in which he lieves and labors. Mortimer Tabue, talking about electronic data processing and information systems, speaks of our man in connection with the problem of gathering data to be retrieved:

> The acquisitiion of information for an IR system is not an EDP problem but a management problem. Management must establish some system to insure that any information which has value beyond the use for which it is initially generated reaches the IR system for coding and storage. This is not always a simple matter, since different types of scientific and industrial operations have different traditions with regard to the reutilization of information generated in the course of regular operations. A chemist, for example, will complete a study of a particular

problem by writing a report, and this report will usually be in a form adequate for storage, coding, and retrieval. On the other hand, an engineer who is concerned with building a structure will generate an enormous amount of paper in the course of his current operations, but when he completes his work, he will have a building instead of a report, and the paper he has generated will normally be filed away uncoded, unanalyzed and in a form suitable only for achieving rather than for information storage and retrieval in any useful operating sense. It is true that no one ever builds the same building or the same industrial complex twice. Nonetheless, there certainly are elements in any engineering operation which are repeated throughout a number of different projects, and the engineer must learn to isolate the various parts of his work into separate and reusable pieces of information which can be coded and stored and thus provide an informational basis for new operations.

In our experience, the data acquisiton problem has usually been largely a matter of convincing the company's engineers that reusing information already on hand will not interfere with the free exercise of their creative ability as engineers but, rather, will release such ability and stimulate it toward genuine advances instead of a laborious rediscovery of the wheel. (6)

In that passage is a question for the information specialist: How easy are inputs to his retrieval system? Can this system accept engineering notes in their raw form? Or are the format requirements for system input rigorous, in the name of efficient storage and effective retrieval? A balance must be struck here. Some information specialists are trying to strike it. But too many are unwilling to sacrifice system efficiency. They forget who owns the information and what his needs are.

At the beginning of the passage quoted above, Dr. Taube pinned certain responsibilities on management. Later, he cited some things that the engineer "must learn." And my question now is, "Is there no task here for the communications specialist? No responsibility?" We'll come back to this question later.

The cases we've discussed to this point have been those of people who view documentation as a by-product of their main work, a side issue, an annoying and stultifying digression from important matters. As much hell, if not more, is raised by the man who perceives his written work as his product, as his image, as his very identity. The opposite of indifference is fussiness. For example: I first broached the subject of this paper to Dr. Weisman two years ago. One year ago we agreed that we would put the topic on this year's program. I immediately started compiling notes and recording an occasional lapidary phrase that came to mind. You haven't noticed any lapidary phrases because I deleted them on Quiller-Couch's advice that when writing one should "murder his darlings." In Mid-February, I started writing the text, one page a day, Monday through Friday. Mid-April, the first draft was done and I began picking at it. And if there were any way I could get out of it, I wouldn't present it to you today, because it isn't right.

Fussiness leads to another kind of system indifference. Were I preparing this for print, I should pick at it until the last possible minute -- the last possible minute being a week or two past the printer's deadline, and then I should arrive at the print shop, red-faced and roaring that it must be printed instantly! And the duplication shop, drawing on the deep well of patience and forgiveness that such shops usually have, and at considerable expense to itself, would print my document instantly.. And then, on looking it over once more, I should find two or three things still wrong and ask that it be done again, and faster. This is a parable for our time, familiar to all supervisors of duplication shops.

The problem here, whatever its real nature, is intensified if the writer has not a year in which to write, and worry, but only a few days. Let me take the fashionable avenue, and say it's a management problem. The task of writing, reviewing, rewriting, and printing takes more time than schedules commonly allow. And schedules are in the purview of management. How easy it is to pass this buck! We'll have more about this in part four of this talk.

We're talking at this point, remember, about what happens to information that the possessor or originator views as something closely related to himself — a main product of his energies and intelligence. If this information gives him a competitive edge in the market place, it becomes a business asset of which he is the proprietor. He says it is "Proprietary" information much as we say information is "classified" if it affords a competitive edge in the military or political arenas. Both are treated tenderly. The DuPont Corporation developed indexing systems for chemical subjects that were kept from the literature of information retrieval for several years because they were thought to afford a competitive edge. For some time, automobile maufacturers have used critical-path or PERT methods for scheduling, yet no mention of this has appeared in the industry's literature. (8) What is the cost here? -- particularly if one (cynically? whimsically?) throws in the cost of industrial espionage meant to expose proprietary information. I suspect that there are several things to be said about proprietary information and its effect on the technical and industrial community, but here, as with the subject of classified information, I find myself insufficiently experienced to throw real light on the subject. But I recommend it as a topic about which worthwhile things might be said, in meetings like this, by persons who are in some way actively involved with proprietary information, whether in generating it, hiding it, or trying to steal it.

It isn't competitive potential alone that sets us to hoarding information; any number of other forces conspire to make us possessive about information and anxious, whatever the cost, to impair its flow. Our wish to get information out to legitimate users is distincly second fiddle to all sorts of other wishes. We may wish, for instance, to control people's actions, if we are a manager or supervisor, and we see the control of information flow as a device of management control. Or we wish to handle information, or to structure it, to suit our personal needs and we learn that this becomes difficult in the face of system rules (such as circulation policies, format requirements, or indexing standards) that are imposed on us if information gets out of our immediate jurisdiction. And so we cloister it and pass it under the table to friends, even though others in a wider circle may plead a legitimate use for it.

Or we perceive that information is power. Someone has hypothesized that within an organization a person's salary, position, and influence are functions of the information to which he has access. There is a chicken-egg relationship here about which I'm not certain. Does our man acquire important information and thereby rise in the hierarchy? Or does he rise and thereby gain access to important information? Doubtless the original text dwelt on this. But whether we openly accept this hypothesis or not, I believe that we intuit something of the sort and are not always as generous as others might wish with information in our possession.

The real effect of these various transmission elements in the Anatomy of Uncommunicativeness, plus an infinitude of others like them, is not to be measured at the transmitting end, but at the receiving end. And so we'll go there, closing Part II of this talk and entering Part III.

RECEPTION ELEMENTS

One of the things we can say about information at the receiving end of the communication process is that it ought to be retrievable. In 1960, the subject of information retrieval, that had theretofore been the arcane hobby of dusty and burrowing souls, such as librarians, programmers, logicians, and linguists, suddenly received the Industrial and Mercantile Stamp of Approval by getting itself written about in Fortune. (2) The article spoke of the "cost of not finding it." That cost -- the cost to American business of not being able to retrieve and use existing information -- was estimated at over one billion dollars a year. This theme has been developed at length by the sellers of retrieval systems and equipment. Mortimer Taube (who sells retrieval systems) spoke, you'll recall, against "a laborious rediscovery of the wheel." Executives, stockholders, and boards of directors, all of whom are concerned with profits and anxious to avoid duplicated effort, have taken up the cry in behalf of better retrieval. The Weinberg Committee pointed out that the technical man must attune himself to this effort, and that the responsiblity for seeing that he does so lies in three places: with the man himself, with the colleges and universities that train him, and with the professional seniors to whose pattern the neophyte shapes his behavior.

I trust you agree that the problem exists, that it is important, and that something must be done about it. Let us meditate upon the man who is to use our retrieval system.

Calvin Mooers, who has worked many years in information retrieval, has formulated Mooers' Law:

> An information retrieval system will tend not to be used whenever it is more painful and troublesome for a customer to have information than for him not to have it.

What constitutes painful possession of information? The news that you have been fired, obviously. Or the clear evidence that airplanes can sink battleships, if your are a battleship admiral. The virtue of effective retrieval is often illustrated by reference to a scientist in, say Baltimore, whose work toward a given goal is blocked in a way that he can't resolve and who learns -- quite by chance, since he lacks a decent retrieval system --that a scientist in Palo Alto has solved the problem in the course of a wholly different investigation. Thus the Baltimore man is enabled to move ahead. This is the euphoric version. In the painful version, the Baltimore scientist learns that the Palo Alto scientist has not only resolved the block but has gone on to realize the goal toward which the Baltimore man was working. Painful, indeed.

I'm not sure about the word "troublesome" in Mooers' Law. Is he talking about the trouble involved in retrieving information or about trouble that may arise from having retrieved it? Perhaps both. Let me talk about the trouble to which man is willing to put himself in order to retrieve information.

When we ask people what they expect of a retrieval system --meaning what balance do they think reasonable between the effort required of themselves and the effort required of the system -- they tell us what we wish to hear. They put on a reasonable face. They call up what they know of the state of the art in information retrieval and voice their expectations accordingly. The man in the street whose experience with information retrieval centers on the public library gives us an answer that is reasonable in terms of what he thinks a library can do. Implicit in his answer is a picture of himself meeting the library halfway. The person who has studied retrieval problems in connection with electronic systems reflects different procedures but a similar attitude. He describes himself going to a place that houses input devices for the main gadget, looking up codes in a thesaurus entering the relevant codes into the mechanism, scanning some initial output in order to shape a second, more precise input, and eventually pushing a button to get a hard copy of the information he wants. His personal contribution of time, energy, and intelligence is not small. All very reasonable.

Ten years ago, only a few workers in electronic data processing had come to grips with language problems, including the problem of information retrieval. If one has not tussled with these problems, he is prone to over-simplify them. And so, ten years ago, it was not difficult to find EDP people who believed that information retrieval was a problem that a computer could lick with its left hand and that this would be taken care of as soon as some more pressing matters could be put aside. For them, there was no such thing as system limitation. They envisioned a system that watched them at their work and that delivered to them the information that they needed, when they needed it, and with no effort whatever on their part.

This, I think, is a truer statement than the others of what man, in his heart of hearts, expects of retrieval systems. In other words, I believe that this no-effort concept is part of the layman's real expectations of a retrieval system -- real expectations, as distinct from the reasonable expectations, that he voices when we ask him about it. An this is one of the expectations that we must contend with if we are going to do anything effective about information retrieval.

Why does this inertia exist? It is there for one thing. I suspect, because of the nature of information that is relevant to one's work. The characteristic by which we know that information is relevant, is that it changes one's behavior. The seeker after information had best be set for change. The pleasantest information, in this view, is information that speeds us towards a desired goal. Much less winning is information that forces us to choose between two or more desired goals. For example, suppose we learn that a new machine will lower our operating costs but require that we surrender positions, perquisites, and traditions to which we've grown attached. What now?

We are sometimes held away from the library by the probability -- real or imagined -- that information germaine to our pursuits isn't there. This probability seems very real to the researcher who envisions himself on the frontiers of human experience. Some indeed are out there. The majority, I suspect, are not. But one can deceive himself very comfortably by avoiding the literature. These days, the responsibility, once taken for granted, of the professional man for searching the literature of his field is largely honored in the breach. This means not only that the wheel is likely to be rediscovered, but that the literature of wheel discovery (since some researchers are indefatigable and unstoppable writers) will likely be one story, often-told. But no matter how voluminous the literature of wheel discovery, it will be only a tiny part of the total literature, wheel and otherwise, that must be confronted in a literature search. The corpus is simply too large for easy frisking. To the value, then, that our researcher attaches to his picture of himself in a coonskin cap, let us add the value of the labor that must be exacted of him if we require that he destory his precious picture. Of course, he shuns the library.

But I'm being unfair. If it is true, as I was insisting earlier, that masses of information are withheld from our information systems (even though other masses are present), then our man at the receiving end is sound in assuming that what he wants isn't there, even if he does not fancy himself an intellectual pioneer. He must, in fact, see himself as one who plugs along three or four years behind the frontier if he is to persuade himself that an information search is worthwhile, for that is about the time it takes for ideas to find their way into open publication. This is especially true in the industrial community (as distinct from the scientific) where news about applications is sought and not about theory.

I think too, about the professional man who, in response to a library user survey, said that if librarians could help him with his information problems, they wouldn't be librarians. That statement is hard to cavil at.

To top it all, reinventing the wheel is fun. Is the excitement in the process somehow less for those who come later? The process is not relivable. Each time, it is a new journey. A friend, a chemical engineer, reminds me that Shakespeare rewrote old plays, poets brood over ancient loves, and composers build from medieval themes. Everywhere, yesterday's fabrics are rewoven with no discredit to the weaver. He may in fact acquire a laurel wreath for his work. The process is everything. Who cares about a duplicated product or two? -- who except a handful of executives and stockholders to whom process consists in motoring to the bank.

Information retrieval is a topic too complex for pat commentary. I have only some general notions about it. For one, I think that the retrieval system of the future must reach the man at this desk, in the form of a booktype catalog, perhaps, or (years away) some mechanical input-output device. The man who will leave his desk to go to the library is not the problem. It is the other that we must be concerned with.

A second general thought: Traditionally, the intellectual processes of cataloging and reference have been held to constitute the professional heart of the library. It is on these very processes that the retrieval problem centers. In respect of these processes, one can't yet expect to produce results that are other than ordinary, at least in the layman's eyes. Until the retrieval problem is solved, or at least lessened, the library can best look to its reputation in the technical and industrial community by sharpening its mechanical procedures -- its acquisiton processes, for example -- to cut waiting time and costs to a clear minimum.

Third, I think that the communications specialist must devise means for educating the scientist, engineer, or businessman for his role in the retrieval process. The man is the central element in that process. He will likely remain so for some time. He must be taught to view himself as a data processor. That will take some doing for "data processor" is hardly a fitting title for one cast in God's image. Much happier descriptive labels are surely available. Susanne Langer, for instance, treats of man as a "symbol maker." (3) That is a more profound view of him, with a golden aura, and it hints of poetry. Now poetry, like information retrieval, does not lend itself to pat analyses. But one of the many things we can say about the business of making a poem is that it is an act of data processing -- specifically, of data reduction and display. But is our man to be won by that? He thinks of his work in mathematics, or business management, or engineering, or physics as an endeavor fit for an angel, while data processing, in the sense that I mean it here, is documentation, not poetry at all, and documentation, whether at the writing or the reading end, is hack work.

He will not be easily won to his role in the retrieval process. But he must be won, and the winning of him is one of those tasks that the communications specialist must undertake. For no one else will.

Now that we're back to the communication specialist, let me stay with him. He professes, to some extent, at least, to be a healer of information hurts. On the other hand, he turns away from certain ailments. What are the rightful limits to his physician's role ? This is the question that I wish to consider in the next section of this paper, about the Hippocratic elements in the Anatomy of Uncommunicativeness.

HIPPOCRATIC ELEMENTS

If we are to talk sensibly about the healing function of the communication specialist, we need to understand how he relates to information processes and to other persons who are involved. These relations are not at all clear. Think back to the "symptoms" of management difficulty that I listed earlier. Such as these:

Support data not developed.

Development of technical data late and of poor quality.

Inaccurate technical order publications.
And so on.

Do these truly represent management difficulties? Or are they documentation problems? Or both? Or neither? Like an optical illustion, they shift from one thing to another, before our eyes. Does this matter? On the conversational plane, the plane of institutes and journal articles, what we call these difficulties is of no consequence. That is possibly why institutes and journals pay so little attention to what in my opinion is a major problem facing the communication specialist in the industrial world. For in organizations, where given problems are assigned to given people, this illusion-effect determines the specialist's grip on information as much as does his professional skill. That is why this chapter on Hippocratic elements will focus less on Hippocrates himself than on Hippocrates in his working world.

We can diagram the communication process by drawing two circles with a line between them. Let the circles represent people involved in the process; the line stands for messages that pass between them. That line is the concern of the communication specialist -- that line plus whatever elements in the two circles or their environment affect the goings-on along the line.

Now let's place this structure in a system with a number of other circles and lines. A system by popular definition is devoted to some task or other, and so we can represent this larger construction like this:

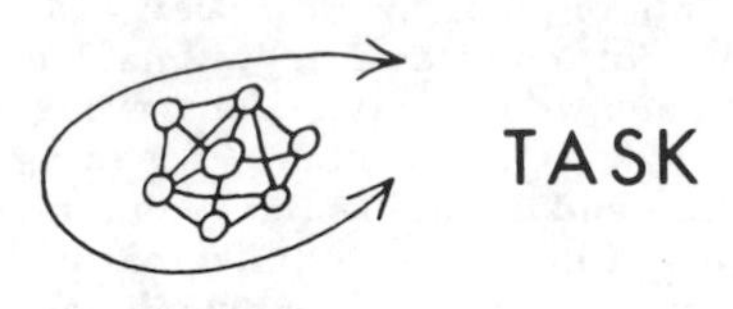

TASK

Now let us further say that this system represents an organization. It is here that the manager enters. He is charged, among other things, with seeing that our various system components interact smoothly. In other words, he too is concerned with the communication lines between the circles.

One can't generalize about the consequences of joint proprietorship, In some cases, the object of concern finds itself doubly well cared for. In others, it finds itself victimized by misunderstandings, cross-purposes, and contradictory instructions. In still other cases, the two custodians will take up separate parts of the object, with their portions abutting as precisely as the stones in a proper wall. (This is the best case and possibly the rarest.) And in still other cases, the object will be picked up at the ends, like a piece of string, with the middle sagging and untended.

I think the latter is a precise metaphor of the handling afforded communication systems. When we communications specialists shove a difficulty out the door, saying that it's a management problem, we do not shove it into the waiting arms of someone else, for that someone else isn't there. He is elsewhere, attending to wholly different matters. And I suspect that if we were to go talk to him about it, he would say that the problem we pushed out the door is precisely the kind of nit that we were hired to take off his back. And he would perhaps add, "Is your job too big for you?"

Is there such a thing as an important nit? Here is a manager (perhaps the one we just talked to) speaking at a documentation conference:

> Documentation is important -- Obviously. Engineers taking part in any project must realize that the project is not completed until it satisfies the group's documentation requirements. Some of our brighter, more creative more industrious engineers must be sold on the importance of documentation work. . . We must make it fashionable to assign a man to documentation. The technology of information handling systems — programs, reports, formal documents, design control, -- has not been sufficiently developed; we must establish the technology on an equal footing with programming, operation or equipment. (6).

I have since kept an eye on that manager. It will be a cold day indeed when he assigns one of his brighter and more creative engineers to documentation. Or takes steps to make documentation a fashionable occupation for an engineer, for he faces a problem in translation. To understand this, break the idea "documentation is important" into smaller ideas. Say to yourself, "Clear writing is important!" But then you must ask how much one should pay a technical writer. Or say, "Retrieval is important!" But then as how much one should pay a librarian. Or say, "Speed in dissemination is important!" And then ask how much one should pay for print shop and mail room personnel.

Through such questions as these, the ringing words "Documentation is important" are translated into people and things. And so the reality is that the writing and editing services, the library services, the printing services, and so on, are established at a point in the organization's hierarchy some distance from the manager's working realm, or the realm of the engineer.

The communications specialist, is in fact, likely to find himself curiously isolated. To understand this, think of the range of questions that in the answering shape our communication processes:

1. What task is to be performed? (Build a bridge, design a computer, paint a wall, launch a satellite, meet a payroll, etc., etc.)

2. Who is to do it?

3. What information does he need?

4. Who should supply it?

5. In what form?

6. How can it best be transmitted?

7. How can we control the communication?

8. What services can be brought in to help?

These questions do not necessarily take this order. Question one, concerning what we do, may well depend on question six, concerning our ability to transmit messages. It is a rare and fortunate case when the answer to any of these questions is clear cut -- that is, when authority and responsibility are unequivocal, when the analysis of information needs is impeccable, when the controls are infallible.

These, to repeat, are questions whose answers, or lack thereof, are basic in shaping communication processes. Yet where does the communication specialist commonly enter? At question eight. And there he labors, as best he can, on matters relating to that question, while the problems that we discussed earlier -- the scientist who won't read, the engineer who won't write, the technician who wont format this thoughts for retrieval -- these problems are embodied in the answers we coin to questions one through seven; and although these problems touch on question eight, it is only a finger's touch, at arm's length. The heart is elsewhere.

But are they thereby management problems? How many enigneers must procrastinate on documentation before a schedule slips, or machines come off the line improperly assembled, or two hundred cutomers learn that if you turn valve A as the maintenance manual says, you burst line D at a cost of twelve thousand dollars? These are management problems. And these are the problems the manager has in mind when he says that documentation is important. But only dimly, if at all, does he see in these problems the individual scientist who won't read, the engineer who won't write, the whole host of particular failures that accrete into the management difficulties I listed near the beginning of this paper.

Who does see them? You. And I. And Mortimer Taube. And Calvin Mooers. In short, the communication specialist sees them. And that, in my mind, makes these problems his.

Well then, what is he to do? Whatever it is, he must

begin by escaping his isolation, and to do this he must understand how he got there in the first place. Let us review:

It was partly by choice. How quick he has been to say, "That is a management problem!" It was partly through the process of interpretation whereby the words "documentation is important" are translated into the dollar amount that one is willing to spend for documentation services. It was partly through the circumstance that separation feeds on itself, so that managers and communication specialists seem increasingly unable to understand one another. Neither sees problems as the other sees them, and so it is only by luck that either in assaying answers will appear at all bright to the other. Parenthetically, let me say that the manager can afford this. The communication specialist can't.

Appear bright? The communication specialist struggles mightily with the billion-dollar problem of information retrieval, and the manager watches, reviews the budget, frets, and finally bursts out, "Good grief, man! Why don't you use a Dewey decimal system, like down at the library?" And the gap widens.

This manager, then, seems to be one who suffers unfocussed hurts from documentation problems, who does not understand the source of hurt, who hires writers and editors and librarians and so on out of a half-formed faith (the Havard Business Review says that there are profits in prose) that these people can help, and who, in general, is vaguely disappointed with the results. And so quite reasonably, it does not occur to him that the communication specialist, who is having trouble enough with matters relating to question eight, has any useful business in the realms defined by questions one through seven.

You see, a whole set of forces has vectored our man like a pool ball, into a side pocket, from whose depths we hear him muttering, "Management problem." If he is to get out, it will be that he takes himself by the scruff. And with what in mind? I believe, as I said back near the beginning, that if information problems are to be solved, the manager must play an active role. The task of the communications specialist, then, is to qualify himself to help the manger see these problems correctly and to find solutions that make sense from the manager's point of view.

How does one stress that last point? It is not enough that the communications specialist clarify problems, as by pointing out that problems of class x are management problems. Nor is it enough that he propose solutions, however correct they be. In life, the most elegant of solutions will not likely be a clear gain on all counts. Improvements are measured on balance; they consist of sets of gains that outweigh sets of losses. The communication specialist must learn to propose his solutions in detail and in terms of the gains and losses, however complex or obscure, that the manager must weigh in evaluating the proposal. To do less, at least until one acquires oracular status, is to suggest that he is neither mature enough, informed enough, nor responsible enough to be taken seriously as a problem solver. This is to say nothing at all about managers who are constitutionally unable to act on even the clearest and most complete proposals. We are talking here about the responsibilities of the communication specialist, not the manager.

Nor are we talking about something that is easily done. The over-whelming difficulty of it may be the largest, most formidable element of all in our Anatomy of Uncommunicativeness. Our ability even to discuss communication problems, much less explain them, is nearly non-existent. It is a skill that is yet to be acquired, and I'm saying that the acquiring of it is one of the important tasks to which the communication specialist must address himself.

The specialist's effectiveness and his esteem in the eyes of line people move hand in hand. As one goes up or down, so goes the other. He must look to his professionalism if he prefers that his progress be upward. He does not do this by pointing to his diplomas or his long years of experience. These ought to be taken for granted. The test of a professional is in other things:

1. Discipline. Our man is disciplined. I do not mean that he is a skilled and diligent practitioner of formalized method, but simply that he can work without supervision. He thinks that the responsibilities of his profession outweigh the privileges. He sees the tasks to be done, and places the doing of them above his immediate personal advantage. Perhaps responsiblity is a better word than discipline.

2. Results. He produces results. He performs the rituals of his chosen field better than those who are not of his persuasion. Is this worth saying? In how many organizations do the technical writers produce the best writing? How many librarians are in the van of inquiry into methods in information handling—or even reasonably informed about the inquiries of others?

3. Vision. He can see beyond his discipline. If he is a librarian, for instance, he recognizes that cost accunting, electronic data processing, and any number of other arts, crafts, and disciplines have meaning for him. Moreover, he can see those places within his discipline where change is called for. Possibly I'm really saying that he is well-rooted in theory, so that he feels secure in adjusting his surface responses to best meet the wind. I repeat: well-rooted in theory. In this quality of vision we have the seeds of the all-important trait I mentioned earlier -- the ability to see communication problems as others see them. And that brings us to the next item.

4. Teaching. He can teach. Here, the need to see as others see, is absolute. Our man can define problems and propose solutions in ways that tie to others' experience. For the manager, he quantifies problems and procedures. He searches out those that he can quantify, knowing that his first problem is to reach this other person.

With a proper concern for these things, our latter-day Hippocrates can begin to lend a hand with those hard-to-fight communication ailments that we talked about earlier.

And with that, we can close Part IV of this paper and attempt to summarize the whole of it.

What I've tried to say is this: Many people do not pull their weight in our communication systems. To change this, some attitudes and techniques must change. The communications specialist, because of his professional insights, and his professional skills, should be a key figure in this process. But to be so, he must first overhaul some of his own attitudes and techniques. His grip on communication problems depends as much upon this as upon his knowledge of technique in writing, librarianship, graphics, or what have you. I believe the communication specialist can enlarge his contribution to the technical and industrial community, if he will only do so.

REFERENCES

1. Air Force Systems Command, A Summary of Lessons Learned from Air Force Management Surveys. AFSCP 375-2. Washington, 1963.

2. Bello, Francis, "How to Cope with Information," Fortune. September, 1960. Pg. 162-192.

3. Langer, Susanne K. Philosophy in a New Key. Mentor Books. New American Library of World Literature, Inc. New York, 1953.

4. Light, Israel. "Communication Problems in the Life Sciences," 1963 Proceedings. Institute in Technical and Industrial Communications. Herman M. Weisman, Editor, Fort Collins, Colorado, 1963. Pg. 69.

5. President's Science Advisory Committee. Science, Government and Information. Washington, 1963.

6. System Development Corporation unpublished conference report. 1957.

7. Taube, Mortimer. "Advances in Information Retrieval and Data Acquisition: I. Progress in the Design of Information Retrieval Systems," Advances in EDP and Information Systems, American Management Association, Inc. Management Report No. 62. New York, Pg. 56.

8. Wattel, Harold L. editor. The Dissemination of New Business Techniques: Network Scheduling and Control Systems. Hofstra University Yearbook of Business, Volume 2. Hempstead, New York, 1964. Pg. 325.

The Gulf between Correctness and Understanding

JAMES LUFKIN

Abstract: The traditional preoccupation with "correctness" and "clarity" in technical writing frequently goes hand in hand with a neglect of the reader's point of view which results in publications of such poor quality that instead of admiring them we should consider them unacceptable. Robert Pirsig's *Zen and the Art of Motorcycle Maintenance* includes an exploration of this problem and a reexamination of the relationship between "objectivity" and quality.

THE GULF BETWEEN CORRECTNESS AND UNDERSTANDING

Robert Pirsig's *Zen and the Art of Motorcycle Maintenance*[1] is an autobiographical essay in which an extraordinarily gifted and sensitive technical writer tells how he gradually discovered some essential truths about himself and about the relationship between technology and human values. Cast in the framework of a father's cross-country motorcycle ride with his 11-year-old son, it is also a moving plea for an end to the present irrational and dehumanizing separation of art and technology, and finally—perhaps above all—it is a searching examination of the concept of quality, beginning with the ancient Greek notion of *arete.*

This is ostensibly a book about philosophy, rather than about technical writing, and yet it is loaded with observations that technical writers will profit from. Since its publication in April of 1974, Pirsig's book has been treated with respect by a number of reviewers including *The* [London] *Times Literary Supplement,* which devoted its entire first page to the book before the British edition appeared. It has also become a standard reference in the field of technical communication, and has been used successfully as a text in college English and composition courses and in courses for technical writers and instructors.

An example of Pirsig's view of technical writing appears early in the book where he describes an experience in a motorcycle repair shop in which some fumbling mechanics nearly destroyed his bike. He is horrified and afterward tries to imagine what went wrong. As he recalls the mechanics' faces, he says:

> ". . . . the biggest clue seemed to be their expressions. They were hard to explain. Good-natured, friendly, easygoing—and uninvolved. They were like spectators. You had the feeling they had just wandered in there themselves and somebody had handed them a wrench. There was no identification with the job. No saying, 'I am a mechanic.' At 5 P.M. or whenever their eight hours were in, you knew they would cut it off and not have another thought about their work.
>
> They were already trying not to have any thoughts about their work *on* the job . . . . living with technology without really having anything to do with it. Or rather, they had something to do with it, but their own selves were outside of it, detached, removed. They were involved in it but not in such a way as to care.
>
> Not only did these mechanics not find that sheared pin, but it was clearly a mechanic who had sheared it in the first place, by assembling the slide cover plate improperly. I remembered the previous owner had said a mechanic had told him the plate was hard to get on. That was why. The shop manual had warned about this, but like the others he was probably in too much of a hurry or he didn't care.
>
> While at work I was thinking about this same lack of care in the digital computer manuals I was editing. Writing and editing technical manuals is what I do for a living the other eleven months of the year and I knew they were full of errors, ambiguities, omissions and information so completely screwed up you had to read them six times to make any sense out of them. But what struck me for the first time was the agreement of these manuals with the spectator attitude I had seen in the shop. These were spectator manuals. It was built into the fomat of them. Implicit in every line is the idea that 'Here is the machine, isolated in time and in space from everything else in the universe. It has no relationship to you, you have no relationship to it, other than to turn certain switches, maintain voltage levels, check for error conditions. . .' and so on. That's it. The mechanics in their attitude toward the machine were really taking no different attitude from the manual's toward the machine, or from the attitude I had when I brought it in there. We were all spectators. And it occurred to me there *is* no manual that deals with the *real* business of motorcycle maintenance, the most important aspect of all. Caring about what you are doing is considered either unimportant or taken for granted."

Pirsig is fascinated with the opposition between what he calls the "classic" and "romantic" views of the world. At the beginning of a long series of passages in which he defines these concepts at length, he gives these examples:

> "The romantic mode is primarily inspirational, imaginative, creative, intuitive. Feelings rather than facts predominate. 'Art' when it is opposed to 'Science' is often

Manuscript received January 29, 1976.
Mr. Lufkin is with Honeywell, Inc., Minneapolis, MN 55408.

Reprinted from *IEEE Trans. Prof. Commun.,* vol. PC-19, pp. 4-6, Mar. 1976.

romantic. It does not proceed by reason or by laws. It proceeds by feeling, intuition, and esthetic conscience . . . . The classic style is straightforward, unadorned, unemotional, economical and carefully proportioned. Its purpose is not to inspire emotionally, but to bring order out of chaos and to make the unknown known. It is not an esthetically free and natural style. It is esthetically restrained. Everything is under control. Its value is measured in terms of the skill with which this control is maintained.

To a romantic this classic mode often appears dull, awkward and ugly, like mechanical maintenance itself. Everything is in terms of pieces and parts and components and relationships. Nothing is figured out until it's run through a computer a dozen times. Everything's got to be measured and proved. Oppressive. Heavy. Endlessly grey. The death force.

Within the classic mode, however, the romantic has some appearances of his own. Frivolous, irrational, erratic, untrustworthy, interested primarily in pleasure-seeking. Shallow. Of no substance. Often a parasite who cannot or will not carry his own weight. A real drag on society. By now these battle lines should sound a little familiar.

This is the source of the trouble. Persons tend to think and feel exclusively in one mode or the other and in doing so tend to misunderstand and underestimate what the other mode is all about. But no one is willing to give up the truth as he sees it, and as far as I know, no one now living has any real reconciliation of these truths or modes. There is no point at which these visions of reality are unified.

And so in recent times we have seen a huge split develop between a classic culture and a romantic counter-culture—two worlds growing alienated and hateful toward each other with everyone wondering if it will always be this way, a house divided against itself. No one wants it really—despite what his antagonists in the other dimension might think.

That is a fair sample of Pirsig's style: plain, familiar, straightforward, but flexible and adaptable, and always lucid. Elsewhere in the book he describes far more sophisticated ideas, but he is always readable.

In the course of the next 300 pages, Pirsig continues his autobiographical narrative (which includes the recollection of parts of a frightening experience of insanity) and develops his concepts of the "classic" or "theoretic" view of the world on one hand and of the "romantic" or "esthetic" on the other. How he brings these antagonistic modes of thinking to an eventual reconcilation is too much to summarize convincingly here. But he does so by means of his concept of *quality*, which involves both. And to the skeptic's question as to whether this *quality* resides in the *object* (and therefore should be possible to *sense* and *measure*) or in the *beholder* (and is therefore subjective and not real) he answers that it is neither, but rather a third entity in its own right, that cannot be measured but must be perceived by intuition.

Thomas E. Pearsall, Professor of Technical Communication at the University of Minnesota, addresses Pirsig's theme in his book, *Teaching Technical Writing: Methods for College English Teachers.*[2] Pearsall sees the dichotomy between the *romantic* and the *classic* view of the world as a frequent obstacle for literature teachers suddenly faced with the task of teaching technical writing. In analyzing this problem, Pearsall says:

". . . . some teachers, used to the heightened, emotional language of literature, may find the objectivity of scientific and technical writing flat and dull. English teachers do have to come to grips with this problem, and I do not wish to minimize it. Because of the self-selection processes that help decide people's careers, a fundamental difference may exist in the way some English teachers look at the world as compared with a scientist's view. . . .

Though I may be overgeneralizing, I'll risk saying that more English teachers are romantics (in Pirsig's sense) than are classicists. Yet the students in the technical writing classes will be largely classicists.

Perhaps simply recognizing the problem is a start on solving it."

We may well ask how a book as serious and as difficult as *Zen and the Art of Motorcycle Maintenance* ever became a best seller and found its way into so many universities.

First, there is the human appeal in the narrative of the motorcycle journey. To be sure, the journey symbolizes both the philosophical exploration and the voyage of self discovery, but it also relates the father's concern for his son, who is struggling with a form of mental illness similar to his father's. Secondly, the book appeals by the wide range of its cultural concerns, which go far beyond the scope of systematic philosophy.

It makes the antithetical relation between the rational and the irrational come alive in a number of vivid examples. It touches upon the irrationality of the presumed "scientific objectivity" of much technical writing. It also exhibits the paradox of our educational system in which the brightest students are frequently the ones that fail. It points out the fatal defects in the traditional ways of teaching freshman English, and the irrelevance of the traditional emphasis on "correctness" in writing. It suggests that the conventional grading system is in some ways counter-productive, that the better students don't want it, and that it is used to cover up inadequacies in teaching.

These observations on education are brief but relevant asides to the main discussion, for the contrast between the rational and the intuitive views of the world is never out of sight.

Pirsig also touches upon the supposed materialism of scientists and quotes Albert Einstein on scientists' motivations. He summarizes Poincaré's argument to the effect that scientists work mainly by means of intuitive, subconscious choices among many alternatives, much as artists do.

The Greeks, Pirsig points out, did not make anything like the distinction between art and technology that we have adopted to our loss. But, he says, "we have artists with no scientific knowledge and scientists with no artistic knowledge and both with no spiritual sense of gravity at all, and the result is not just bad, it is ghastly. The time for real reunification of art and technology is really long overdue."

In a short passage about what in technical writing is called an "analytic" description, Pirsig exhibits such a description of a motorcycle. The text is beautifully complete, in both concept and function of the machine; the writing is perfectly logical in both classification and sequence. It is economical to a fault, and it uses the plainest possible English. And of course all of it is perfectly *correct.* In fact, it could serve as the very model for all such descriptions, for a textbook or a beginner's manual.

But, says Pirsig, it is a failure on several counts. First and most obvious, it is so dull that no one will ever read it unless he absolutely has to. Then, for all its clarity and correctness, it is impossible to understand unless you are already familiar with the whole machine. And the observer has been left out, so that there is no human point of view for the reader to look from. In fact, Pirsig says, "there are no real subjects in this description. Only objects that are independent of any observer." And finally, there are no value judgments of any kind, not even carefully qualified comparative ones.

And yet the machine so described was built by human beings for the use of human beings.

Pirsig is asking us to reconsider the state of mind that has made it possible for us not only to accept but even to admire that kind of writing.

---

[1] New York; Morrow, 1974, 412 pp. $7.95; paperback, Bantam, $2.25.

[2] Society for Technical Communication, Inc., 1975, 23 pp. $2.00 members; $3.00 non-members.

This article is based on a contribution to the panel discussion, "The Critics' Circle: An Analysis of the Literature on Writing," at the 22nd International Technical Communication Conference, Anaheim, California, May 16, 1975. In my absence, the presentation was made by Kathleen Block, of Design Data Laboratories, Arlington, Virginia, who added a number of ideas of her own for which I am grateful.

THE WRITING PROCESS AS A DESIGN TOOL

Erno R. Bernheisel
The John Hopkins University Applied Physics Laboratory

The purpose of technical writing is seen as more than just the creation of a product that stands in a dependent relation to the system. Rather, it is seen as a valuable process that can aid in designing the system. To secure the benefits of that process, the designer must free himself temporarily from certain inhibiting forces and then concentrate on writing precise explanations.

Writing maketh an exact man--Bacon

Technical writing is properly taken to be the creation of a secondary product that describes, explains, reports on, or stands in some relation to a primary product--where the primary product is a nonlinguistic entity such as a unit of equipment, a research activity, or a procedure. For simplicity, we shall refer to the primary product as the "system," though it is by no means restricted to an assemblage of physical components, and to the secondary product as the "document." We shall also refer to those who create the system--scientists, engineers, and programmers--as "designers."

That the document depends on the system for its very existence seems so obvious as to be taken for granted. Though there are exceptions, we do not in the course of our day-to-day work write documentation for systems that will never exist (i.e., we do not ordinarily write manuals for equipment that will never be built, reports on research that will never be undertaken, or instructions for procedures that will never be carried out.) Proposals and specifications are exceptional cases in which the documentation precedes the system (if the system comes into existence at all), but they are a class of documents different from what we have in mind here. In most cases it is fair to say that the relation of the document to the system is much like that of the map to the territory.

Given this dependency of the document on the system, it is not surprising that the value of the technical writing process is usually seen as residing exclusively in the result of that process; it is the document as such that is valued, not the process that created it. With that focus on the end-product, the process of writing, especially when undertaken by the designers themselves, is usually seen as nothing more than the necessary (and perhaps unpleasant) means to the desirable end; in the view of many designers it is simply an onerous task to be done after the "real" work (the design of the system) has been completed.

Since the system is often designed before the writing begins, the most the document can do is faithfully mirror the system, and that is the standard by which the entire writing effort is adjudged successful or unsuccessful. What else might the writing process do? I would suggest that it can be a useful tool in designing the system.

The view that the document can only mirror the system, as the map does the territory, fails to exploit a phenomenon that the designer can use to his advantage; namely, that the mere act of sitting down and writing about a subject causes the writer to see that subject in a new light. Somehow, a writer's understanding of his subject is not the same after he has written about it, but is in fact increased. Subtle flaws in the design of the system or the reasoning on which it is based are frequently revealed when the designer writes up his work. Or a technical writer, if he has an understanding of the subject approaching that of the designer, sometimes discovers minor "technical" errors or design shortcomings that escaped the attention of the designer. Ordinarily, little practical use can be made of these discoveries since the system design is usually frozen or the investigation completed by the time the writing gets under way.

Before turning to ways in which the writing process can be used as a design tool, we should give some consideration to just how this phenomenon works. Two factors seem to be important. One is that the discipline of writing sound English sentences to explain precisely and logically all the system functions causes the designer to think about certain details of the system that he may have overlooked. The second factor is that certain inhibiting forces interfere with his (or anyone's) ability to write such sentences. We shall here give somewhat greater attention to the second of these factors.

The relation of language and thought and the question of what goes on in the writer's mind when he is composing have been matters of speculation for philosophers, poets, psychologists, and linguists--but they have found little to agree on. Writers themselves are not much help, for as Hersey says, "The testimony of most writers about the process of creation is vague."[1] However, there does seem to be some little agreement on the notion of two contending forces at work: one operates at the unconscious level (the Muse, ectatic inspiration, or in Hersey's terminology, the supplier); the other, a censor, operates at the conscious level. Both the supplier and the censor have desirable and undesirable characteristics. The supplier is the necessary source of all verbal behavior, but without the censor the supplier would produce a message meaningful only to the person generating the message. On the other hand, the censor, while shaping the message into something meaningful, may also be a severe inhibiting force that exalts form at the expense of content. The supplier and censor appear to be very closely related to Skinner's notion of the ecstatic generation of verbal behavior and its subsequent, euplastic self-editing.[2]

This may sound like so much mysticism, of interest perhaps in "imaginative" writing, but hardly appropriate in technical writing. Yet it seems undeniable that many designers, who clearly are in the best position to say something meaningful about their systems, have great difficulty in communicating that information to others. We shall return to this idea shortly.

To use the writing process as a design tool, one must begin the writing at the earliest possible moment and continue it faithfully along with the system development. Quite apart from any advantages to be gained in system design, there are a number of secondary benefits to be obtained from such a practice: There is a greater chance the documentation will always be up to date; it will be completed at the same time as the system; and there will be some protection against the project being damaged by the sudden departure of a designer.

To get the writing started early and continue it as the system develops is no easy task. To begin with, the designer is too busy creating the system to take time to document it, especially if he sees documentation only as a valuable product, not as a valuable process. The situation is not much improved if a technical writer is assigned to the task since the designer is also too busy to talk to him. Moveover, documentation during the early stages of the project seems premature and wasteful to many because the system is still in flux and documentation produced then may soon become obsolete. Add to these problems the aversion that many people have to writing, and you have a situation in which it is almost impossible to get anything down on paper.

While all those problems are serious, perhaps the one most difficult to deal with is the designer's aversion to writing. In that connection, one frequently hears that scientists (or engineers, or programmers) do not like to write; indeed it has been suggested that there is something antithetical between the way their minds work and the way a writer's mind works. That strikes me as sheer nonsense. What is probably closer to the truth is that most people (including, but not restricted to designers) do not like to write. Dewey would attribute this aversion to that part of the early education process that emphasizes criticism of mistakes, thereby replacing creativity with self-consciousness and inhibition.[3] This clearly sounds like the contention between the supplier and censor with the censor being the dominant force. Dewey's theories are, of course, unpopular these days, and it is not our intention to attack or defend them. But regardless of the original cause and subsequent development of this aversion, it is real and must be dealt with.

The waste of effort from premature documentation may not be so great as first imagined. If the writing helps in the design process, then some savings will be realized there. Further, at many steps in the documentation process, the designer can expedite his work by using an audio tape recorder, provided a secretary is available to transcribe the tapes. Of even greater value, though less likely to be available, are word-processing systems, especially the more powerful (computer-based) systems that facilitate the adding, deleting, or moving of text.

In using the writing process as a design tool, the designer must produce two kinds of documents; the first is a collection of design notes and the second is a draft of the system document itself. The design notes serve as a repository for information about the system design, and the system document serves as a vehicle for the designer to test the soundness of his design. The system document will subsequently become the input for the technical writer or editor, if such help is available; otherwise, it will have to be revised for publication by the designer himself.

The designer may organize the design notes in any way convenient for him, though it is useful to have them in a loose-leaf notebook so that they can be re-organized if necessary.

In the design notes, content is emphasized over form. The designer is to be free to write in whatever style is most comfortable to him, thereby reducing his aversion to the act of writing. To the extent possible, the supplier should be in control and the censor's role minimized. An important mechanism for accomplishing that is to ensure that all personnel involved in the project understand that the design notes are for the eyes of the designer only. The purpose here is to effect a transfer of information from the designer's mind to paper as easily as possible. Anything that interferes with that transfer is to be avoided. The writing may be very informal with little regard being given to the niceties of English style. Abbreviations and the designer's own form of shorthand may be used freely with no concern for whether anyone else understands them. The designer may want to develop an artificial language of his own, something rather like the higher level programming languages, if that allows him to write more rapidly.

All pertinent information about the system or that portion for which this designer is responsible should be included. For example, the design notes should state the specifications for the system; they should describe any problems the designer anticipates in meeting those specifications, how he plans to overcome those problems, and what alternative solutions are available. For hardware systems, the design notes should include the characteristics of the inputs the system will receive, the outputs it must produce, and the interfaces between it and any other systems. For software systems, such information should be included as file formats, programming conventions, and subroutine calling sequences.

Also contained in the design notes should be information on decisions to be made and the implications of those decisions for other decisions. Assumptions that the designer has made should be stated as such, along with information on how and when those assumptions can be verified. Any tests, experiments, or data needed to verify the design should be described. Finally, the designer should not hesitate to include such elements as reminders, questions, guesses, doubts, and opinions.

The design notes should be continually updated as more information becomes available. Information that no longer seems useful should be flagged in some way but should not be removed, for such information may later be valuable. In brief, the design notes should encompass all the designer knows about the system, and in preparing them he should be unencumbered by any restrictive forces that tend to interfere with his getting the information on paper

Early in the project, as soon as the system's overall function has been defined exactly, the designer must begin work on the system document. This work will be done concurrently with designing the system and writing the design notes.

For any document, the writer needs some overall concept of its structure to obtain a sense of direction, and to achieve that sense he needs at least a rudimentary outline. If a technical writer is available, he should be called upon to help with the outline, but writing the document remains the responsibility of the designer alone. In the early stages of the project, when information is minimal, precious time should not be wasted in attempting to develop a detailed outline. The designer should make a sketchy one, recognize that it is not inviolable, and then start writing. As the writing proceeds, the outline must be refined continually to a level of detail at least one step below the writing. From time to time, the structure and logic of the outline should be compared with those characteristics of the system itself. Disparities between the two may be caused by flaws in either.

The writing will consist principally of descriptions and explanations of how the system works. That task will have been made easier by the designer having prepared the design notes. The designer begins by writing an explanation of the overall system, treating each subsystem as a "black box." Then as he designs the subsystems, he must concurrently document them in a similar manner, down to the lowest level of design detail. Thus the system document is continually supplemented by explanatory information in a top-down manner as the design progresses.

In his writing, the designer must test the soundness of his design at each level by explaining what he has done and how it works with all the precision and exactitude he can muster. He should go through every step in the functional operation of the system without skipping over those that are "obvious" or "self-evident." There are several reasons for requiring that discipline. Those steps may contain subtle flaws that he has thus far overlooked and would not be revealed until a rigorous system test was performed. In writing an absolutely logical explanation of the operation, the designer may discover such flaws. Moreover, while certain characteristics of the system may now be obvious to him, because he is deeply involved in the system design, they may not be so obvious to others. If he has reasonable doubt about whether or not to include certain information in this document, chances are it should be included. It is considerably easier to delete excess information in the final document than it is to supply necessary information that is missing.

Since this document must be comprehensible to the technical writer who will use it as input, the designer must here be more concerned with form than he was when writing the design notes. Nevertheless, the emphasis is still primarily on content.

If technical writing or editing help is not available, then the designer must prepare the final draft for publication. Here the emphasis clearly shifts to casting the information in a form that is comprehensible to the reader. The writer must impose upon himself a kind of mental dualism, as it were, forcing his mind to commutate rapidly between the position of the reader, who will have only the final text and illustrations from which to catch the exact meaning, and the position of the writer, who alone can supply the appropriate words and complementary illustrations. He must search for ambiguities and other possible sources of misapprehension, anticipating any questions that might occur to the reader. He must assume that the reader will be critical and will expect a logical, cause-and-effect explanation of the system. At the same time he must question whether or not the system itself exhibits that relationship. If it does not and is still amenable to change, then the process that discovers that discrepancy is to be valued.

Overall, this technique requires a change in attitude on the part of the designer, who must come to see the document not simply as a reflection of the system but as an aid in designing the system. It is intended to free him in the early stages from the inhibiting forces that may make writing a threatening undertaking, but ultimately it requires him to test the soundness of his design by determining whether or not it can be explained in straightforward English. The exactness he seeks in his design may be achieved by seeking that same exactness in his writing.

NOTES

1. John Hersey, The Writer's Craft (New York: Alfred A. Knopf, 1974), pp. 7-9.

2. B. F. Skinner, Verbal Behavior (New York: Appleton-Century-Crofts, 1957), p. 382.

3. John Dewey, How We Think, 1st Gateway ed. (Chicago: Gateway, 1971), pp. 245-46.

Clear Thinking and Its Relationship to Clear Writing

Charles H. Lang,
Methods and Systems Analyst,
Engineering and Research Staff,
Ford Motor Company

Once upon a time there was an engineer, a highly talented young man, who was loaned by his company to the government of a small country of the free world. His job was to help that country solve a particularly vexing drainage problem.

The government was a kingdom. Eventually this highly talented engineer hit upon the solution to the problem, and the king in his gratitude presented the engineer with a bag of precious stones: rubies, emeralds, diamonds--a fabulous collection.

But the engineer, being a highly ethical fellow, and knowing that such a gift might raise eyebrows among congressional committees, refused the present.

"Why, I can't take that," he protested. "There must be a fortune there. The job wasn't worth all that, and anyway, my company pays me well; they sent me here to do a job and I did it. Thanks very much, but no thanks," he said.

The king, however, kept insisting that he would like to show his appreciation, and, in order not to appear ungrateful, the young engineer finally said,

"Well, OK--tell you what. I like to play golf in my spare time on week ends, and my golf clubs are getting pretty well beat up--why don't you just buy me some new clubs and we'll call the whole thing even?"

The king protested that such a gift wouldn't nearly show how indebted the entire country was to the young man, for his wonderful job, but agreed. Since golf clubs were not to be bought in that country, the engineer was assured he would hear from the king shortly after returning to his job in the United States.

You can imagine the surprise of the engineer a few weeks later when he did indeed get a message from the king. "I have purchased in your name," the message said, "six of the finest golf clubs I could find. I am very sorry, however, that only two of them have swimming pools."

I tell this story here for two reasons. One reason is that tradition holds that speeches are supposed to begin with a story. (This is contrary to what I used to tell my speech classes back in the days when I taught college freshmen.)

But the other, more important, reason is that the story demonstrates a very important point that I want to make here today.

My topic is entitled "Clear Thinking and Its Relationship to Clear Writing." My thesis is simply stated: it's merely that clear writing is impossible without clear thinking preceding it (and conversely, that if we are thinking clearly on a given subject, we can write clearly about it.)

If we do not think clearly we cannot achieve the basic purpose of writing--communication. Our writing becomes fuzzy, ambiguous, proposes invalid conclusions, makes unsound recommendations.

Unclear thinking is evidenced in writing chiefly in three ways:

> confusion of fact with non-fact,
>
> failure to assign precise meanings to words,
>
> drawing conclusions which are unwarranted on the basis of facts presented.

CONFUSING FACT WITH NON-FACT

Let's look at each of these in a little more detail, then see what, if anything, we can do about them. First of all, let's take up this item of confusing fact with non-fact.

One big cause of communication breakdown--and incidentlly, of ill-will between persons--is our tendency to accept inferences and assumptions as facts, and especially the tendency to behave as though they were facts.

Take a simple example: Each of you entered this room and assumed, among other things, that the seat you now occupy was solid. Since you assumed the seat would hold you, you then behaved as though your assumption were a fact and sat down.

Now here is the point: accepting that assumption as a fact had this important result--you stopped looking further; you stopped observing <u>and</u> you stopped thinking. If the chair had collapsed you would have been taken completely by surprise.

Another example, less obvious, but more practical: Some time back I was called on to help prepare an SAE paper for one of the engineering supervisors. He was to talk to a group of engineers like many of you on one of the company's product lines.

In the rough he had prepared--and I use the term loosely--I came across the words "ample" and "adequate" with painful frequency. I pointed out that such words as ample and adequate were really not very informative, especially to a group of engineers. "Those words, " I said to him, "represent your opinion. I should think an audience of SAE people would like you to be more specific. "

His answer? "Everybody admits the cab space is adequate. Why bore them with the figures on it?"

Now I want you to note three things about this statement:

1) he confused fact with non-fact, i. e., "adequate" with 4 x 6 x 8;

2) he made an assumption, an inference, in thinking that "everybody admits so-and-so;" and

3) he then behaved as though his assumption were a fact, that is, he gave his speech on the assumption that everybody admitted the cab space was adequate.

One result? He was criticized, not only by some members of the SAE, but by his boss who was present in the audience, for not being--as they put it--technical enough. In other words, for not being factual.

A final example of this tendency to confuse fact with non-fact, that is to make an assumption and then to behave as though that assumption were a fact:

Early last season the management of the Detroit Tigers, like the managements of all other teams in the American League, had to give up in a draft a certain number of their players to the two new teams, Washington and Los Angeles. But while they had to allow some players to be drafted, at the same time each club was allowed to protect a certain number of their players, that is, they were allowed to say, "You may draft three of our players, but you may not pick them from this group. "

Now, Detroit assumed that both Washington and Los Angeles would want seasoned veterans for the nucleus of their clubs, and on the basis of that assumption, placed as many of their seasoned players as possible on the protected list.

Manager Rick Ferrell then expressed great surprise when Los Angeles drafted three young Tigers who had just the previous winter been promoted from Detroit's farm clubs. Because Ferrell behaved as though his assumption were a fact, he was surprised when it turned out to be an assumption after all. Worse, Ferrell then had to re-do a lot of the ground work that he had already spent up to three years doing.

PRECISE MEANING

Let's look now at another way in which unclear thinking reveals itself in writing: failure to assign precise meanings to words. As an example of how this particular aspect of unclear thinking turns up in writing, let's look at a policy letter published by one of the nation's automobile manufacturers.

Policy letters, you may or may not know, are the official declaration of what the company stands for and how it operates--the bible to which all its management turns for guidance. So here at least one should expect to find clarity of thought and expression.

The policy letter of this automobile company attempts to fix responsibility, and it says, in part...

> The Vehicle Product Engineering Office will have final responsibility for the complete vehicle.. The Basic Manufacturing Groups will have basic responsibility for body, engine, electrical, transmission, etc...

A continuing source of confusion has been the meaning of the word "final, " the meaning of the word "complete, " and the meaning of the word "basic. "

That is, if the one office is to have final responsibility for the complete vehicle, then what basic responsibility does the other office have? What does final mean? What does basic mean?

You may ask, "Why not simply straighten the thing out and republish the Policy?" This has not been done, because the two groups cannot agree among themselves on the meaning of the words; that is, they cannot express their thinking in words precise enough to satisfy each other.

Another example: What meaning would you assign to these words in a letter of recommendation? "I have employed this man for the past two years and I cannot recommend him too highly. "

And failure to assign precise meaning to one word may have changed the course of the lives of every one of us here today. Stuart Chase tells us about the Japanese word "mokusatsu. " It seems it has two shades of meaning which are similar, but which do have important differences. It means 1) to ignore, and 2) to refrain from comment. The release of a press statement using the second meaning in July of 1945 might have ended the war then.

The Emperor was ready to end it, and had the power to do so. The cabinet was preparing to accede to the Potsdam ultimatum of the Allies--surrender or be crushed--but wanted a little more time to study the terms.

A press release was prepared announcing a policy of mokusatsu, with the "no comment" implication. But it got on the wires with the "ignore" implication through a mix-up in translation. It came out "The cabinet ignores the demand to surrender."

To recall the release would have entailed an unthinkable loss of face. Had the intended meaning been publicized, the cabinet might have backed up the Emperor's decision to surrender, in which event there would have been no Hiroshima, no Nagasaki, no Russian armies in Manchuria, no Korean war--due to one word that had no precise meaning attached to it.

A more practical example of how failure to assign precise meanings to words confuses readers and listeners can be found in any campaign oratory. The Nixon and Kennedy campaign is the most recent nation-wide example. Both these men discussed at length the presence or absence of a recession, American prestige abroad, and military preparedness. Could there have been such a difference of opinion between them if they had assigned precise meanings to those words?

Let's end these examples on a lighter note. After an irate military man had complained to an editor about a line in a society note, the editor stormed into the writer's office and demanded to know the meaning of this line: "Among the prettiest young ladies present at the garden party was Colonel John Bloodstone."

Well, said the writer, that's where he was.

IMPROPER CONCLUSIONS

A third way unclear thinking shows up in unclear writing is in our practice of drawing conclusions which are unwarranted by the facts we have presented.

In one issue of Communications is an article called "The Changing Advertising Symbol" which draws this conclusion: "Stocks which sell at many times their book value do so because those companies have readily identifiable pictorial symbols on all their products."

This may be a valid generalization, though I doubt it, but the point is, the facts presented in the article do not warrant such a generalization, because too many facts refute it. Many stocks soar without any product, for instance.

In a similar case, I recently witnessed one committee report, which, after making a recomendation regarding the use of engineering drawings, made this generalization. "Intangible benefits," the committee wrote in one part of its report, "will accrue from having a complete record on one drawing..."

Now, you would have to read the report to determine whether or not the generalization was sound, but you don't have to read the report to see that this particular bit of writing is unclear simply because it does not convey any information--the basic purpose of a report. (Just as a bit of incidental intelligence here, you will be interested in knowing how that conclusion was handled by an analyst asked to comment on the report. The analyst commented on each of the points in the report; then, in referring to that point, said with tongue in his cheek, "Certain intangible faults in the proposed system tend to balance the benefits."

A third example of how unclear thinking shows up in the drawing of a conclusion which certainly is not warranted by the facts presented: This one is from the Detroit News, and it is not on the sports page, but on page one.

"...The Detroit Lions can really look to a happy new year. The Lions usually win the championship the following year after being lucked out of the title." That's the conclusion. The facts the writer presents to support that conclusion are two.

1) "That's the way it was four years ago when Detroit lost the final game of the season to Chicago... The following year they broke through for the National Football League championship. 2) In 1951 the Lions similarly were beaten by half a game. They surged to the title in 1952."

FACTS

Well, now let me make an assumption of my own and then procede to behave as though my assumption is a fact. Let me assume that with the use of these examples I have convinced you that unclear thinking does result in unclear writing. I may, in fact, be drawing a conclusion here that is unwarranted by the facts. However, assuming you are convinced, what can be done about these things? What can we do, as writers, to make our communications meaningful, concise, and possess all of the other desirable qualifications that have been discussed and will be discussed from this platform?

Let's take up the points one at a time. First, fact and non-fact. Obviously, if we are to distinguish between fact and non-fact in our writing we must first be clear about the difference in our thinking. So let me define--let me assign a precise meaning to the word fact.

Facts, first of all, are observable--they can be verified. They are agreed on by numbers of people. The number of chairs in this room is a fact; the quality of them is a non-fact. We can count the number of seats and chances are that we will all agree on the result. We would not agree, however, on how comfortable they are.

Second, facts exclude judgements. Statements of facts contain no "loaded" words.

In the area of non-factual statements are non-observable items, or those statements which are unverifiable. This includes assumptions, opinions, inferences, and the like. We cannot verify, that is, observe a man's thinking, so a statement that "John is angry" is non-factual because we cannot observe what is going on in John's mind.

It is factual to say John is shouting and pounding his fist on the table and when I ask him what is wrong he throws the telephone at me--all these things are observable. But when we say he is angry, we are looking into his mind. Whenever we do that we are inferring, that is, we are looking at some things (pounding, shouting, throwing) and putting meaning into those actions.

The practical application of this principle in writing is that the statement about John's "anger" fails to suggest any specific means of improving John. On the other hand, if we report the observations that led us to the opinion, the situation is not only more clear, but more hopeful. We might not eliminate the "anger"--if that's what it is--but there are fairly definite steps we can take to remove John's incentives or opportunities for pounding desks and throwing telephones.

Another way of putting this principle might be this: we can end an argument with facts; we cannot end an argument with opinions. S. I. Hayakawa puts it this way: We cannot determine the truth or falsity of the statement that "Angels watch over my bed at night," because there is no way of factually proving or disproving it. However, he says, whether we believe in angels or not, knowing in advance that any argument on the subject is both endless and futile, we can avoid getting into fights over it.

How do we recognize fact and non-fact? One way is to ask ourselves questions about the statements we make in our writing. Could anyone--Protestant, Jew, Catholic--disagree with this statement? Does the statement contain "loaded" words like "awkward", "smooth", "gawky", "stylish", etc. ? These kinds of words reflect the writer's opinion. Dr. Irving Lee analyzed factual statements this way:

Statements of fact:

> made after an observation;
> are limited to what can be seen;
> we can make only a limited number (since we can't see all there is to see);
> the more who see, the more who will agree.

Statements of inference, on the other hand:

> can be made anytime--before, during, after observation;
> go beyond what can be seen;
> we can make an unlimited number;
> have more room for agreement;
> are predictions, guesses, tell about another's feelings, attitudes, motives, etc.

Wendell Johnson sums it up another way. He says that a fact is an observation agreed on by two or more people situated, qualified, and equipped to make it, and the more agreeing the better.

Skill in recognizing this distinction between fact and non-fact is basic to clear communication. As I mentioned earlier, one reason communication breaks down is that we often tend to accept inferences as facts, and then behave as though they were facts. If our way of talking and writing does not make this distinction, we must expect disagreements. A constant awareness of the difference can have these practical results:

> we will be less surprised when the things we expect to happen don't happen;
>
> we will try to get more information--ask more questions--in an effort to verify or deny our inferences;
>
> we will find it easier to change our minds when changing the mind is desirable;
>
> we will be able to review a decision with less embarrassment.

Now I want to make this thing clear: I'm not saying that our writing should contain nothing but factual statements. I'm not suggesting that opinions, assumptions, inferences, and judgments must be excluded from writing. My point here is that we must always be aware of when we are using non-factual statements.

The use of opinion, point of view, and imagination can be constructive--indeed, we are often expected to use these things in writing. But it is the deliberate or careless or unknowing misuse that causes trouble.

ON BEING PRECISE

Now let's look for a few moments at our second point of confusion--failure to assign precise meanings to words. How do we go about, in writing reports, news stories, features, etc. being sure that our reader knows how we are using the words we use? This matter of definition is more complicated than it seems at first glance. Because how can we talk about the "meaning of words" when words actually have no meaning?

Let me repeat that for those of you who feel you couldn't have heard me correctly: words themselves have no meaning. In the words of Dr. Lee, words don't mean--people mean. The only meaning attached to any word is the meaning we ourselves put into it when we hear it. And we may change that meaning everytime we hear it.

To demonstrate this point rather simply, I'm going to ask you to play a little game with me. If you think games are childish, I'll ask you to humor me--go along with a gag just to prove my point that words have no meaning.

Has anyone in here had any experience with diplopia? In other words, if I were to ask anyone of you, in ordinary conversation outside the scope of this discussion, "What does diplopia mean?" you would have to say "I don't know." Fine.

Now you see the point. One minute ago that word meant nothing to you. Now it does. What happened in the meantime? This is still the same word it was before. It is spelled the same way; it looks the same on the blackboard; for all practical purposes its physical characteristics are the same as they were one minute ago. Yet somehow or other you now attach a meaning to it whereas you didn't before.

What happened? The answer is simple: nothing happened to the word, but something happened to you. In that span of time you had an experience, and you are now able to make a noise to describe that experience--the noise you make with your tongue, lips, teeth, lungs when you say "diplopia." That is, if you wanted to bring to another person's mind the experience you just underwent, you could do so by making the noise diplopia.

Because that's what we do when we speak or write. We use words to label experiences, to picture what we have in our minds, and we expect these words --noises or scribblings--to recall similar experiences --to call up similar pictures--in the mind of our reader or listener. Similar, but not identical.

Why not identical? Because the experiences being summed up by the one person are never exactly the experiences which the word sums up for the other person.

Take the word "diplopia." The very next time you hear the word, a slightly different picture--a recollection of an experience--will come to each of you. To one it will be a picture of a bored audience; to another it will recall a fascinating new concept of meaning; to a third it will be a hazy recollection of new acquaintances; to still another it may even be associated with something completely divorced from this meeting, as, for instance, if the word serves merely to trigger, next year sometime, a picture of a bar in downtown Fort Collins. You mean some things by the word; an eye doctor means much more by it, because he has had more experiences which may be recalled when he hears it or uses it.

So words, then, have no meaning. Their purpose is to recall for us a series of experiences. So it is that we change, every so slightly, the meaning of a given word every time we use it or hear it or read it, because things have been happening to us since the last time the word was used.

Now, perhaps you begin to see why the confusion, and why the necessity for definition. Words symbolize experiences, and experiences differ. Even such a common-place word as "chair" must symbolize different experiences--hard, soft, broken, stuffed, folded, and so on. Therein lies a second source of confusion--we must use the same word--chair--to symbolize many different experiences with chairs.

In other words, there are many, many more things to be spoken of than there are words to speak of them, so the same word must be used to talk about different things. You see, then, the folly of trying to "freeze" the meaning of a word, as some advocate. "Why not establish a hard and fast meaning to every word," they say, "and thus there will be no confusion?"

Well, one sees why this is impossible when one recognizes the function of words, which is to symbolize experiences. What these people advocate, in effect, is passing legislation which would require every human being to have exactly the same experience, the same number of times, and to react in exactly the same way as everybody else, every time.

Let me state this same concept another way, a way used by far more able semanticists than I. Many authors use the analogy of a map to a territory. Maps, they point out, symbolize a territory. They are not the territory, but they stand for a territory. A curved line represents a river, parallel lines joined by short dashes represent railroad tracks, certain types of circles stand for cities of certain size, and so on.

Now, the curved line, of course, is not a river, but you and I have agreed to let it stand for a river for purposes of communicating with each other. By means of this type of communication, I can guide you to Texas without actually having to lead you there in person.

So with words. The noise "diplopia" is not double vision, but you and I have agreed to let it stand for double vision for purposes of convenience. The noise "chair" is not something you sit on, but you and I have agreed to let it stand for a certain action--an experience--and thus we can communicate our thoughts to each other without the necessity of actually physically handling a chair.

Words, then, symbolize reality, just as a map symbolizes a real territory. Now, let me take this concept a little further; again I'm going to ask you to humor me and play a little game with me. When I make a noise, when I say a word, I want you to point to the reality I am symbolizing.

COAT CHAIR PENCIL CEILING MAN

Before going on I want to point out one thing: I've used only very common words, yet many of you pointed to different realities--so there's one possibility of confusion. In a written report you as the author aren't around to tell the boss, "No, I didn't mean that pencil; I meant this one," meanwhile pointing to the reality--providing him with an experience, as I did for you with diplopia.

Now let me get to a second important point about definition. Let's play some more:

Democracy Love Management Labor Union

Why can't you point to something? Aren't these words realities? Haven't you had experiences with them? When you answer those questions you are well along the road to an understanding of the problem of definition.

Democracy most definitely is real; we here most assuredly have had experiences with democracy. But democracy isn't A thing or A reality; democracy, like all of the other so-called abstract words, is a collection of many experiences, many sets of circumstances, of relationships to which you and I have put a collective label.

We use that one label to recall for a reader or listener dozens, scores, hundreds of experiences, each our very own. Is it any wonder, then, that sometimes confusion arises in our communication process?

So much for why definition is necessary. Now: how can we go about being reasonably sure the reader of our communication knows how we are using our words? We've already taken the first step when we realize how easy it is to confuse. That makes us cautious. We then try to use words that have roughly similar meanings for all of us, although we recognize our writing can't be limited to a vocabulary of words like pencil and cat and desk without raising eyebrows among the brass.

Thirdly, when we do get into words that have varieties of meanings, we define in terms of experiences, not in terms of other words--we try to show real people doing real things. We try to recall for our reader experiences he has had which are roughly similar to the experiences we have had and which we are symbolizing with the use of the particular word in question.

I got a lesson in this from my daughter when she was seven years old. She asked me what the word "vie" meant. I explained (I thought) that "vie" meant to compete. She looked doubtful for a few seconds, then brightened. "Oh," she exclaimed, "you mean like when you and I raced to see who could get to the corner first."

Real people doing real things--definition in terms of experiences--is called operational definition. Walpole says definition is the use of a certain road to take your reader from a common referent to one which is new to him. I provided you with a common experience a few minutes ago to enable you to put meaning into a strange word.

Fourth, we must realize that words are not fixed in meaning, which is to say that the meaning you and I bring to the words we hear shifts from time to time. At one time "cowboy" in America meant traitor, to use but one of hundreds of examples.

There will be little difficulty accepting the concept of meaning change when we realize that words symbolize experiences, and as experiences build up, slightly different pictures are called to mind upon hearing or seeing the word.

Fifth, note that context helps determine what experiences a reader will recall. Context determines the sense in which a word is used. When we set up certain conditions in our writing we are helping our reader decide which sets of experiences to recall when he sees a given word.

The old joke about "If a rooster lays an egg on the peak of a roof, which way will it roll," and "Which is correct: yolk of eggs is white or yolk of eggs are white?" point up the idea. Our king and his engineer friend whom I talked about when I began this address demonstrate contextual confusion also.

To sum up, it's not enough that we know what our words mean--we must be sure they mean roughly the same thing to our reader.

ON DRAWING CONCLUSIONS

Now let's move to the third of the three items we have listed as causing confusion in writing--drawing conclusions that are unwarranted on the basis of facts presented.

This is the most difficult of the three concepts to grasp, mostly because all of us seem to recognize the habit in others, but fail to see that we ourselves are often guilty of drawing invalid conclusions.

It reminds me of a cartoon that appeared in, of all publications, the Wall Street Journal. The artist pictured a lady in a bookstore who was saying to the attendant, "I particularly like to read things that will help me make better persons of everybody I know."

We make inferences--draw conclusions--we generalize--I use the terms synonomously here--we make inferences on the basis of observations. We note what has happened in the past, and we predict then what will happen in the future. We note, for example, that every time we walk into an auditorium the seats are solid enough to hold us. We then draw the conclusion that if we were to sit down in a perfectly strange auditorium the seat would be solid.

We note that the union has asked for more money at each of the past contract negotiations, and we infer that it will ask for more money at the time the present contract expires. Our sports writer friend noted that twice before the Detroit Lions finished one-half game out of first place and then won the championship the following season, and he concludes that next year Detroit will win the title because this year they finished second.

A generalization or conclusion in the sense I use the words here is a statement about the unknown based on the known. It is a statement about members of a group or class--women drivers, for example--based on observation of part of that group or class--our wives, perhaps. It is a statement that says that different things are somehow similar--that pencils, for example, are made of wood. It is a process of leaving out details, in this case every detail except the detail of wood-ness.

In that a conclusion goes beyond the observable, it may be identified with the non-factual as we discussed that concept earlier.

How do we go about checking the validity of our conclusion or generalizations when we attempt to communicate? There are a number of questions we must learn to ask and, especially, to answer:

On what evidence (observations)do I base this conclusion?

Have I made enough observations?

Do these observations actually relate to my conclusion? (I note that everytime I forget my overshoes, it snows --is the forgetting related to the snowing?)

Have I unwittingly selected my evidence to support the conclusion I would like to see accepted?

Have I considered negative evidence?

Could someone else possibly draw a different conclusion from the same set of observations?

Does my conclusion suggest a finality? ("Adopting this method would improve efficiency 50%.")

THE METHOD OF SCIENCE

We would do well to adopt the method of science to our language habits, particularly our language habits in this area of generalizing. The method of science, says Wendell Johnson in his People in Quandries consists of

a) asking questions that can be answered on the basis of observations;

b) making the relevant observations, or using those made by others;

c) reporting the observations accurately so as to answer questions asked; and

d) revising conclusions previously held in accordance with the answers obtained--and asking further questions that are prompted by the new conclusions.

I have tried to point out to you today some of the pitfalls that await the unwary. I've tried to show that language, our chief means of communication, has serious limitations. My main theme has been that whenever we don't think clearly, we cannot possibly write clearly, and thus, of course, cannot achieve the basic purpose--communication.

I've indicated that unclear thinking shows up in writing chiefly in three ways:

in confusion of fact with non-fact;

in failure to assign precise meanings to words;

in drawing conclusions which are unwarranted by the facts.

And I've tried to give you a few specific ways to counteract these three weaknesses. I couldn't, of course, go into great detail on any of these three concepts, let alone all three of them, in a one hour talk. My purpose has been rather that of opening up new lines of thought for you to pursue on your own.

I hope I haven't left the impression that this is the only thing that's important about communication in general and about writing in particular. Certainly the messages of the other speakers must find their way into the communication process as well: organization, visualization, form, grammar, spelling, outlining, analysis, revision and editing--all these things are important to clear communication.

A DISSEPIMENT ON DISSERTATING

by

Raymond P. Perkins*

The form, format, and function of a research paper, and the sequence of activities involved in completing one, have been evolved over a long and carefully analyzed process. Apparently many who approach this type of activity for the first time are so overwhelmed, overawed, or over-impressed by the form that the logic of the structure escapes unnoticed.

Five steps, generally, constitute not only a sound procedure but also a logical one which assures avoidance of oversight and promotes the most efficient use of time and resources. At first blush this outline may seem overly simplistic, but this results mainly from the fact that the underlying principles are not complicated. Even the largest forest is composed of trees.

Step One: This either begins with, or soon evolves into, a question. The genesis is recognition of a problem or "wonderment" about how or why something is or should be. It frequently begins as, "How did that happen?" Or, "Are these things related?" At other times first emergence takes the form of "Every time A happens, B occurs—I think." What action follows this initial discovery or question will be dictated by the real or imagined importance of an answer—the strength of the need for an answer, based on the seriousness of the problem or the breadth of value, e.g., the number of people, places, or situations affected.

Unless the phenomenon being investigated is isolated and unrelated to anything else (a unique situation), the result is most likely to be useful and complete if examination is undertaken within a logical structure. At one level, logical structures are called theories. Since theories find their greatest usefulness in generating hypotheses, an existing structure should be adopted or adapted, or a new theory may be required. One could say that the origin of the "wonderment" represents the discovery of a new truth; what follows is some experiment to support or disprove that truth.

A clear statement of the question, derived from insight or observation, consideration of the importance of an answer, perspective obtained through a theoretical structure, and, behold Chapter I!

Step Two: Once the problem or question has been defined and some measure of importance has been derived, if it appears that further effort is warranted, all available sources should be consulted to determine: (a) Has a solution/answer been provided by someone else, directly? (b) Is there a transferable or substitution solution/answer available? (c) If there have been unsuccessful attempts in either a or b, what were the deficiencies? (d) What approaches have or have not been used successfully on this and similar problems?

If an acceptable explanation or solution has already been provided, the problem is solved, and one would probably turn to an extension of this solution or move to another problem.

The most stable body of evidence is likely to be found in recorded reports of previous efforts—applicable literature, for the most part. An essential product of this review should be a summarization of previous studies which impringe on the question at hand. Along with results and methodology, other pertinent aspects should be reported; your reader should be provided with enough information to "judge your judgment" about the efficacy of these efforts.

This step could be entitled, "Chapter II, Review of Literature."

Step Three: Assuming that a thorough review of pertinent literature produces no viable existing solution, one can now develop some final form of hypothesis as a basis for quantifying or testing suspected relationships. This hypothesis will usually suggest factors to be considered. The task then becomes one of determining ways to control, measure, record, compare, isolate, eliminate, identify, or examine variables which contribute, counteract, or compound. To wit, a strategy or "Methodology" is derived.

It may become necessary at this point to augment the previous step by consulting similar sources for guidance and assistance in locating and evaluating techniques, measuring devices, and other "tools" with which the procedure can be implemented.

---

*The late Dr. Perkins was Associate Professor of Industrial Education, Texas A&M University.

Reprinted with permission from *J. Indust. Teacher Education,* vol. 14, pp. 36-37, 1977.

Step Four: All of the new data, raw and analyzed, produced by the methodology should be sorted, sifted, studied, grouped, compared, and examined internally and in relation to information garnered in Step Two. Here, only examination is important—and recognizing implications.

Not only must the data be examined, the methodology should also be scrutinized for gaps and weaknesses—oversights as well as errors. It is important that the soundness of the foundation influence subsequent judgments about the value of the superstructure.

This Step could be labeled "Data," or Chapter IV.

Step Five: Now, all that has emerged is interpreted. From immersion in the data, one hopefully will spring forth able to pull the information into cohesive units, draw definitive conclusions, project the worth of these conclusions into their respective meanings for application and better practice. Suggesting ways the information can be used, identifying further questions or dimensions needing answers, and prescribing improvements for subsequent efforts round out this portion. Why not call it Chapter V; "Summary, Conclusions, and Recommendations!"

Many more trees are in the forest, mostly trees like these; but the original idea was to see something besides the forest.

Rhetoric Applied to Scientific Method*

Daniel Marder

Phenomena are observed every day, but unless they can be formulated in terms of a problem, they remain aspects of our primeval world. Once man has asked the questions "What causes the sun to rise?," "What causes rain?," he was well on his way to an understanding of his universe. A clear perception of a problem is the first step toward scientific inquiry.

Isolating the problem lends unity to the investigation. A man doing chemical research does not research chemistry. He studies some aspect, such as carbohydrates, and studies this aspect with some purpose in mind, according to his interest. "What are all the kinds of carbohydrates found in natural petroleum?" The general subject plus the particular interest limits the problem. The interest limits and fixes the subject so that it is manageable. The problem is the subject brought to focus on a single question within some sort of limit.

ISOLATING THE PROBLEM

Often a group of phenomena is recognized as comprising a situation, but a problem cannot be formulated because the constituents in the situation do not hang together. The scientist is often in the midst of such things, and the problem itself is to formulate the question for scientific inquiry. Defining the problem, then, gives the researcher a purpose for his investigation.

Technical papers and reports also begin with a formulation of the problem which is often called objective or purpose.

Basically, the activities to be reported fall into two catgories, investigations and operations. Introducing reports of investigations, the writer includes the circumstances from which the purpose or problem arose, that is, the background leading to its formulation. These introductions may also include definitions of unfamiliar terms, very general descriptions of equipment and operation, if applicable, and the scope or range of the investigation. Reports on operations are often intended for instruction. They usually introduce the subject by telling what it is, what it is for, how it was developed, and how it is used. But the essential item in every introductory section, either in the investigation or operational report, is the purpose, or problem.

Here is a worthwhile introduction to a report of an investigation:

> The advances in the technology of missile guidance and propulsion have led to a number of proposals by agencies of the Department of Defense for small, unmanned earth satellites. Some of these proposals are based on techniques available under the current state of the art, while others look toward developments of the near future to provide means of placing larger and more useful satellites in an orbit around the earth. Although current technology would severely limit the size of the satellite and perhaps its use as a scientific tool for geophysical research, a minimum satellite project can be justified on the basis of experience to be gained in launching and tracking the satellite. The valuable data obtained could lead to a better understanding of certain geophysical phenomena. This Laboratory has conducted a study of the problems involved in instrumenting a minimum satellite vehicle.
>
> The prime objectives of the study were to determine (1) if tracking could provide proof that the satellite had been actually placed in an orbit, (2) if tracking with sufficient accuracy to meet the scientific objectives of the program was possible, and (3) if an electronic tracking beacon for the satellite vehicle could be designed.

The background is fully rendered so that the reader knows where he is in the scheme of things. The problem — the justification for a minimum satellite project — evolves from the background. The problem is then broken into three smaller problems, each stated explicitly as an area of inquiry. The reader knows where he is and so does the writer. In each of the three minor problems, the subject is brought to focus on a single question within some sort of limit.

The observations which lead the scientist or engineer to formulate the problem also suggest a tentative answer which we call a hypothesis or working hypothesis. The word comes from an old Greek verb meaning "to put

*Presented at the Army Chemical Center, Edgewood, Maryland, June 7, 1957.

Reprinted with permission from *Soc. Tech. Writers and Publ. Review,* vol. 7, pp. 22-24, July 1960.

under," that is, to support. Hypotheses, then, are placed under a collection of observed facts as a means of ordering them for investigation.

Before he arrives at the hypothesis, the investigator sifts through all available information pertaining to the problem so that all the work previously accomplished can be utilized, but not repeated. Perhaps someone has already solved the problem. Sometimes this preliminary search changes the nature of the inquiry. For example:

> For three decades, carbon monoxide combustion has intrigued investigators because of its extreme sensitivity to very small quantities of water vapor. The burning velocity of carbon monoxide is accelerated not only by water but also by hydrogen or by hydrogen-containing organic compounds. It is postulated that H atoms and possible OH radicals participate in the reaction. The presence of these H and OH species increases the reaction rate not only by heat conduction but also increases it by diffusion. Diffusing rapidly ahead of the flame, they act as individual ignition sources.
>
> W. Bone,[1] in his earlier studies, found H_2 more potent than H_2O in carbon monoxide combustion. He suggested that the relatively inert CO might be resolved into a more active state by the H atoms produced during combustion. Bone did a remarkable job in drying his CO over phosphorus pentoxide for 500 days, but in most instances, reliable data on the $CO + O_2$ system have been hard to obtain because the gases used were not dry enough. Usually a small amount of moisture is tolerated and more water added as a controlled variable.
>
> Many investigators, including Slootmaekers and van Tiggelen,[2] have collected data and postulated mechanisms for the $CO + O_2 + H_2$ or $CO + H_2O$ reactions, and most agree that the reaction mechanisms are analogous to those of $H_2 + O_2$ because the CO does not enter into the first stages of ignition. Moisture in the CO, therefore, is a possible source of error in the data. Studying the kinetics of the reaction of stoichiometric CO and O_2 mixtures containing H_2, Buckler and Norrish[3] proposed a mechanism for the reaction at lower temperatures than those attained by flames. The moisture problem in the CO was not considered.
>
> In our investigation, the carbon monoxide, which was generated by formic acid decomposition, and the oxygen were carefullly dried and distilled. Beginning with the driest gases possible, effects of hydrogen on the flame velocities were compared with those of deuterium. The substitution of D_2 for H_2 gave velocity ratios, which are compared with the thermal or the diffusion theory, and a mechanism for the reaction of H_2 with dry carbon monoxide and oxygen has been postulated.

ATTACKING THE PROBLEM

After the hypothesis is formed, the investigator tests it by experimentation, which is observation under controlled conditions. The investigator carefully tries to avoid any bias in favor of the hypothesis. He never underemphasizes negative evidence. In fact, he tries to disprove his hypothesis by testing his own experimental methods of control. The results, or data, are then grouped, or classified. Unnecessary data are weeded out, and gaps usually appear which mean the investigator must return to the laboratory for more experimentation. The method is oversimplified here because we wish to examine the skeleton of the idea, not the complex details and frustrations of scientific method in operation.

Procedures or Methods. When the investigator turns writer, he again follows the method of his investigation. He has stated his problem and knows what he is looking for. The hypothesis is a statement of inquiry, regardless of its form.

The writer may know the answers when he sits down at his desk, and the whole process of technical writing would be greatly simplified if he could merely state them. But every reader comes from Missouri. The writer is required to prove his answer to the reader as he has proved it to himself. He presents the experiment or test which proved or did not prove his hypothesis. The reader who is also a professional engineer or scientist demands details of the procedure that was followed, of the experimental equipment employed, and of the control measures instituted.

Results. Finally, the writer presents the results — the data — and the reader should have all the evidence for evaluating them so he can determine for himself if the data are accurate and complete and if they actually bear on the problem as the investigator intended.

SOLVING THE PROBLEM

The last part of the scientific method is the interpretation of results. Do they support the working hypothesis or not? If no working hypothesis was formulated at the beginning, then the results cannot be applied. Quite often, however, scientific reports never state the hypothesis, or purpose, and therefore the results can have no specific meaning. Every reader wants an end to the story. If the hypothesis is clearly stated in the beginning, everything in the report can be related to it. In this way, the hypothesis, whicb is the statement of problem, assures unity, coherence, and the correct emphasis for the reader. It creates a flow of action which appeals to the reader's logic.

Evaluating Data. The proof or lack of proof for any scientific inquiry resides in the data. The ending of the investigation is the evaluation of the data in terms of the hypothesis. Here is where the problem is solved. Sometimes the data are simple, and the writer can proceed from the results directly to his conclusions. But usually interpretation is necessary. In reports of investigation, the interpretations of data are presented in sections called "Analysis of Data" or "Discussion." If the data substantiate the working hypothesis, then a new fact has been added to our store of knowledge: the hypothesis becomes an accepted generalization and may be dignified with the term "theory," although this term is reserved for generalizations which have a great weight of observed evidence to support them.

Drawing Conclusions. Whether the evaluation of data substantiates or disproves the hypothesis, the observed facts are generalized as a conclusion to the inquiry, and the writer inserts such statements in the concluding section of his report. When the data do not substantiate the hypothesis, the investigator may wish to reexamine the problem, to modify, restrict, or expand it. Rather than conclude the report, he may decide to postpone it until he has completed a second, and perhaps a third, investigation. Eventually, he either discards, retains, or restates his hypothesis to correspond with the data he has obtained. The statements in the conclusion are direct answers to the questions in the introduction. The ending of the inquiry and the report describing it are like the completion of a circle.

Recommending Action. In applied research and in most engineering projects, the investigation is often prompted by the need for action. In reports of such investigations, the ending includes a section for recommended action. The recommendations are always based directly on the conclusions. When the conclusions of an investigation do not verify the hypothesis, the recommendation may be to discard the project.

STICKING TO THE PROBLEM

The methods of science and exposition are parallel as we have seen. They both have a beginning, a middle, and an ending. In the beginning, the investigator conceives a problem from his observations and forms a hypothesis, a tentative answer; the writer states these observations as the situation or background from which the problem stems, and then he presents the hypothesis as the purpose of his investigation. In the middle, the investigator attacks his problem through controlled experimentation; the writer describes the experimentation step by step, giving procedures, equipment used, and results obtained. In the ending, the investigator evaluates his results and applies them to substantiate his hypothesis which may suggest a source of action; the writer presents the evaluation as an analysis or discussion of data and states the general conclusions stemming directly from the evaluation. He may also offer recommendations which are projections of the conclusions. The writer is always bearing on his problem. If the problem is the constant point of focus, then the writer can see which elements do not belong in his report, or in certain sections of his report, and he can also see how to fit the parts together. By emphasizing the problem, therefore, the writer achieves unity and coherence — the basic principles of rhetoric.

These principles of rhetoric are the natural means of organization in all mental work — in the systematic working of the human mind — whether that mind is operating within a scientific method or within the craft of writing. Ω

Rules, Context, and Technical Communication

CAROLYN R. MILLER
Department of English
North Carolina State University

The concept of "rule," derived from linguistics and anthropology, provides a way of understanding the relationship between context, purpose, and message production and interpretation. "Rules" are shared expectations which structure situations and guide individual action. This paper reviews some of the concepts that have come out of rules theory in communication research and suggests their particular relevance and utility to understanding the problems and situations in technical communication.

We've all heard a lot about how important context is in determining the meaning or evaluating the effectiveness of a message. Context, frame of reference, situation, intention--these and other terms are used to suggest that meaning is not contained in a sentence or utterance but is somehow related to the way the sentence or utterance is used.

But what exactly *is* context? And how does it help us to interpret communication? Loosely understood, context is the surrounding situation, especially previous related statements. These provide clues which help us, for instance, determine the referent of a pronoun or resolve a grammatical ambiguity. But just *how* does context enable us to make sense out of such potentially confusing information? What are the clues and what do we do with them?

In a more general sense, context provides us with a structure, into which we can fit problematic statements. Usually, they will fit only one way, and we take that to be the "right" way. Thus, context will constrain our expectations about what a sentence will most likely mean. For the sentence, "I saw a man eating shark," context will supply clues as to whether a shark or a man is being discussed. Out of context, the syntactic ambiguity is impossible to resolve.

The structure of context is normally loose; otherwise it would predict the entire message for us and there would be no need for further communication. It provides a set of expectations and likelihoods which may in fact be violated. This view of context suggests a particular perspective for the study of communication, a perspective quite different from the mechanistic or scientific one which has informed us about channel capacities and information quantities. This perspective will not permit the prediction and control of communication effects, as empirical science might, but it can help to explain how effects occur and thus provide a basis for judgment and choice in communication design. Science seeks to establish laws to which natural phenomena conform; this alternate perspective seeks to elucidate rules to which human beings conform. In this paper the concept of "rule" will be used to explore the way context operates in communication. The particular relevance of this approach to the study and practice of technical communication will then be suggested.

RULES AS SOCIAL AGREEMENTS

The study of rules as the basis for most human action has its origin in anthropology and linguistics. The structure of human cultures and languages can be understood as due to sets of rules which people learn and conform to, not by mechanical necessity but by some degree of choice, or habit formed by choice.

Rules, in effect, are social agreements about the nature of reality and the place of human beings in it. They show how to conform to an established order--to the everyday world of commonsense reality just as much as to a formal or ritualized situation like a court of law.

In the court of law, for instance, rules are likely to be explicit, and we are likely to be conscious of them. We stand when the judge enters; we keep silent; we answer questions in certain ways. If we do not, we may be censured. Rules of etiquette are another kind of explicit social rules. But many rules are implicit; we become so habituated to the actions they specify that we find it hard to believe they are really matters of choice, not necessity. For instance, anthropologists have found great cultural variations in concepts as fundamental as those of time and space. If we grow up Hopi, we learn to conceive of time as an event rather than, as Americans do, as quantifiable linear extension. Time does not *necessarily* exist in one form or another; rather, we learn to construe it as those around us do. A person who construes it differently may be thought of as mentally ill.

Many of the rules which provide structure to our language are also implicit, ones which we learn by example rather than by precept. Given a list of modifiers and a noun, most native speakers of English will find that the group of

Reprinted with permission from *Proc. 25th Int. Tech. Commun. Conf.*, 1978, pp. 60-64. Subsequently reprinted in the *J. of Tech. Writing and Commun.*, vol., 10, no. 2, pp. 149-158, 1980.

words "sounds right" in only one order, although they have never encountered that particular group of words before: "polished first libation highly six brazen the bowls" will resolve itself into "the first six highly polished brazen libation bowls." We have internalized a set of rules about the relative positions of adjective types that helps us sort them into an agreed-upon order. But few of us were ever taught that adjectives derived from nouns (brazen) precede noun adjuncts (libation) yet follow verb participles (polished).

At the level of etiquette, then, rules provide reasons for doing one thing rather than another; at the level of subconscious habituation, rules provide motives for doing what may seem right or even necessary. They help us construct, codify, and anticipate the confusion of sensation and the multiplicity of possible random actions. They create meaning out of potential chaos. So meaning is neither inherent in words or situations and thus absolute and unarguable, nor inherent in people and thus unapproachable and capricious. Alice and Humpty Dumpty had a famous debate on this point. As Alice put it to Humpty Dumpty, "The question is whether you *can* make words mean so many different things." "The question is," said Humpty Dumpty, "which is to be master--that's all." The concept of rules helps us compromise this paradox: words are both our masters and our slaves; meaning grows in the social mediation between the inscrutable external world and the isolated internal consciousness. Rules help us find a way to achieve our own purposes within a set of constraints, within a larger social order which supports those purposes.

Rules arise in human society because of the conflicting human impulses for cooperation and competition, according to some anthropologists. Social situations are thus rather like games, in which we cooperate in order to compete; in communication we must begin with common, public conventions in order to convey our private intentions.

If rules are merely social agreements, it follows that they may be adhered to or not. Since common rules are what enable us to interpret the actions and utterances of others, noncompliance may also be interpreted. In fact, we may have an entire set of rules which help us interpret the violations of other rules or the incongruities of actions referrable to two or more conflicting rules. For example, if I say, "This is a fine piece of work, O'Malley," with heavy stress on the first word and a snarl on my face, you will interpret the incongruity between my facial expression and my words as indicating sarcasm; you will interpret my violation of a basic rule by which we expect all aspects of a message to agree as indicating that the intended message was not the literal one.

CONTEXT AS A HIERARCHY

A rule violation may have socially comprehensible reasons behind it. Consequently, noncompliance is not a matter of incorrectness (or wrongness, which implies a kind of moral depravity) but (if it is not due to ignorance) an invocation of another rule at a higher level, a rule for interpreting rules. The rules we know are not arranged in a lengthy list but in a hierarchy. This arrangement provides for both greater flexibility and greater economy. We can systematically eliminate many possibilities for interpreting an unusual action or message by considering, for example, socio-economic class, sex, tone of voice, the surrounding environment, other people present, etc. A given situation or environment, like the court of law, will automatically eliminate the likelihood of many kinds of utterances: jokes, prayers, etc. The judge is more likely than the witness to give orders. A statement uttered by the judge may be a charge to the jury; the same statement uttered by the defense lawyer is more likely to be a plea to the jury.

In fact, what we have been talking about here is what we began with--context. But now we see that context can be defined as a set of rules understood as applying to a statement. The set of rules is a branch, so to speak, of a hierarchy. The upper levels of the hierarchy provide rules for determining which sets of rules at lower levels will apply. Figure 1 is a version of what this hierarchy might look like. At the top, action may be seen as ultimately grounded in human biology, which makes certain types of actions necessary. But the next level, culture, is so various and so influential that it greatly overpowers the ultimate effect of biology. Within a culture there are various types of situations and certain ways of handling aspects of those situations, which I have called genres. These three levels may be seen more generally as a "semantic environment." I have borrowed this term from Neil Postman's book, *Crazy Talk, Stupid Talk*. He defines the semantic environment as "an ordered situation in which messages can assume meaning," and it includes people, purposes, applicable rules of discourse, and the actual messages being used (page 9).

The next general level I have called pragmatics or speech acts. At this level the conscious intentions or subconscious motives of a person are linked to the social structure or semantic environment. An episode may be seen as a constituent of a culturally recognized genre; for example, the episode of greeting is a standard constituent of the genre, interview. The interview takes place within a situation, as for example, employment recruiting or news gathering. A person's moment-to-moment intentions are revealed partly by what I have called here the "superfix," a term borrowed from descriptive linguistics. In speech, the superfix consists of stress, pause, tone, facial expression. In print, the superfix can be indicated by punctuation, type style, layout, and other details of format.

The grammatical structure and lexicon of the language used constitute the third general level of context. Together these are interpreted as symbolizing aspects of common experience.

Several points about this hierarchical context are noteworthy. First, it not only helps us interpret what others are doing or saying, it also helps us formulate our own actions and messages.

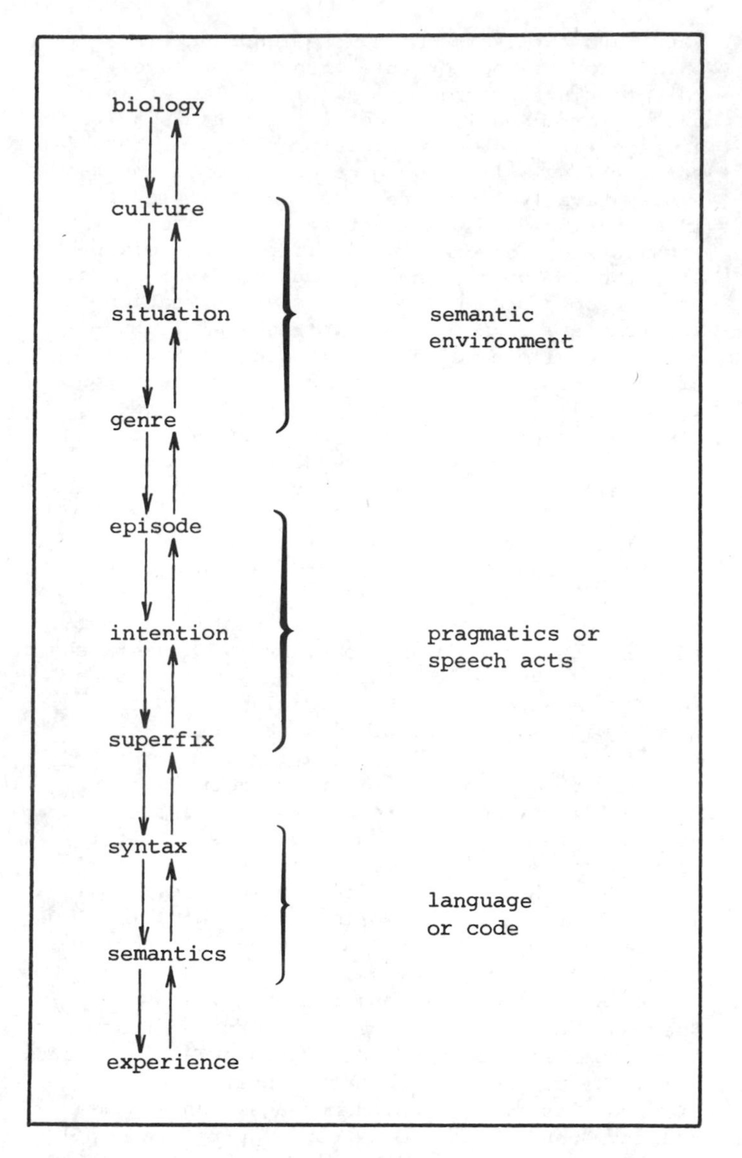

FIGURE 1. Rule-Based View of Context Hierarchy. Rules of formulation and interpretation connect the levels. The eight levels in the middle may be grouped into three major levels, as shown.

Communication occurs most efficiently and precisely when both formulator and interpreter are using the same version of the context hierarchy at all levels and the same sets of rules to connect the levels. Second, the many levels I have suggested here provide some degree of redundancy to increase the chances of the parties understanding each other. And third, the levels are not necessarily used only in the order shown in the figure but in a recursive process which feeds back interpretation from one level to the next.

One more point must be made before I come to the relevance of all this for technical communication. There are two basic kinds of rules, that is, two basic kinds of understandings about the relationships between levels: constitutive and regulative. Constitutive rules are agreements about what actions will "count as" an interpretable action on that level. To return to the game analogy, the constitutive rules of, say, checkers, specify the layout of the board, the number of checker pieces, and the types of moves permitted. If someone sits down to play checkers and proceeds to move every piece as though it were a knight in a game of chess, we would agree that what is being played is not "really" checkers. Violation of a constitutive rule, like this, usually results in confusion on the parts of other participants. There is no appropriate response--there are no game rules to tell us what to do with this kind of opponent. The other kind of rule, the regulative rule, tells us what to do to play well or effectively. Regulative rules would specify opening strategies in checkers, for instance. Failure to follow regulative rules usually results in censure or sanctions, such as that of losing your checkers to the other player if you play badly. In Neil Postman's terms, failure to observe constitutive rules results in "crazy talk," talk which other people cannot recognize as belonging to a recognizable semantic environment. Failure to observe regulative rules results in "stupid talk," talk which is ineffectual in accomplishing its recognizable purposes.

Consider one more example, which will bring us to our concern with technical communication. If I purchase a fancy new electronic calculator and request an instruction manual from the manufacturer, what I receive in response to that request may be unsatisfactory in two possible ways. If it looks to me more like a price list for accessory parts, I will say, "Why'd they send me this? This isn't an instruction manual!" I will be confused. The manufacturer has violated my understanding of what constitutes an instruction manual. If it looks like an instruction manual but is full of errors or is arranged with the most difficult operations first, I will say, "This is a lousy manual. That company is a bunch of incompetents. I'll never buy anything from them again." I will invoke censure and perhaps apply economic sanctions. The manufacturer has violated my expectations of what makes a good manual by making the material difficult to learn and understand.

RULES AND CONTEXTS IN TECHNICAL COMMUNICATION

What we call technical communication occurs in a highly ordered context or set of related contexts. Many of the situations or genres which seem concrete and compelling to someone in the field are not widely recognized semantic environments. Many of the rules which seem compelling (that is, seem to be constitutive) are just the only way we've ever tried it (that is, are unchallenged regulative rules). Constitutive rules supply us with a set of choices. Regulative rules specify which choice is best, for a given situation. Much misunderstanding has occurred and much stupid talk is committed because a regulative rule was mistaken for a constitutive rule or because a constitutive rule in a given semantic environment was mistaken for an absolute and unconditional constitutive rule. Our understanding of English grammar and usage has been plagued by this confusion. What is "correct" English depends upon the dialect, subculture, situation, and intention--depends, in other words, upon a complex contextual structure. The growing use of cartoons in training manuals must at one time have violated somebody's sense of what constitutes a manual. Both constitutive and regulative rules change as cultures, situations, and entire contexts change. I believe that the detailed view of context that rules theory gives can be useful in designing and evaluating technical communications.

Figure 2 shows a fairly simple example of how a statement "belongs to" a technical communication context and how both constitutive and regulative rules may be seen to lead to the statement. The entire semantic environment consists of a complex of people and purposes. At each level, constitutive rules specify general understandings of what will "count as" belonging to that level. Regulative rules then help us choose a specific way of fulfilling that requirement. What is chosen by the regulative rule becomes the next level of the context. Thus, a technical manual must be orderly, relevant, and complete if the manual is to be recognized as a manual by people in Western technological culture. One effective way of achieving those requirements is to write sequential instructions. Similarly, we write short imperative sentences in a certain way not because some stern authority said we must but because, for the people and purposes involved, standard English is the most effective choice of dialects. In other contexts it may not be.

Context and its rules operate during the processes of formulating and interpreting communications. Problems in technical communication can often be better understood (and prevented) by analyzing this process, rather than by inspecting the product. A company I once did some consulting for provides an example of the differences between these two approaches to a problem. The company was having difficulty with its contract

CONTEXT LEVEL	A TECHNICAL COMMUNICATION CONTEXT	RULE TYPE	CONSTITUENTS OF CONTEXT
subculture	technology-based industry		
situation	production for market	*constitutive* →	specific people and purposes
genre	customer relations	← *regulative*	
episode	technical manual	*constitutive* →	order, relevance, completeness
		← *regulative*	
intention	sequential instructions	*constitutive* →	emphasis, directness, action-orientation
		← *regulative*	
superfix	short, imperative sentences	*constitutive* →	various surface structures and dialects
		← *regulative*	
syntax	standard English	*constitutive* →	(you) + verb + complement
		← *regulative*	
semantics	next action	*constitutive* →	semantic content
	↓ "Set the selector on C-mode."		↓ imperfect agreements on what aspects of common experience are being symbolized

FIGURE 2. Constitutive and Regulative Rules in a Technical Communication Context. Constitutive rule: "If you want to play this game, do r_1, r_2, . . . or r_n." Regulative rule: "If you want to play this game well, do r_x."

proposals. The managers had hired me to teach some engineers and supervisors "how to write," to improve the product. I was to instill what they saw as absolute rules (or laws) at the levels of syntax, superfix, and possibly intention. A few questions, however, revealed that solving the difficulties with the product would require elucidating and evaluating rules at the level of the company's semantic environment, rules which affected the entire communication process. The company was organized in such a way that the sales staff, who were the only employees ever to talk with customers or potential customers, and technical staff, who were the people who designed the systems for the customers and wrote the proposals, were in two different main divisions of the organization and never talked with each other. At this level, what we might call the situation of the organization, the company structure determined the rules about who talks with whom. At another level, the level of semantics, the proposal writers were having to symbolize experience for customers, many of them international, with whom they had little in common--and even that little was difficult to find at such distance. The constitutive rules at these two levels were in conflict and further in conflict with commonly understood regulative rules at the level of episode or intention that specify both brevity and completeness of detail in technical descriptions. By trying to get effective proposals out of such a situation, the entire company was engaging in a form of stupid talk.

This rules-based view of context can, I believe, be a useful perspective for both the study and the practice of technical communication for the following five reasons:

1. It sees the source of authority for rules in those who use them, not in some impersonal or absolute authority. A clear understanding of the constituent audiences and readerships for technical communication will help determine what expectations are relevant and useful and how those expectations can be met effectively and efficiently.

2. It sees communication as a process in a dynamic situation. Because so much of what we call technical communication is written rather than spoken, we may neglect this view. But writing, as well as speech, can not be interpreted (or effectively formulated) without constant reevaluation of evolving context and changing agreements on applicable rules.

3. It looks for meaning in the violation of rules as well as in their observance and thus provides an entire dimension of new resources for the formulation and interpretation of messages. If technical communication aims especially at precision and efficiency, it ignores those resources at its own risk.

4. It makes a distinction between constitutive and regulative rules. Ignoring this distinction limits us to single, "correct" ways of doing things and discourages us from freely adapting means to ends. Much of technical communication is constrained by highly conventional techniques because of the highly structured organizational (industry or government) environments in which so much technical communication takes place.

5. It organizes communication contexts as a hierarchy. Just realizing that context at one level provides a comment or set of directions for interpreting content at lower levels explains, for example, the widely recognized advantages of standard formats. By deliberately holding context at the pragmatic and environmental levels constant, attention is efficiently focused on language code and the attempt to elicit meaning from common experience. Technical communication usually has no place for jokes, sarcasm, or surprise because these involve rule conflicts at lower levels of context and require recourse to the more abstract higher levels of context for comprehension. They are thus inefficient, and consequently inconsistent with the overall semantic environment of technical communication.

In summary, since technical communication is what we might call "context-heavy" (that is, it occurs in highly structured situations), an understanding of what context is and how it affects communication is particularly important. This rules-based approach suggests both the complexity and the flexibility of context. It may provide ways to both study and achieve the efficiency that technical communication seeks. In addition, because of its very obvious structure, the technical communication context provides a fruitful arena for the study of the types of rules which underlie all human interaction.

BIBLIOGRAPHIC NOTE

I will be happy to supply reference citations for the technical works which contributed to my understanding of rules theory in communication. An excellent non-technical source which I highly recommend is Neil Postman's *Crazy Talk, Stupid Talk,* Dell, 1976.

Structure, Content, and Meaning in Technical Manuscripts

Herbert B. Michaelson

IN CONTRAST TO THE USUAL CRITERIA *found in the literature on technical writing (i.e., readability, effective expression, and logical development of ideas), this article offers the special viewpoint of a journal editor. It analyzes, from an editorial standpoint, the over-all qualities of a good technical report or paper of the type intended for scientific or engineering readers. The merit of any manuscript is shown to depend on its structure, content, and meaning, and on the complex relationships among them.*

The literature on engineering and scientific writing concentrates on three kinds of quality criteria:

1. *Readability and intelligibility*: The favorite target of the communication specialist, who frequently favors word economy and condemns jargon.
2. *Effective expression*: The objective of the English professor, who often stresses sound literary style and clear exposition for maximum communication to readers.
3. *Logical development of technical concepts*: The main approach of those writing experts who have engineering or scientific training.

All discussion and advice on technical writing does not, of course, fall neatly into these three categories. Whatever the viewpoint, however, the usual admonishments on how to write will generally start with the bland assumption that the work to be reported is eminently sound and worth writing about. We will not make such a brave assumption here because, as is well known, the glitter of today's technical literature is not all pure gold. Indeed, the problem is not how to swell further the flood of engineering reports (which no one can keep up with anyhow), but rather how to direct the flow of the better papers into useful channels. Many technical manuscripts are too lengthy for their purpose, and the labor of the writing and heavy rewriting is not always justified by the quality of the work reported. On the other hand, many fine potential contributors to the literature are never written—not because the work is not worth publishing, but because the author's skills are undeveloped. For him the job of composing an acceptable report, getting it referred and published is too time-consuming and demanding to be worthwhile.

When we consider that tremendous effort is expended on writing about routine and trivial topics and that some papers on excellent work never materialize, we must reconsider the usual advice of experts on technical writing. In this article we will try a new approach by examining three properties of a manuscript:

1. *Structure*: The way a technical report is put together to show the relative importance of various aspects of the author's work.
2. *Content*: The exact nature and purpose of the work reported and the author's interpretations.
3. *Meaning*: The significance of the manuscript to a given class of readers, i.e., the efficiency of the author's communication of technical ideas.

Abstract Introduction Theory Analysis Experimental Results Conclusions References Appendix
Summary Statement of Problem Design Approach Derivation of Design   Equations Applications Illustrative Examples Conclusions Bibliography
Abstract Contract Requirements Historical Background Design Principles Development of   Prototype Model Properties Life Tests Reliability Conclusions Appendix

Figure 1 — Representative outlines of technical manuscripts.

These three properties are by no means separate and independent. The way they interact and the effect on the quality of a technical manuscript are the subjects of our discussion.

STRUCTURE

By the "structure" of a manuscript we mean the way it is put together—its division into sections and the disposition of its tables and illustrations. The structure, in its truest sense, is the foundation of a manuscript and the author's choice of a structure determines at the start the whole character of the paper.

The first structural detail is the beginning section of the manuscript. Without a sound framework the manuscript will mislead the reader in several ways. If it lacks a good orientation of the problem in the early sections, the average reader is at a disadvantage to understand the real purpose of what follows. When one section of a paper does not logically follow another, or tabular material is badly designed, or charts and photographs are poorly chosen, the popular remedies for poor technical writing don't help the reader. Short words and short sentences do not improve the situation. Avoiding jargon accomplishes nothing in this case. Precise grammar, spelling, and punctuation are of no avail. The use of well-turned phrases and a style of language meticulously chosen to suit the level of the readers will not solve the inherent problems here. The precision and style of language, of course, are important but are secondary to a careful overall shaping of information and the exposure of technical concepts in appropriate portions of the manuscript.

A sound structure tends to show logic and merit in the work being reported. The structure is sometimes revealed in the abstract. If the abstract comes directly to the point, showing the purpose and content of the paper, the writer is off to a good start. On the other hand, a confused, vague, or wordy abstract will frequently be followed by a poorly constructed paper.

An ideal type of brief, informative abstract, adaptable to any kind of engineering or scientific manuscript, may consist of only three sentences: (1) a pithy statement of the problem, (2) an identification of the author's approach to the problem, and (3) a short statement of the essential results. The structure of such an abstract is similar to the classic building block of a manuscript: Introduction-Body-Conclusion.

As for the manuscript itself, the "fine structure" depends a great deal on the topic and on the purpose of the work. A few specimen outlines are shown in Figure 1. There are, of course, many other varieties, and the choice of an appropriate structure depends on the author's sense of order and balance. These forms of technical manuscripts appear stereotyped only to the novice. To the experienced author an imaginative approach to a basic outline, proportioned to emphasize the author's more important contributions, is the basis of an effectively written manuscript.

A sound structure, however, does far more than provide proper order and emphasis in a paper. In subtle ways, *choosing a structure actually affects the content of the manuscript.* For example, the standard type of Introduction in a piece of technical writing demands a statement of the problem; an explanation of what has been previously done in this field, and briefly who has published the important developments; some remarks about how the present approach differs from what others have done; and perhaps a brief advance view of what will be discussed in the manuscript. This scheme of writing the introductory portion is suggested in Figure 2.

Following such a conventional routine will require the author to assemble several classes of information for his Introduction. He may well have to go back to the library for more information, or he might have to check with his colleagues to ensure that he is reporting the correct background information. In this sense, the *content* of the introductory paragraphs can be affected by a set of conventional structural requirements. Of course, we are not suggesting a stereotyped Introduction that answers the items in Figure 2 in monotonous succession. These general types of information, however, tend to find their way into most manuscripts.

Another effect of structure on actual content sometimes occurs when an author is attempting to arrange data in some logical way. Redesigning a table or chart for clarity can unexpectedly reveal new trends and inspire new interpretations; and rewriting for better literary structure involves rethinking. All of us at one time or another have gained a new insight into the nature of a technical concept by having to explain it in writing.

By the same virtue, outlining a manuscript and building it according to a plan will sometimes reveal an unintended gap in the work. Many an author, while writing a manuscript, has suddenly sensed a missing piece of information and has rushed back to the laboratory for more data! And many an author has acquired a new understanding of what he had been doing in the laboratory when he struggled to shape up an evaluating Conclusion for his report. In these subtle ways, then, the attempts to structure a paper can actually contribute to its content.

Structure has another obvious but important property—it also affects the over-all significance to the reader, i.e., the meaning. Since any scientist or engineer finds large quantities of literature in his field of interest, he appraises the over-all meaning of any paper by scanning it before he actually reads it. In scanning, he observes its structure, i.e., he looks at the headings, senses the relative length of the various sections, glances at the illustrations, and forms an opinion of what is in the paper and what is significant about it. The way the blocks of information are fitted together affect the over-all meaning. The sequence of sections, for example, shows the author's line of thinking and indicates how his concepts or descriptions will develop. The relative length of each section gives an impression of its relative importance, i.e., lengthy sections imply a heavy percentage of the author's effort and may lead the reader to attach undue importance to those portions. In this sense, the structure of the manuscript actually colors the meaning that gets through to the reader.

We can see, then, why the structure of a report is a terribly basic consideration—one that unfortunately may be cast aside by the author who claims that he doesn't need a written outline or that he can't work with one because it continues to change as the paper develops. The fact is, especially for those who have difficulty in writing, that an outline carefully thought out will establish the desired structure and become the key to the character of the manuscript. In the over-all structure, perhaps far more than in the author's skill with language, is evidence of an author's judgement and his ability to highlight the significant aspects of his work and to deemphasize the less important ones.

Abstract
A. The purpose of the paper
B. A very brief summary of the results
C. General statement of the significance of the work and its applications

I. Introduction
A. A statement of the exact nature of the problem
B. The background of previous work on this problem, including published work
C. The purpose of this paper
D. The method by which the problem will be attacked
E. The significance of this work; explanation of its **novel** features
F. A statement of the organization of the material in the paper

II. Body of the Paper
(The organization of this main part of the paper is left to the discretion of the author. The information should be presented in some logical sequence, the major points should be emphasized with suitable illustrations, and the less important ideas subordinated in some appropriate way. This portion of the paper should be styled for the specialist and should not be "watered down" for the general reader.)

III. Conclusions
A. A statement of how the original objectives were met
B. Summary and evaluation of the work done
C. The limitations of this work
D. Advantages over previous work in this field
E. Applications and general significance of the results

Figure 2 — Suggested outline for journal's papers.

CONTENT

Our definition of the content of a paper was "the exact nature and real purpose of the work reported." We use the adjective *exact* because the real nature and intent of any given manuscript are usually subject to interpretation. Clarity in technical writing can be a surprisingly elusive quality—what is clear to one class of readers might not be at all clear to another. For this reason the first requisite of a good engineering or scientific paper is an adequate statement of the problem at hand and a lucid explanation of the background of the work. Without this kind of preparation for what is to come in the main body of the report, the average reader might have to peel off layer after layer of detailed exposition before he gets to the core of the author's real contribution. But in technical writing, unlike other forms of literature, it is considered bad form not to come immediately to the point.

Somewhere early in the manuscript, then, the precise nature of the work should be identified. In the whole realm of scientific and engineering literature there are really only a few kinds of content. These might be classed in general as follows: the original contribution (either theoretical or experimental); the application or analysis of known concepts; the description of a device, system, or methods; the properties of a material or process; a report of progress or completion of a project; and the tutorial review.

Confusions as to actual content arise in several ways. If an application paper, for example, is written in the flavor of an original contribution, with the vague implication that the information is being published for the first time, the reader can be misled. A writer seldom intends to be misleading but can neglect to place the proper labels on his work.

Another type of confusion arises in whether the work reported is complete or fragmentary. An example is a paper describing a device, which appears at first glance to be a completed and practical design but on closer examination turns out to be merely a progress report, with many gaps and unfinished areas of development.

Technical papers should give some clear indication of the status of the results. If the work can be called complete (and few papers are really "complete"), it is a well rounded investigation, including a discussion of all the pertinent aspects, with analyses and proven results. If the work is fragmentary, it has gaps, inadequacies, or tentative results, or is merely one phase of a segmented work project. Such manuscripts need a section devoted to the limitations of the results. Any technical paper that lacks this self-imposed critique implies that it has adequately covered everything there is to say about the subject and, therefore, that the study is complete.

The real purpose of a manuscript may differ from its *apparent* purpose. For example, the actual purpose might be the survival of the author in an academic world of publish-or-perish! Or it might be to justify the expenditure of project funds. Or at the other end of the spectrum of motivation the purpose might be to expose a new concept to the world of science. Since the actual purpose of the work is included in our definition of content, we might list here a variety of the true objectives of technical reports and papers:

1. *To reveal* a new concept or application for the edification of the scientific or engineering community.
2. *To evaluate* work that has been previously reported: to compare, to analyze, to determine relative importance, and to investigate feasibility of ideas or processes.
3. *To recommend* a solution to a given problem; usually involves an assembly of pertinent technical information and an analysis of unresolved questions.
4. *To inform* on the progress of project work and how the objectives have been approached.
5. *To instruct* in technical principles, showing the extent of existing knowledge and methods.
6. *To support marketing* by providing information about a product or service.
7. *To justify work or funds* on a specific project, or to request new funds or other support.
8. *To get on record* so as to establish a reputation for competence, or merely to obtain credit for work done.

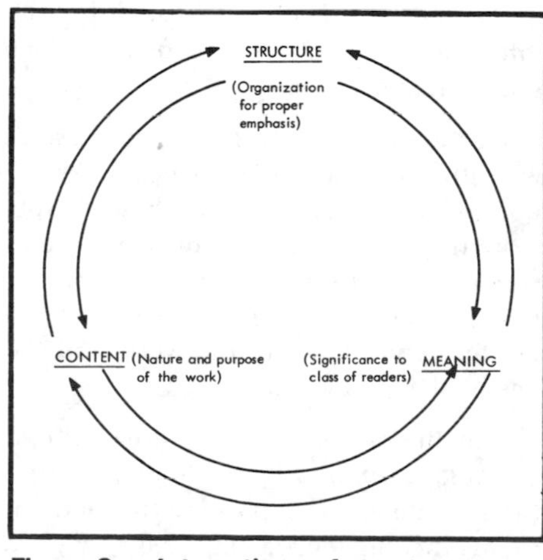

Figure 3 — Interactions of structure, content, and meaning.

These eight reasons for preparing a technical report are not entirely distinct from one another, and manuscripts are often written for several such reasons. The manuscript will usually show evidence of how one of these reasons predominates the author's thinking. For example, a paper on the engineering design of a new system might have a long section explaining the difficulty of the original problem if the author's main purpose is to bolster his reputation for competence. Or the report might have a lengthy, detailed section dealing with the background of related work if his chief purpose is tutorial. Or it might contain an extensive analytical section if his main purpose is to evaluate the merit of the design. Indeed, an author's real motives and the main emphasis in the paper can sometimes be judged better by observing the relative length of its sections than by reading the abstract!

We pointed out previously that the structure adopted by the author has certain subtle effects on the content of a paper. The intended content, on the other hand, also influences a writer's choice of structure but in a way that is anything but subtle. Some of these effects become rather obvious when one considers that a heavily theoretical paper will have one or more sections showing how the ideas are used in practice. And a report that seeks the solution of an engineering problem will generally have a Recommendations section, and so on. Moreover, each engineering or scientific discipline tends to demand its own structural form, and so attempts to prescribe a standard outline for all reports and papers are usually futile. Content dictates structure, which then develops according to the taste and judgement of the author.

The content of a paper also has a rather drastic effect on what we have called "meaning," i.e., the significance to a given class of readers. Here is a crucial point that is sometimes underestimated even by the most experienced technical authors. Unless the writer can get a clear idea beforehand of his audience, he is not in a favorable position to decide on the main content of his paper. What is technically meaningful to the specialist may not always be meaningful to management who must base decisions on a technical report. Another aspect of technical content concerns today's overlapping fields in science and engineering. For example, when a report will be read by a physicist, a chemist, and a device engineer who are working on the same project, the technical content can be more meaningful to one class of specialists than another. In the long run, the judicious author finds himself making a compromise in his choice of content to suit his various readers.

MEANING

The most controversial and troublesome aspect of technical literature is not the structure or content of manuscripts; rather, it is how efficiently a manuscript communicates information to readers—a quality that we have called "meaning." The most frustrating single problem for the technical reader is ambiguity. There are three kinds: *linguistic ambiguity*, which concerns multiple word meanings and connotations; *logical ambiguity*, consisting of gaps or contradictions in technical concepts; and *structural ambiguity*, an uncertainty of over-all significance due to poor organization of the manuscript. These faults, of course, are the favorite targets of editors and referees; any manuscript free of such ambiguities has a great deal in its favor. As we have already seen, however, structure and meaning alone are no assurance of quality—that third essential element of *content* is as important as the first two. Indeed, a paper can be structured in fine style and written in the clearest language and with sound logic—but if the nature of the work reported is trivial, the manuscript is trivial!

We have already suggested how structure and content can affect the meaning of a manuscript. There are reciprocal effects, too. Most technical work, particularly research and development, involves tackling a problem and finding a solution. Sometimes the problem is relatively simple but demands involved or elegant methods for solving it. Or the problem may be terribly complex but can be solved in a dramatically simple way. Whatever the situation, the evidence of technical skill eventually appears in a manuscript of some kind—either in laboratory notes or in formal writing, such as a report of a paper for publication.

In these circumstances, the real "meaning" of an author's findings is frequently in a fluid state. Laboratory work or analytical studies can be routine for long periods of time, but results may come in fits and starts. New directions and new thinking come unexpectly in any technical project—sometimes even after the project is supposedly finished and while a report is being written. The alert and imaginative author finds that "meaning" grows with a manuscript and that refining his ideas involves reshaping his writing. The author may not be fully aware that he is actually reinterpreting his months of work in the laboratory. But, aware of it or not, the author finds new shades of meaning in his work as he writes and rewrites. In this way, meaning affects the structure and content of his paper.

CONCLUSION

We have attempted to show that the over-all quality of a piece of technical writing does not depend on such simple criteria as its literary style or the technical merit of the author's work. Even if a paper is "clearly" written and polished in the best expository style, it will not be considered a high-quality paper if its content is trivial or outmoded. On the other hand, an excellent piece of technical work might be buried in an avalanche of poor, confused writing. But even if a paper has good content and literary style, any attempts to define its quality are further complicated by the different viewpoints and needs of various sets of readers.

Our approach to this over-all question was to suggest three benchmarks of quality for any technical manuscript: structure, content, and meaning. This combination of three criteria accounts for the questions we have just posed on expository style, merit of the work, and the needs of various readers.

The interactions of the three criteria are suggested in Figure 3. Since structure, content, and meaning depend on each other in several important ways, a weakness in any of the three will certainly affect the other two. For those who appraise manuscripts, the diagram might offer a new and useful way of looking at a paper and judging its ultimate worth. The diagram can have another purpose, however: It can illustrate for authors and editors the real-life process of technical writing. Although the time-honored recommendations of "clear" writing, sound literary style, and logical development of technical concepts have their special value, Figure 3 offers a broader view of the writing process.

The first question that occurs in any interpretation of Figure 3 is: Where does the writer enter the situation and where does he leave it? This would depend on the individual, but the likely place to enter the circle of interaction would be at Content, i.e., a consideration of the nature of the work to be reported and the type of manuscript to be written—a short, informal note, a formal report, or a full paper for journal publication.

The next step would be a careful shaping up of the Structure, or outline, according to the main purposes of the manuscript. The last consideration should probably be Meaning—the choice of proper emphasis and flavor for the manuscript, according to the class of readers who will primarily be concerned with it.

The only practical way of testing the rough draft for Meaning is to send it for review to some competent person in the intended class of readers. Later the author can revise the draft on the basis of at least one such evaluation.

Our viewpoint here has been that of writing a high-quality manuscript in any of the disciplines of science or engineering, and the writing process summarized in Figure 3 is our special view of how a good paper matures. Needless to say it does not apply to the manuscript that is mechanically constructed by a plodding, unimaginative author who feels that he has finished his work in the labaratory and proceeds to "write it up for publication," tossing off the first draft and then abandoning further effort or interest in his work. The readers of such a manuscript inevitably sense the routine flavor and the lack of inspiration or excitement, even though the manuscript may boast of technical accomplishment. But the paper of quality, written in the heat of the interactions shown in the figure, will have the earmarks of fresh insights and of the inspired thinking that characterizes good work.

Part III
Tricks of the Trade

THIS COLLECTION OF ARTICLES represents the nuts-and-bolts section of the book. These articles offer a compendium of proven devices designed to streamline the process of writing a paper or article.

Michaelson leads off with an unconventional approach to writing papers. He combines ideas offered in the Bernheisel and Lang articles of the preceding section and suggests that writing should be done in increments. This system of writing helps integrate the writing and designing processes and thereby makes writing a more natural, interesting, and productive part of the engineer's work. Hawes and Harkins outline a procedure for the engineer faced with a different problem: writing the abstract before the actual paper is written. In a six-step procedure, they show how to solve the problem while at the same time providing an effective means of getting started on the paper.

The next four papers examine aspects that can be considered as both writing and editing functions. Jarman demonstrates that the use of punctuation, rather than being an arbitrarily applied system of dots and dashes, should be dictated by common sense and logic. Once we master this principle, we can use punctuation as a communicative aid. Mathes offers a follow-up on Miller's essay in the preceding section. Using a concept called "vertical thinking," he proposes a three-step procedure to focus attention on the interrelationship of sentences within a paragraph; this approach guides us in choosing the most appropriate form for each sentence. Colby adds to our understanding of the paragraph; he explains about types of paragraphs, how to organize them, and how to create smooth transitions within and among them. Emphasizing the value and creation of logical coherence, Marshek discusses the various types of transitions and how they can be used effectively.

Winkler's article offers further insights into the concept Marder introduced in the preceding section. She closely examines similarities between problem-solving and report design and shows how we can make use of this correspondence; her analysis demonstrates how technical communication, like engineering, can, indeed, be an applied science.

In the next four papers we find practical information on how to evaluate the report. Farkas details the steps in writing the paper; leading us through the successive steps from inception to submittal, he provides examples of each step and argues for the logical development of text and thought. Douglas informs us how writing that is grammatically, logically, and stylistically correct can still be bad. He explains that writing should not be too "compact," too much like a tedious catalogue of uninterrupted facts. He tells how to avoid this "cobblestone writing" by being "polite" and "generous" to the reader. Greif provides a simple-to-apply checklist for clear copy; these nine questions are intended to give your article "at least a fighting chance for survival in the jangling jungle of articles clamoring for reader attention." This checklist is complemented by the nine-point checklist offered by Gould and Shidle who employ a point-counterpoint format as a forum for voicing their often-differing opinions. Yet they jointly echo a central theme of this volume: "writing can be an art, but it also can be a science."

Carroll concludes this part with an overview of how to increase the chances for your article's acceptance. He explains how articles are selected by trade publications and offers useful information on the "do's" and "don'ts" of writing technical articles. His advice, coupled with the other "Tricks of the Trade" included in this part, should increase the chances of any article's acceptance—whether by a trade publication, a manager, or by a colleague.

The Incremental Method of Writing Engineering Papers

HERBERT B. MICHAELSON, SENIOR MEMBER, IEEE

Abstract – A method is proposed for writing journal papers in increments during the progress of the engineering project in a way that will actually aid the development work. Papers written in this manner are not merely a report of work done in the laboratory; instead, they combine the discipline of formal writing with the discipline of engineering effort. The proposed method applies to any kind of engineering manuscript: theoretical, analytical, experimental, or developmental. Specific guidelines are offered for preparing a manuscript in this way and some unique advantages of the method are shown.

The value of writing manuscripts in this manner lies in the interactions among the engineer, his work and his manuscript. The excitement of current accomplishment in the laboratory becomes reflected in the manuscript, the insights acquired while writing contribute to the progress of the work, and the engineer finishes his paper far sooner than he ordinarily would.

INTRODUCTION

WRITING engineering papers for journal publication is frequently a highly inefficient process. The purpose of this paper is to show how manuscripts that describe development work can be written by the engineer with a real economy of time and effort.

The inefficiencies in engineering writing are due to a variety of well-known problems that confront the development engineer. Some engineers dislike to write and prefer instead to devote time to laboratory work. Others may have poor writing skills and tend to delay the writing job for indefinite periods. Or because of the press of new work assignments, an engineer might not easily find time to write a journal manuscript about his completed work. As a consequence, many manuscripts are not published until long after the work is finished and others are never written at all. There is certainly no panacea for the solution of these various writing problems. But the usual way of writing a paper, i.e., after the laboratory work is completed, can be improved upon by other more efficient methods.

THE NORMAL PROCEDURE

The conventional procedure is to make entries in an engineering notebook during the progress of development work, to write progress reports (and perhaps phase reports) throughout the project, and subsequently to construct a manuscript from these sources. The engineer's enthusiasm, however, tends to cool down after months or years of work on a development project! His writing tends to be unimaginative when he grinds out the same tired phrases and explanations he has already used repeatedly in progress reports and other informal descriptions of his work. The resulting paper lacks spontaneity and does not always display fresh, imaginative thinking. The boredom can be eliminated from such a manuscript by rewriting and by extensive editing; this, of course, demands additional time and effort.

THE INCREMENTAL PROCEDURE

A different way to write a manuscript is to start at the beginning of the development project. From the project plans, and from decisions as to how the development work should proceed, an engineer can construct a short, tentative abstract of his future paper. By the same virtue he can then draw up, in a short time, an outline of his proposed paper. The plan for formal writing then becomes a part of his work and, in a limited sense, even a *guide* for his work. The insights that come during the writing process can actually improve the development work and add to the engineering accomplishments on the project [1].

Writing in increments is done in the following manner. After the development project is well under way, as indicated in Fig. 1, the engineer is fully prepared to write the Introduction to his paper. He is equipped by this time with all the necessary information that appears in a good Introduction: the purpose of his work, a clear view of its relation to previous literature, a statement of the novelty of his engineering approach, and a preview of the content of the paper. Later on he can devote a short time to writing successive sections of his paper, which are then typed in draft form and set aside. For example, fairly early in the project, he can write a description of his experimental laboratory setup if his project concerns hardware or circuit development. As various phases of his work proceed he can develop corresponding portions of the manuscript from his outline, as implied in Fig. 1. Near the end of his project, he is ready to write a Conclusions section. Finally, as his development program draws to a close, he is not in the customary position of planning to "write it up for publication." Instead, he has already accumulated the typed sections of a complete draft ready for final revising and polishing.

Based on paper presented at 1966 International Convention, Society of Technical Writers and Publishers, Fort Worth, Texas, May 13, 1966.
The author is with the IBM Corporation, Armonk, N.Y.

Reprinted from *IEEE Trans. Prof. Commun.*, vol. PC-17, pp. 21-22, Mar. 1974

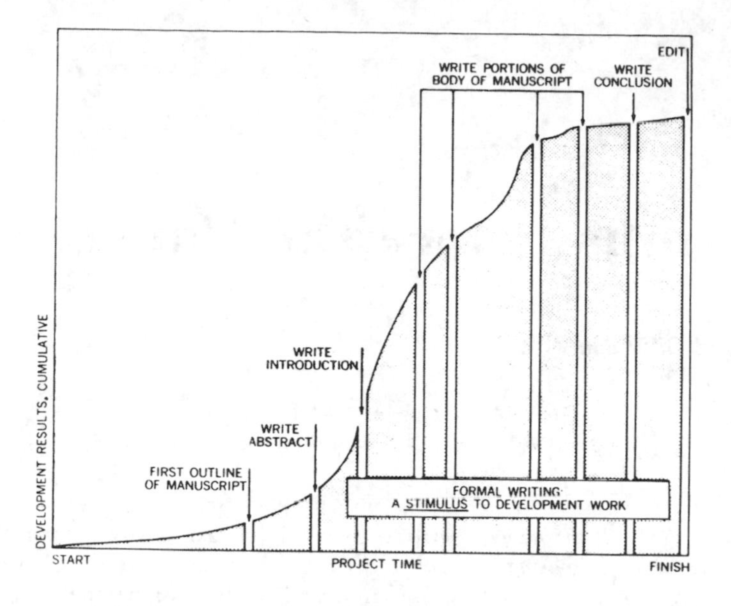

Fig. 1. Writing a manuscript in increments during an engineering development project.

DISCUSSION

I have been experimenting on this incremental method with some 25 authors of engineering manuscripts. The initial results of the experiment are most encouraging. Engineers like the method because the writing becomes an interesting and productive part of their work rather than a mere task of recording results.

This approach has several unique advantages over the traditional manner of writing a manuscript. In contrast to the conventional way of tackling the whole writing job at once, the engineer instead breaks down the writing task into easily manageable pieces, i.e., one section at a time. In a recent IEEE paper, Tracey [2] discussed another aspect of preparing one section of a report at a time, in which authors are encouraged to divide a manuscript into several topical sections, each roughly 500 words long and consisting of perhaps three or four paragraphs and a graph chart. Each section is typed on a two-page spread and is introduced by a topic sentence. This tends to produce straightforward, clear exposition, since all the material following the topic sentence becomes an expansion of the theme for that section. This is an excellent way to draft a paper for publication and can well be adopted in the incremental writing scheme of Fig. 1.

In addition to dividing the writing job into easily manageable increments, the technique recommended here has other distinct advantages over conventional methods. Writing as the work proceeds promotes interactions among the engineer, his work and his manuscript. All three stand to benefit from the process: *the author* will experience the psychological advantage of rapid progress on his manuscript; *the work* will reflect the fresh insights that normally come during the writing process; and *the paper*, written during the excitement of accomplishment, will be far more interesting to read and much easier to write.

REFERENCES

[1] H.B. Michaelson, "Creative Aspects of Engineering Writing," *IRE Transactions on Engineering Writing and Speech*, EWS-4, No. 3, 77-79 (December 1961).

[2] J.R. Tracey, "The Effect of Thematic Quantization on Expository Coherence," Paper No. 9.4, *1966 IEEE International Convention Record*. Part II, page 17.

Writing an Abstract Months before the Paper

Clinton Hawes and Craig Harkins

SINCE TECHNICAL CONFERENCES *generally require the abstract of a proposed paper months before the actual meeting, the abstract is usually written before the paper. This poses a problem for most engineers, leading many to lose interest in participating in the conference. A six-step method of writing the abstract before the paper is written is given in this article. You may find it helpful in dealing with prospective conference participants in your organization.*

The word *abstract* implies that a complete document has been written, and most technical communicators would agree that it should be. The abstract, then, truly *abstracts* the significant information rather than being merely a prose table of contents. Robert Rathbone in *Communicating Technical Information* defines what a useful abstract should be:

> Abstracts should be written *after* the main body of the communication. . .not before. The original communication will thus shape its own image, not vice versa. . .A well-written, informative abstract is a replica, in miniature, of the original. It, too, has a beginning, a middle, and an end, with emphasis on the key ideas and/or results.*

His advice is logical—but often impossible to follow. Take this situation: An engineer is working on a project that he knows will be completed by the end of the year. During the first quarter of that year a conference publishes its "Call for Papers" with abstracts due by May 1. This engineer would like to go to the conference and present a paper on his project. He is far enough into the project to have tentative results. He knows the project will be of interest to the audience at the conference, but he isn't really in a position to write the complete paper before May 1.

How does he get to that conference? Generally, he writes his abstract before he writes his paper. As a result, he frequently writes a bad abstract, which the conference turns down.

We have found that this is a very real situation. Many professional meetings and conferences *do* require that abstracts be submitted months in advance. For example:

–IEEE Region Six Conference – Electronics Serving Mankind; May 20-22, 1968; Portland, Oregon. . .500 word abstracts should be submitted by *January 5, 1968* . . . Final manuscripts will be required by March 11 . . . .

–1968 International Electronic Circuit Packaging Symposium; August 19-20, 1968; Los Angeles, California . . . Abstracts of 200 to 500 words should be submitted by *March 15, 1968* . . . Complete papers will be due by June 15 . . . .†

In spite of the obvious advantages of writing the abstract last, many – if not most – of the abstracts submitted to these and similar professional meetings will be written *before* the actual paper.

Many will be accepted by the program committee for the meeting. Then the author is committed to making sure his paper follows the abstract. Unless he has given considerable thought to the original abstract, he may find himself committed to writing a paper he doesn't really want to write.

We have devised a short communication, titled "Writing an Abstract," to help authors avoid this dilemma. It is a six-step method for writing abstracts *before* writing the paper (Figure 1).

†Information from Technical Meetings Information Service, 79 Drumlin Road, Newton Centre, Massachusetts.

WRITING AN ABSTRACT

As an aid to the writing of good abstracts, we suggest using the following six-step technique:

(1) *List the major topics to be described in the paper.*
(2) *Select between 4 and 6 of the most significant ones.*
(3) *Write a sentence or two on each — explaining or reporting results.*
(4) *End with one or two major conclusions.*
(5) *Add a sentence describing the number and type of drawings and/or photos.*
(6) *Go back and add a beginning sentence or two which discusses the context, scope, and significance of the paper.*

Remember:

An abstract should be a miniaturized version of the main report which *highlights* main topics.

An abstract should *not* be — any more than necessary — an expanded table of contents, indicating main topics to be covered.

Papers often are selected on the basis of their abstracts. The abstract deserves careful attention.

Figure 1 — Abstracting "how-to" sheet for engineers.

*Addison Wesley Publishing Co., Reading, Massachusetts, 1966, p. 25.

STEP 1

List of Major Topics

a. Specifications of system
b. Use of two probes
c. Operation of system
d. Advantages of casual part placement
e. Adaptive principle
f. Description of a measurement
g. Description of a correction
h. Use of a stored program
i. Creating own parts program
j. Example of use

STEP 2

b. Use of two probes
c. Operation of system
d. Advantages of casual part placement
e. Adaptive principle
i. Creating own parts program

STEP 3

b. Two different probes are used for measuring two different types of components -- conductive parts and printed circuitry.

c. The operator need place a part at only the approximate location required for measurement. The system automatically calculates the skew and makes the necessary corrections to all measurements.

d. This unique approach saves considerable alignment time, minimizes the need for expensive setup fixtures, and -- most important -- eliminates an important source of measurement error.

e. Unlike with conventional measuring machines, no attempt is made to make the machine's table, cross slide and spindle slide travel in precisely striaght paths or to set these paths square to each other. Instead, the amount of roll, pitch and yaw is measured by 14 sensors and its effect calculated out by the on-line computer.

i. By manually controlling the system to measure one part, an operator can create a program which allows subsequent parts to be measured automatically.

STEP 4

The Precision Measuring System uses lower-cost hardware to obtain high accuracy, high productivity and a meaningful inspection report at the conclusion of a series of measurements.

STEP 5

Three slides will show: the actual machine; a portion of an inspection report; and a line drawing which illustrates the principle by which roll, pitch and yaw corrections are made.

STEP 6

A computer-controlled measuring system has been developed as an experimental project by IBM, Kingston, N.Y. The system automatically measures parts and processes measurement data under the control of a stored-program computer.

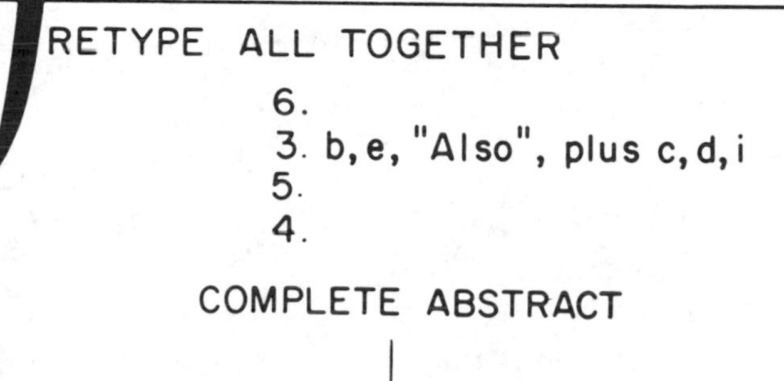

A computer-controlled measuring system has been developed as an experimental project by IBM, Kingston, N.Y. The system automatically measures parts and processes measurement data under the control of a stored-program computer. Two different probes are used for measuring two different types of components -- conductive parts and printed circuitry. Unlike with conventional measuring machines, no attempt is made to make the machine's table, cross slide and spindle slide travel in precisely straight paths or to set these paths square to each other. Instead, the amount of roll, pitch and yaw is measured by 14 sensors and its effect calculated out by the on-line computer. Also, the operator need place a part at only the approximate location required for measurement. The system automatically calculates the skew and makes the necessary corrections to all measurements. This unique approach saves considerable alignment time, minimizes the need for expensive setup fixtures, and -- most important -- eliminates an important source of measurement error. By manually controlling the system to measure one part, an operator can create a program which allows subsequent parts to be measured automatically.

Three slides will show: the actual machine; a portion of an inspection report; and a line drawing which illustrates the principle by which roll, pitch and yaw corrections are made. The Precision Measuring System uses lower-cost hardware to obtain high accuracy, high productivity and a meaningful inspection report at the conclusion of a series of measurements.

Figure 2 – Example of an abstract written by the described system.

It has proved to be a valuable tool. In addition to providing authors with a mechanical step-by-step process for getting the abstract written, we have found that "Writing an Abstract" helps us to control the *type* of abstract we submit to conferences.

Before we started using this tool we were getting too many descriptive (or indicative) abstracts, rather than the preferred informative abstract. While the descriptive abstract only tells what the paper is about, the informative abstract provides some of the information–at least the most important points–of the complete paper. To include in an abstract a statement that "the operation of the machine is to be described and test results will be presented" is hardly enlightening; furthermore, it could almost be assumed.

However, if the length or subject matter makes an informative abstract impractical or incomplete, then a combination of the two types should be used. When the intended reader is the program committee of a professional conference, the combination abstract frequently helps to provide information on the actual presentation, such as whether or not slides or other audio-visual devices will be used. In this way the program committee not only knows the subject matter of the proposed paper, but also something about the type it will be and how it will fit into their program.

Reaction to our "Writing an Abstract" has been favorable. Since introducing it earlier this year, we've had comments such as:

- "It is a real good way to come up with an abstract and an outline at the same time."
- "It helped me to focus on the audience and what the audience is interested in."
- "It helped me to separate the wheat from the chaff."

Figure 2 follows an abstract through each of the six steps in the process.

It's not a panacea, of course–it's only a tool. In many instances, such as the writing of long and highly technical journal articles, its usefulness might be limited. But even in these cases, it can provide a good starting point.

We have found that "Writing an Abstract" is a good device to pull out of our bag of tricks when a potential author says to us, "I'd love to write a paper on this subject–but I just don't have time." We ask him to follow just these six steps (he usually has that much time).

When he does this, although he has taken little time and written down only a few words, he often has committed himself to publication. More often than not, he winds up on a podium delivering a paper to his colleagues.

THE FORGOTTEN ART OF PUNCTUATION

Brian Jarman
Lawrence Livermore Laboratory
University of California

The current trend to minimize punctuation has gone too far. It has become a refuge for the ignorant and the lazy. The forgotten art of punctuation is destroying technical communication, and the reader is left holding the bag. Why is there an indifference to an art that originated in the 15th century to help clarify the printed word? When fulfilling its purpose, punctuation clarifies meaning and, in so doing, justifies itself. We should be on intimate terms with punctuation: it is integral to expression. I explain how we can utilize certain punctuation marks to make our meaning clear in what we write.

Early Beginnings

If we stylize a sentence of modern technical prose in the unpunctuated form of very early writing, it appears like this:

thisisamatterofmathematicalinver
sionusingradiativetransferequations
andspectroscopictransmissionfunc
tionstodeducetheverticaltempera
tureandcompositionalprofiles

Before systematic punctuation appeared, sentences did not exist; words were not separated; letters were not capitalized; periods were not used. Scribes of early English manuscripts tried to separate words with space, but they were careless about it. Also, they threw in periods here and there, but only for decorative purposes to improve the look of their manuscripts.

Evidence of man's first attempts to punctuate his written thoughts appear in the Moabite stone (850 B.C.). This artifact records, in Semitic script, King Moab's war against the Israelites. Its punctuation takes the form of vertical lines between phrases. If we apply this style to the previous example, we get some notion of what it may have looked like:

thisisa|matterofmathematicalinver
sionusing|radiativetransferequa
tions|and|spectroscopictransmission
functionstodeducethe|verticaltemper
atureandcompositionalprofiles

It is here, perhaps, that we get the first glimpse of the writer trying to express his message in distinct segments for easier readership. It is here, perhaps, that we see the first evidence of a writer's conscious effort to separate abstract thoughts as he scribed them on the available media. His motivation brought some clarity to the meaning of his message - a clarity that was not there before. This motivation still stands today, although many of us have forgotten it or choose to ignore it.

Formal Acceptance

The earliest books printed in English reveal three marks of punctuation: a stroke (/) to mark off word groups or phrases, a colon (:) to indicate a distinct syntactical break or pause, and a full stop (.) to signal the end of a sentence or a brief pause.

These punctuation marks helped the reader of those times. But manuscripts and books were still a matter of study by scholars. Laymen were excluded from readership. The average person had been taught neither to write nor read. But with the advent of schools for common people, the need for educative books dictated a standard form of punctuation - one that could bring general clarity to the written word and make it easier and faster to read. Punctuation at this time separated words and structures to avoid misinterpretation.

The man to whom systematized punctuation is attributed is the Italian printer, Manutious Aldus, who lived in the 15th century. He contributed most to the invention of the modern printed book as distinct from a printed imitation of a medieval manuscript. His system of punctuation, which was adopted by other printers, forms the basis of our present system. But it has evolved since then to fit the needs of the English language over the centuries.

Opinions on Evolution

Punctuation may have begun as a breath-taking device, or it may have emerged as an integral part of sentence pattern.

Some of us consider it to be a leftover from speech. We maintain that an inter-relationship exists between pauses in oral language and punctuation in writing. This seems a logical approach. After all, writing is an extension of spoken expression. This is a good way to understand punctuation. When we talk, we pause and drop or raise our voice to mark off the smaller word groups - the clauses, phrases, modifiers, series, etc., which a complete sentence includes. When we write, we have no way of indicating such pauses and reflections, except by punctuation.

Some of us hold that punctuation is entirely grammatical. We discredit the relationship of punctuation to elecutionary performance or to oral reading. We decree that writing is to be seen with the eye and not heard with the ear. As purists, we maintain that punctuation functions where the positioning of written words,and the little residual inflection the language retains on paper, do not show or indicate grammatical relationships.

Common Purpose

Punctuation cannot be pigeonholed. We cannot assign it a well-defined box in our minds. No two writers or editors can expect their styles of punctuation to be the same. Similarly, no two readers' interpretation of a given piece of prose, through its punctuation, will match.

Basic rules of punctuation have been handed down to us, which we are persuaded to abide by. And the writer as well as the editor whould be acquainted with them. But some rules are illdefined and invite differing interpretations. Only two rules of punctuation appear inviolate. One is the use of a period at the end of a declarative sentence. The other is the placement of a question mark at the end of a sentence that genuinely asks a question. The remaining rules are up for grabs when it comes to understanding and applying them.

The common purpose of punctuation is to make writing as clear as possible in its meaning. In technical writing, the effective use of punctuation should be dictated by common sense and logic. We should make punctuation work for us, as a craft and not a science. We should use it as a communicative aid, to improve the reader's acceptance of our work.

Modern Trend

Today's trend in formal and informal writing is to use the "open" system, which minimizes punctuation. At one time, in the 19th century, punctuation became very elaborate. The "close" system prevailed then, in which the comma, semicolon, and colon were climactic in value. Eventually, this gave way to a simpler, more practical style to meet the demands of creative, informal writing.

James Joyce has been blamed by many for the downgrading of punctuation in modern literature. His "Ulysses" and "Finnegans Wake" are notable examples of the disuse of punctuation. Joyce excluded punctuation as part of his creative technique, which was copied by his contemporaries and has influenced modern authors ever since.

Minimal punctuation is a literary device. It belongs to informal writing. It has no place in technical writing, which is formal. In our business, we write to inform not to entertain. And we need punctuation to help us make our meaning clear.

The blame for the present lack of punctuation in technical writing should be placed at the feet of the editor. He is the watchdog of the written language. He is the liaison between the writer and the reader. It is his responsibility to make sure that what the writer expresses will be clear in meaning to the reader.

Editors who minimze punctuation, either through ignorance or laziness, are inviting writers to style their own work accordingly. The lack of punctuation increases the chances of unitentional obscurity and ineffective structure, since the principles of punctuation are unlearned or ignored in the process.

Some of today's published technical writing is gibberish. Here's an example:

> So far we have dealt with the problem of reduced dependent assignment of a machine having the next-state function as a one-to-one mapping for at least one input and we have derived the reduced dependent assignment for that input from the CCM representation of the next-state function for the input and then by checking which of them also gives reduced dependency of the internal variables for other inputs also.

A sentence of this length without a pause or mark is an insult to a reader. Writing of this kind has no business getting published. Whatever happened to the writer's and editor's obligation to the reader?

Effective Usage

Let's now turn to those marks of punctuation whose absence or misuse is prevalent in the technical literature.

The first mark, which is conspicuous by its absence, is the period. The most effective communicative use of the period is to separate complete thoughts. When it appears at the end of a single thought, it indicates to the reader that the writer wants him to stop and digest the thought. The period acts as a red light.

Technical sentences, today, are crammed with too many ideas for the reader to swallow in a single visual pass. They are overpacked. The reader finds himself going back over them to uncover the heart of the writer's message. Because we deal with complex subject matter, we should strive to embody one idea in one sentence. This dictum is designed for writing in which instant clarity and swift reading are desirable. The effective use of periods to isolate complete thoughts will help us to put this principle into practice. It will relieve overpacked sentences and lighten the load for the reader.

An infrequent mark, within a sentence, is the colon. This pause has great communicative value: it emphasizes information that follows it. It can draw the reader's attention to two appositional or strongly contrasted statements. When we see a colon, we pause a little longer than we would at seeing a semicolon or a comma. In fact, we almost come to a full stop. The colon announces that the promise implied in what precedes it is about to be fulfilled. Its appearance creates a telegraphic brevity and conciseness that are appealing as communicative aids. We should take more advantage of its function.

The colon's nearest relative is the semicolon, but the latter is the weaker sister of the two. We should use more semicolons in technical writing. They are effective devices for linking or uniting thoughts that relate to a common topic. Also, they can separate long, involved sentences into manageable portions for the reader to digest.

Why do we neglect the question mark? This punctuation mark can add variety to our expression. Its interrogative nature will pique the curiosity of the reader. The significance of a sentence ending in a question mark is that the sentence begs to be answered. The question mark is a stimulant. It is direct in its function. It puts the reader on the spot, since it carries with it an implied obligation for the reader to respond.

Finally, we come to the comma, the briefest pause in written expression. When misused inadvertently, or when not used at all, it can create havoc in a writer's flow of written thought. The comma is the reader's natural pause that refreshes. But its application needs studying, because it has many functions. The main purpose of the comma is to cut off, for the sake of clearness, a distinct part of a sentence from the other parts. This we should remember, because in fulfilling this purpose the comma can prevent vagueness or confusion.

In the following examples, I try to show you how to recognize the need for effective punctuation. I have chosen the examples at random. They reflect the lack of the punctuation marks I have discussed above.

EXAMPLES

Use Periods for Offloading Overpacked Sentences

Before:

> Since the ultimate objective of the work is the development of a mathematical representation of the reduction process which can be readily used for modeling the behavior of moving-bed systems, we shall select the simplest model that is consistent with measurements over the range of variables encountered in industrial operating practice.

After:

> The ultimate objective of the work is to develop a mathematical representation of the reduction process. This representation can be used to model the behavior of moving-bed systems. Consequently, we will select the simplest model that is consistent with measurements over the range of variables encountered in industrial operating practices.

Use Semicolons to Combat "Anditis"

Before:

> If the correlation is disturbed at any distance within a critical radius an exchange of energy between the two electrons sets in and the out-come of the process is that the faster electron escapes while the slower is captured by the ion into an excited state of the atom and hence no ionization has taken place.

After:

> If the correlation is disturbed at any distance within a critical radius, an exchange of energy between the two electrons sets in; the out-come of the process is that the faster electron escapes, while the slower is captured by the ion into an excited state of the atom; hence, no ionization has taken place.

Use Question Marks to Spark Interest

Before:

> The cause of the rapidly increasing degradation rate as R increases is not known but there are several intriguing possibilities.

After:

> What causes the degradation rate to increase rapidly as R increases? We don't know. But there are several intriguing possibilities.

Use Colons to Achieve Conciseness

Before:

> Defining the volume and nonradiochemical composition of high level waste that we are going to concentrate is complicated by the fact that different processing schemes may be used and that different combinations of process wastes may be made to generate the high level wastes.

After:

> Defining the volume and nonradiochemical composition of high level waste that we are going to concentrate is complicated: differing processing schemes may be used and different combinations of process wastes may be made to generate the high level wastes.

Use Commas to Avoid Confusion

Before:

> In this model absorption process are assigned to the lowest electric dipole transition. At very low temperature migration energy does not occur and thus the same site is responsible for both absorption and fluorescence.

After:

> In this model, absorption processes are assigned to the lowest electric dipole transition. At very low temperature, migration energy does not occur; thus, the same site is responsible for both absorption and fluorescence.

CONTEXTUAL EDITING: THE FIRST STEP IN EDITING SENTENCES

Prof. J. C. Mathes
University of Michigan

Of the many "correct" forms a sentence may take, the form it takes derives first from the context within which it exists. Contextual editing requires analysis of this context to edit sentences in terms of their single sentence editing. Contextual editing requires <u>vertical thinking</u>.

Most technical writers could write their own manual on sentence structures and sentence style, but can they explain how to revise and edit sentences in a report efficiently? We all are familiar with the discussions in technical writing texts and handbooks of grammar on simple, compound, complex, and complex-compound sentences, on subordination and parallelism, on natural and periodic forms, on varying sentence structures and on the use of the active and passive voices. Yet with all of this knowledge we may not be able to choose the most effective sentence form quickly and appropriately as we revise and edit technical communication.

The purpose of this paper is to introduce a technique for editing sentences that must have priority over traditional sentence editing techniques. The first stage in editing sentences should be contextual editing, that is, editing sentences in terms of the contextual patterns provided by the paragraphs.[1] I will explain what a contextual pattern is, then I will explain a three-step procedure for editing sentences systematically.

<u>Context Has Priority</u>

Why does context have priority? Let's look at some variations of a rather long sentence from a technical report and ask ourselves, which form is most effective?

> The naval architecture characteristics of "weight controlled" destroyers, such as buoyancy and stability, were those chosen to carry the weight of hull, machinery, ship systems, people and payload.
>
> For "weight controlled" destroyers, the naval architect formerly chose characteristics such as buoyancy and stability to carry the weight of hull, machinery, ship systems, people and payload.
>
> "Weight controlled" destroyers were those formerly designed to carry the weight of hull, machinery, ship systems, people, and payload.
>
> Formerly, destroyers were "weight controlled," and the naval architecture characteristics such as buoyancy and stability were those chosen to carry the weight of hull, machinery, ship systems, people, and payload.

Considering principles of subordination, variety, voice, etc., we can generate numerous variations of this (or any) sentence. Writers with a particular flair for expression certainly could improve upon any of the above sentence forms if that sentence were to exist independently. However, we must subordinate such considerations to the context within which that sentence exists.

That sentence exists in a paragraph of comparison and contrast and is paired with another sentence[2]:

> Modern destroyers are "volume controlled," so the significant design features are those relating to utilization of volume. Formerly, destroyers were "weight controlled," and the naval architecture characteristics such as buoyancy and stability were those chosen to carry the weight of hull, machinery, ship systems, people, and payload.

Here the form of the second sentence is determined by the form of the first--both main clause and second clause--as the following figure shows:

Modern	destroyers	are
Formerly	destroyers	were
"volume controlled"	so....	
"weight controlled"	and....	
...the significant design features	are those	
...the naval architecture characteristics such as....	were those	
relating to utilization of volume		
chosen to carry the weight....		

Figure 1. Context In A Paragraph of Comparison and Contrast

The forms of these sentences derive from the comparative purpose of the paragraph, which requires the sentences to be parallel in order to clarify the relationships between the two concepts being compared and contrasted. If the form of the second sentence had been, "For 'weight controlled' destroyers, the naval architect formerly chose...," the pattern would have been broken and therefore the ideas obscured.

Thus context has priority when you come to revise, edit, and polish sentences. Your purpose is to enable the reader to read efficiently and with comprehension; you edit sentences to clarify the context and enable the reader to do so.

Pattern Provides Context

Implicit in every paragraph should be a purposeful pattern that establishes the context within which the individual sentences take form. The pattern thus is at the interface between the paragraph as a structural unit and the sentences as subordinate structural units. The pattern of a paragraph is neglected in almost every textbook discussion of paragraph development.

Discussions of paragraph development dwell on the familiar concepts of the topic sentence, paragraph unity, paragraph development, and paragraph continuity. The topic sentence provides the central idea to which the rest of the sentences must be related for the paragraph to be unified. Paragraph development is explained in terms of the order and relationships of the ideas in the sentences. Paragraph continuity is the use of transitional words and phrases to establish a "*recognizable* flow of thought."[3]

Reconsider these discussions of paragraph development, however. Notice that they do not clarify the structural context within which the sentences exist. Although they may clarify what each sentence is about, they do not explicitly clarify the basic form each sentence should take. They do not focus your attention on the implicit pattern that will establish the basic forms of the individual sentences.

A conceptually unified paragraph will have an implicit pattern appropriate to the purpose of the paragraph. This pattern establishes the context for the individual sentence forms. For example, an effect-cause paragraph will state an effect and then explain the causal process leading to the effect. Given an effect E, the paragraph will trace a causal process through a series of stages, A→B→C→D→E. The paragraph will then have the pattern illustrated in Figure 2.

E←(A→D)	Core sentence sets pattern
A→B B→C C→D D→E	Supporting sentences follow S-V-O pattern established

Figure 2. Pattern For An Effect-Cause Paragraph

Here is a paragraph in which the causal process pattern provides a context for the individual sentences:

> During a storm in the North Pacific Ocean in the early hours of 19 December 1969, the SS MICHIGAN sank because its cargo of bombs exploded. The storm caused the vessel to experience heavy rolling (up to 52°). The rolling created excessive loading forces on the cargo securing system from the palletized bombs. The cargo securing system failed, permitting a complete row of palletized bombs to shift. When the palletized units shifted, numerous 2,000 lb. bombs broke loose and slid and rolled in number five hold. This resulted in an impact detonation which ruptured the starboard side shell plating and damaged the forward bulkhead. Pro-

gressive flooding occured. The SS MICHIGAN was abandoned by her crew and eventually sank.

In this paragraph the causal process pattern provides a context within which the subject-verb-object relationship in each sentence takes form. Although each sentence still can assume various forms and several ideas can be combined or separated in various ways, the pattern has priority and establishes a basic form underlying any variations. The verbal focus which relates subjects and objects in a consistent sequence must control any editing of the sentences in this paragraph.

Contextual editing therefore is a systematic procedure of clarifying the context and then editing sentences in terms of their interrelatedness with other sentences before editing sentences individually. The procedure consists of three steps:

(1) Determine if there is a core sentence to establish a pattern for the paragraph.

(2) Determine if the pattern is appropriate for the purpose and content of the paragraph.

(3) Determine if each sentence clearly follows the pattern established.

A technique I label *vertical thinking* will enable you to edit efficiently.

Core Sentence Provides a Pattern

First, determine if there is a core sentence to establish a pattern for the paragraph. I use the term, "core sentence," to distinguish it from the related term, "topic sentence." Both "core sentence" and "topic sentence" refer to that sentence in a paragraph stating the central idea of the paragraph. Both concepts mean that this sentence provides paragraph unity and development. That is, this sentence presents an idea that the other sentences must relate to and develop logically. The concepts also imply, especially for reports, that this sentence should come at the beginning of most paragraphs. The concept of "core sentence," however, unlike the more familiar concept of "topic sentence," in addition means that this sentence also should signal the pattern or structure of the paragraph. In this first step of the contextual editing procedure you identify or establish a sentence that provides a structural as well as conceptual key to each paragraph unit.

This step should be simple for most paragraphs in a rough draft with any coherence at all. If core sentences are missing, or if two core sentences appear in a paragraph, more extensive revision than sentence editing will be required. This step usually requires you to emphasize, clarify, or state core ideas that are left implicit or are buried in clauses and structurally subordinated sentences in rough drafts. Most rough draft paragraphs are written around core ideas; many of these core ideas, however, do not get stated explicitly in core sentences. You do this during the first step of the contextual editing procedure.

The core sentence in the paragraph quoted above was identified and edited contextually: "During a storm in the North Pacific Ocean in the early hours of 19 December 1969, the SS MICHIGAN sank because its cargo of bombs exploded." This sentence underwent several modifications during editing, the most significant occuring when the subordinate clause was added, and another when the stronger conjunction "because" replaced the weaker conjunction "when."

Here are two similar core sentences, one after editing and the other before editing:

> Three sources of nutrient load in West Bay must be identified before we consider how these nutrient sources may be reduced.
>
> Based on these assumptions we identify three potential sources of guest patronage for a motor hotel at the proposed site: business travellers, tourists, and group meetings and small conventions.

The first core sentence clearly established an enumerative pattern of particularization, "The first source....," "The second source....," and "The third source...." The second core sentence, however, came at the end of the paragraph which enumerated several assumptions but which lacked a core sentence at the beginning. The core idea is contained in the phrase, "Based on these assumptions." Contextual editing would require this phrase to be expanded to a core sentence and placed at the beginning of the paragraph:

> We had to make certain assumptions about the design of the motel in order to identify the potential sources of patronage.

This revision establishes a core sentence which provides a structural as well as conceptual key to following sentences, which explained the assumptions rather than the sources of patronage. ("We assume that the design..., the site..., the accommodations..., the amenities...," etc.).

Pattern Must Be Appropriate

Second, determine if the pattern is appropriate to the purpose and context of the paragraph. In this step you examine the core sentence to make certain that the structural signal complements the conceptual signal. Does the implied form reinforce or undermine the content? Does the core sentence, for example, signal a narrative sequence of events with "when its cargo shifted" where a causal process signaled by "because its cargo shifted" would be more appropriate?

To determine the pattern you closely examine the subjects, verbal concepts, and objects or modifers in the core sentence. You analyze this S-V-O(M) relationship to determine what the focus of the sentence is and what structural pattern it implies, and whether these are appropriate to the rhetorical purpose of the paragraph. What pattern does this core sentence indicate, for example?

> Here, the poor sampling technique is evident.

This sentence clearly focuses on the subject concept, "poor sampling technique." Thus, in this form the core sentence implies an analysis of the sampling technique, either in terms of a stage-by-stage process or in terms of a flaw in one stage of the process. (Since analysis involves the breaking up of a whole into its constituent parts in order to examine either the relationships among those parts or individual parts discreetly, another sentence may be needed to signal which analytical pattern is being invoked.) This paragraph topic, however, yields other core sentences which signal different patterns. For example:

> Here, the proportion of N_2 to O_2 deviated significantly from the expected proportion because of the poor sampling technique.

This version of the core sentence focuses on the effect, and implies a contrastive pattern to particularize the actual and expected proportions of N_2 and O_2. (The clause, "because of the poor sampling technique," implies that this paragraph is part of a larger pattern.) Or:

> Here, the deviation from the expected proportion of N_2 to O_2 indicates poor sampling technique.

This version of the core sentence focuses on the cause, and implies an effect-cause pattern to explain the cause of an observed effect.

Your purpose in the second step of the contextual editing procedure is to make certain that the pattern signalled is appropriate to the rhetorical purpose of the paragraph. This step contributes significantly to the editing of a report because it forces the writer to clarify the intent of a paragraph during rough draft stages of report writing. The consequences are some paragraphs with no clear rhetorical purposes and others with several rhetorical purposes. The rough draft paragraph for the "core" sentence discussed above had two purposes--to particularize the effect of the poor sampling technique and to analyze the sampling technique. Consequently, the rough draft paragraph inefficiently mixed these two patterns. This step of the editing procedure enabled the writer to separate these purposes and establish two paragraphs, each with an appropriate core sentence and pattern.

<u>Pattern A Guide For Editing</u>

Third, determine if each sentence clearly follows the pattern established. In this step you use the pattern as a template to inspect and revise each of the subsequent sentences in the paragraph. The pattern will suggest some meaningful sequence of subjects, verbal concepts, and objects and modifiers in the sentences following the core sentence.

The pattern for the causal process paragraph presented above, $E \leftarrow (A \rightarrow B \rightarrow C \rightarrow D, \ldots)$, determined the forms of the sentences in that paragraph, as illustrated in Figure 3.

The SS MICHIGAN sank.	$E \leftarrow$
The storm caused heavy rolling.	$A \rightarrow B$
The rolling caused securing system to fail.	$B \rightarrow C$
...	...
Progressive flooding occured. The SS MICHIGAN sank.	$D \rightarrow E$

Figure 3. Pattern For Editing An Effect-Cause Paragraph

If the subject-verb-object sequence of any of the sentences in the paragraph had been reversed, the flow of the paragraph would have been interrupted. This step of the editing procedure would alert the writer to edit that sentence to fit the pattern.

Similarly, the pattern for the paragraph contrasting "volume controlled" design to "weight controlled" design determined the forms of the sentences:

A is B, thus C is D
but A^1 is B^1, thus C^1 is D^1

This pattern enabled the writer to edit the cumbersome second sentence of the contrast so that it could be read efficiently in context.

The third step of the contextual editing procedure would enable a writer to identify very quickly the sentence in this contrastive paragraph that needs editing:

> (1) The steel system can be constructed more rapidly than the masonry system. (2) Steel columns can be erected much faster than concrete columns, which must be allowed to cure in their forms for two or three weeks. (3) Steel wall panels also can be placed much faster than block walls can be erected, so the prospect of delays caused by the masonry union can be eliminated. (4) Insulation can be sprayed more rapidly than styrofoam panels can be glued in place.

(5) Thus, a steel system can be constructed in one-third to one-half the time a masonry system can be erected.

This paragraph is not ineffective. The writer has maintained rather consistently the S-V-modifier pattern appropriate for the contrastive purpose signalled by the core sentence (1) of the paragraph. However, the writer has varied the pattern with sentence 4, and thus forces the reader to hesitate to decide whether insulation is sprayed on steel walls or on masonry walls. With contextual editing the writer will eliminate this inconsistency by revising sentence 4 to fit the context: "Steel walls can be insulated more rapidly by spraying than masonry walls can be insulated by gluing styrofoam panels in place." (Conversely, the writer who would argue that the original sentence 4 did not need editing would do so on the basis that the pattern is so clearly established by the other sentences that the reader would not hesitate. This argument also reinforces the premise that sentences are edited in terms of the pattern.)

Vertical Thinking

The contextual editing procedure explained in this paper is particularly effective because it requires you to change your habit of thinking. When you write a rough draft, you grope along one sentence at a time. You are trying to formulate your ideas at the same time you are trying to put them into sentence form. You get one idea onto paper; then you move on to the next, hoping the connection between the two will be clear. You write linearly--one sentence at a time in a seemingly endless linear sequence. You are thinking "horizontally."

To edit contextually, you must think "vertically." Vertical thinking is thinking holistically, grasping all the parts simultaneously. To think vertically you establish frameworks and boundaries; you examine interrelationships in two dimensions rather than discrete bits arranged along a single, linear dimension. You take a unit of prose--a cluster of paragraphs or a single paragraph--and mentally arrange the ideas in outline form and place sentences one below another so that subjects, verbs, and objects fall into vertical columns. A two-dimensional pattern emerges. To edit contextually you examine the sentences within the context of this pattern and determine if each sentence maintains the focus of the pattern.

Here is an example of vertical thinking:

(1) steel system	can be constructed	masonry system
(2) steel column	can be erected	concrete columns
(3) steel wall panels	can be placed	block walls
(4) insulation	can be sprayed	styrofoam panels
(5) steel system	can be constructed	masonry system

Figure 4. Vertical Thinking for Paragraph Editing

Notice how vertical thinking for editing this paragraph involves examining sentences in terms of their relationships to other sentences rather than as self-contained units. Think vertically and you instinctively think in terms of interrelationships.

Think vertically and you will effect that difficult change from writer to editor. You then can proceed to single-sentence editing because you no longer will be reading your sentences within the subjective context of your first writing them. You will be reading them within a fresh context, and will be able to examine them and manipulate their forms objectively.

A final observation can be made about vertical thinking and contextual editing. Now that you have established a pattern for the individual sentences, you are free to consider other editing suggestions, such as those concerning variety and continuity. Certainly any pattern has the flexibility that allows for significant sentence variety, as a consideration of the many possibilities for each sentence within the causal process pattern of "the sinking of the SS MICHIGAN" will indicate. The consideration of paragraph continuity is particularly enhanced by the contextual editing procedure. You are now able to establish appropriate paragraph continuity because the pattern will suggest the words and phrases to use to relate the sentences to each other within the context of the whole rather than just back-to-back. You will edit for continuity not just to eliminate the bumps in the road, but to illuminate the entire route from beginning to end.

Notes

[1]Paragraph. I use this term to denote a basic unit of thought. In print, especially in technical writing, the unit of thought may appear as a cluster of short "paragraphs."

[2]This example and several others in this paper appear in a chapter on sentence editing in J. C. Mathes and Dwight W. Stevenson, *Designing Technical Reports: Writing For Audiences in Organizations*, The Bobbs-Merrill Co., 1976.

[3]John B. Colby, "Paragraphing in Technical Writing," *Technical Communications*, 18:2 (March/April 1971), pp. 13-16; quote p. 15.

Paragraphing in Technical Writing

John B. Colby

GOOD PARAGRAPHING *is characteristic of professional writing. Knowledge of types of paragraphs and paragraph movement, plus use of the summarizing topic sentence, will help a writer understand and use the tools of his trade to do a better job. Mr. Colby gives a review and analysis, with examples, that should be useful to many communicators.*

A paragraph, my new desk dictionary tells me, is "a subdivision of a written composition that consists of one or more sentences, deals with one point, or gives the words of one speaker, and begins on a new usually indented line." My Perrin's *Writer's Guide*[1] goes a bit further into the subject, devoting 46 pages to it (including exercises). Somewhere between the dictionary's impoverished definition and Perrin's abundance of thought must lie a manageable area that will enable the practicing technical writer to review the subject quickly and put it to use. For it appears to me, and possibly to some other editors, that good paragraphing is a vanishing – but useful – art, like good cabinet-making and blacksmithing.

CHANGING STYLE

Perrin, Weisman,[2] and countless others who have written on the subject point out that a paragraph is a basic unit in the development of a subject. Let me add that the paragraph is peculiar to writing; we can readily perceive that speakers use sentences just as writers do, but it is difficult to recognize the spoken paragraph. I deliberately inject this comment because I think I have stumbled across something new. Nowhere have I seen anyone write this down. I stake out a claim on it, therefore, and I hope to soon find references to it in the literary efforts of others.

Students, neophyte writers, and sweaty engineers laboring over a draft of a report worry about the proper length of a paragraph. Like the length of women's skirts, the length of a paragraph depends a great deal on the current style. A hundred years ago paragraphs were quite long; Victorian novels featured paragraphs that ran a page or more. In *The World of Mathematics*,[3] James R. Newman reprints a lecture by William Kingdom Clifford ("The Postulates of the Science of Space") written about 100 years ago. This early technical writer opened with an 1100-word paragraph. The abnormal length fits Perrin's observation that philosophical works and journals tend to have longer paragraphs than the more popular writings do. Newman says that Clifford had a singular power for making hard concepts understandable; if what he says is true, then we see at once that paragraph length has little to do with ease of understanding. Of course, we should recognize two things about this example. The first is that if someone took Clifford's lecture down in shorthand and then converted it to writing, he may not have had the slightest idea where to paragraph. The second is that regardless of his ability to explain, very few of us could ever understand much about Clifford's subject, which was really non-Euclidean geometry.

In contrast to Clifford's cross-country paragraph, modern writing, including technical writing, is characterized by paragraphs that run to 100 to 150 words. Newspapers and magazines use even shorter paragraphs, often only one or two sentences (20 to 50 words). When I apply the criterion of units of thought to modern paragraphing it seems to fail. I believe the real reason for the 150-word paragraph in technical writing is that most writers nowadays get their material typed double-spaced. Now, a full page of double-spaced typing looks bad; it looks like there ought to be a paragraph in there. So the author or editor paragraphs, usually about the middle of the page. This works fine as long as one uses identation to indicate a paragraph. If, on the other hand, block paragraphs are used, a problem comes up when a paragraph ends at the bottom of the page near the right-hand margin. In this case, the editor cannot tell that a paragraph was intended at the top of the next page. This aggravates him, especially if he is marking copy for a printer.

UTILITY OF PARAGRAPHS

Some claim that paragraphs help break up text into units of thought that readers can assimilate more quickly and readily than if they weren't used. This may be true, but I know of no evidence that it is. Who has methodically studied the absorption and retention of paragraphed versus unparagraphed writing?

The real utility of paragraphing, it seems to me, accrues to the writer. What he wants to do is arrange his work in an orderly fashion, and good paragraphing helps.

TYPES OF PARAGRAPHS

To find an orderly arrangement, it helps to look at the types of paragraphs used by most writers in the trade. A block diagram may be the best way to indicate different types of paragraphs and how they are used to develop a section of a report. (Actually, there are probably better ways, but block diagrams are fashionable and are widely accepted by engineers. So I contrived this diagram.) As Figure 1 shows, a report or a major

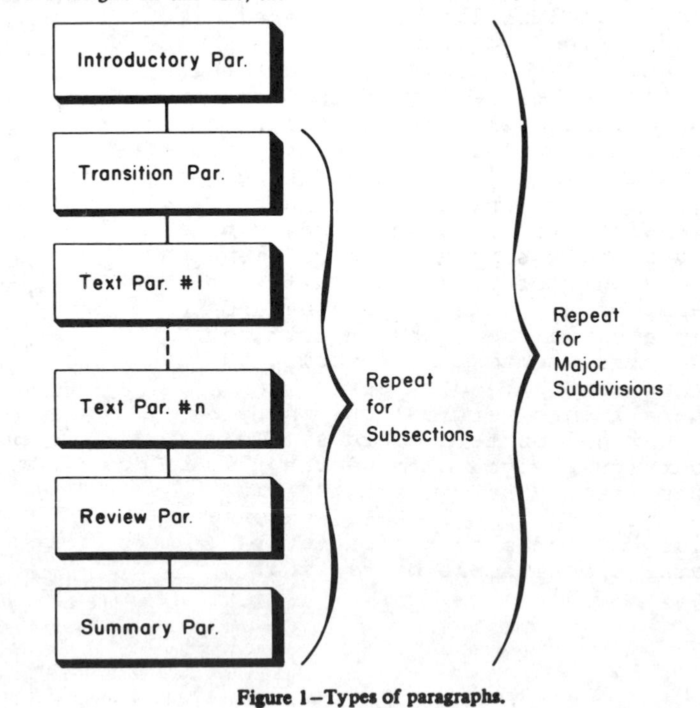

Figure 1—Types of paragraphs.

Reprinted with permission from *Tech. Commun.*, vol. 18, Mar/Apr 1971, pp. 13-16. Copyright © 1971 by Society for Technical Communication, Inc.

section can be led off with an introductory paragraph. A transition paragraph, probably very short, can be used to link the introductory paragraph with the first text paragraph. Then follow several text paragraphs, as many as needed to develop the section. Finally, it might be useful to review what has been described to this point, especially if the content is complex, so a review paragraph might be written. This can be followed by a summary paragraph that contains conclusions drawn from what has been presented so far. This might also serve as an introductory paragraph for the next major section. As the diagram indicates, this sequence is repeated until the subject or the writer is exhausted.

PARAGRAPH MOVEMENT

A paragraph should move, Perrin says, so that the reader at the end of it is better informed than he was at the start. Perrin lists five main types of movement: narrative, descriptive, support, climax, and pro-and-con. A narrative paragraph depends on movement in time for its development. In technical writing, we do not often follow a chronological development, so we do not often write narrative paragraphs. However, an introductory paragraph often has characteristics of time:

"This project was initiated in 1965, when the sponsors agreed to fund it to the extent of $100,000. Early in 1966, the project was staffed, space rented, equipment purchased, and implementation began . . . ."

A text paragraph might be descriptive:

"The assembled space platform is about 10 feet high, 40 feet long, and 10 feet wide. Viewed from the front, the living quarters are on the left, the laboratories in the center, and the service module on the right."

Most technical writing makes frequent use of the support paragraph. Often, a writer makes a general statement (a topic sentence) then follows this with detailed statements giving the evidence that supports or particularizes his lead:

"The search for commercial deposits of oil and gas is becoming more difficult. In recent years, the average depth of exploratory wells has increased from 5000 to 8000 feet. More wells are being drilled offshore, some in waters more than 1000 feet deep. The reason for this is that in the continental United States big pools of oil (greater than 10 million barrels) are found less and less often. The explorationist is forced to search for large deposits in remote areas and at greater depths."

The pro-and-con paragraph is often used in technical writing, usually to give both sides of a question or argument, or to discuss the advantages and disadvantages of something:

"Color photography is the best way of recording the differences in the samples collected. The yellows, greens, and browns contrast vividly, enabling the analyst to classify and quantify the particulate matter present. The cost of color photography is much higher than black-and-white, however, and reproduction in printed reports is expensive. Furthermore, the shades of color vary from time to time, apparently because of differences in batches of film and in batches of processing chemicals. Consequently we cannot depend on color matches to infallibly indicate a particular element. What we must do is decide whether the advantages of color photography more than offset the disadvantages."

Climax paragraphs are probably used more often than they should. This type of paragraph starts with particulars and builds up to a general statement, or climax. The writer may choose this method deliberately, or he may resort to it subconsciously when he feels that the generality would not be acceptable to the reader (perhaps because of prejudice). So he backs into it. Only when he feels he has presented enough detailed evidence does he spring the generality.

Let us suppose, for example, that a researcher was employed by the smog industry to study the effects of smog on health. Obviously, the investigator would not want to antagonize smog industry leaders at the very outset of his report:

"Extensive investigations have been made of the effects of smog on lung, heart, and eye diseases. Experimental and control groups of rats were used for subjects, and concentrates of 25 different types of smog were administered in controlled breathing experiments, by injections, and by eyedrops. Statistical techniques plus thorough medical examination of the subjects were used. Carcinogenic and coronary symptoms were developed by rats in both groups. Whether deviation from the norm is significant is a matter of some debate, and the final analysis of the results has not been completed—indeed, further tests are warranted. Nevertheless, we feel we can only conclude at this time that further inhalation of or even exposure to smog should be avoided where possible by young, healthy people, should be minimized by adults, and will be absolutely fatal to debilitated, elderly people."

ORGANIZING THE PARAGRAPH

Most authorities agree that the modern paragraph should be organized around one topic, and many teach that a good paragraph contains a topic sentence. By topic they mean the immediate subject—and this should be gotten clearly in mind. For example, the basic subject of conversation for a group of males taking a coffee break may be women, but the topic may be the blonde at the next table.

The topic sentence may appear as the first sentence, it may pop up somewhere in the middle, or it may be the last sentence. In the pro-and-con paragraph given earlier the topic sentence appeared first; in the climax paragraph, it came last. It takes some skill to organize good paragraphs in which the topic sentence appears in the middle or at the end. Consequently the unskilled writer would do well, in my opinion, to start most of his descriptive, support, and pro-and-con paragraphs with a topic sentence. This helps him keep his mind on the immediate subject; he is not as apt to wander into momentary digressions. Then, having disposed of the topic, he can write another topic sentence and so on. Even better than the topic sentence is the summarizing topic sentence.

SUMMARIZING TOPIC SENTENCE

Mrs. Melba W. Murray, who has written an excellent book called *Engineered Report Writing*,[4] teaches the use of the summarizing topic sentence. In brief, the summarizing topic sentence gives the essence of the topic to be discussed; it gives information that is immediately meaningful, and the paragraph is then developed to support the statement made. For example, let us use the smog paragraph we offered as a climax paragraph, rewriting it for our new purpose.

The *topic* is:

Smog

A *topic sentence* would be:

Inhalation of smog should be avoided.

A *summarizing topic sentence* would be:

Investigations of the effects of smog on health force us to conclude that it causes irreparable damage to the heart, lungs, and eyes.

The summarizing topic sentence is a good lead sentence for a paragraph. The author can develop his paragraph by bringing in information needed to support his opening statement, or assertion. When he has brought in enough detail to be convincing (this is a judgment), he can go on to the next paragraph.

CONTINUITY AND TRANSITION

Inside a paragraph, continuity can be established by using a *recognizable* flow of thought. Recognition is helped by repetition, i.e., using the same words or phrases, especially key words. Contrast helps (using words like *however* and *but*). Continuity-establishing words (such as *therefore*, *hence*, *because*, *consequently*, and *subsequently*) are essential. Reference pronouns help. For example:

"Mathematical and data-processing techniques that employ digital computers have entered more aspects of our lives than most vigorous proponents dreamt of ten years ago. We all expected *computers* to be extensively employed in science and technology. *But* few of us anticipated that *they* would bill us for air travel, for meals and lodging, for purchases in stores. *Few of us* realized that *computers* would become slaves of the Internal Revenue Service, checking and cross-checking our tax returns with absolute accuracy at inhuman speed. *Computers* and their associated *mathematical techniques* are such basic technical tools today that little attention is attracted by *their* use in designing aircraft, space ships, automobiles, ships, buildings, systems, and molecules. *Computers* set type, direct machinery, write music, check spelling, and control traffic. Those who do not comprehend the vast usefulness of *these electronic calculators* fear *them*. *Those* who do, however, appreciate that *because* of *them*, man is capable of making more progress—real *progress*—in the next few generations than he did in all history."

The italicized terms are key words, pronouns, and conjunctions that were deliberately used or repeated to establish continuity. Most paragraphs with continuity are sprinkled with words like these; they are characteristic of a skilled writer. An unskilled writer would do well to practice their use.

Some authors have difficulty developing sections in which one paragraph leads readily and smoothly into another. My experience, however, is that one well-constructed paragraph leads readily into another, so the problem is solved from the outset by constructing a good opening paragraph.

Let us consider the summarizing topic sentence given for the paragraph on smog. The writer might proceed with this paragraph by writing sentences concerning the investigations, the effects of smog, and what it does to the heart, lungs, and eyes. Since very little detailed information can be packed into a sentence, the writer might then write whole paragraphs and sections giving progressively more detail on the investigations, the effects of smog, and so on.

Two devices help in tying paragraphs together. These are the forward and backward references. If he wished to go from his opening paragraph to the next one using a forward reference, he might close his first paragraph with a statement like this:

"Since our experimental techniques are new in this type of work, we shall discuss them briefly before discussing the specific effects of smog on health."

Then he goes on with his next paragraph. On the other hand, he might have decided that he did not want to spoil the effect of his climax, so instead he opens the next paragraph with a backward reference:

"The importance of inhalation and exposure was shown clearly by the investigations conducted, which are new to this type of work and so shall be discussed briefly before we go on to the specific effects of smog on health."

Notice also that both the forward and backward reference indicated that after the description of the investigations there will appear a paragraph (or maybe section composed of several paragraphs) on the specific effects. Simple as they are, these devices are of great help in organizing a document that has continuity.

IMPORTANCE OF THE SUMMARIZING TOPIC SENTENCE

The importance and utility of the summarizing topic sentence cannot be overemphasized. Its use has been taught and encouraged at Esso Production Research Company for several years. Those who employ it report that it helps in expanding an outline into paragraphs and paragraphs into a readable report. In fact, as Mrs. Murray points out in her book, it leads writers into a new way of looking at their material, and it is an effective way of designing reports. Feedback from recipients of reports is that they are more readable and more useful when they employ summarizing topic sentences used as lead sentences. In addition, expanding an outline with summarizing topic sentences cuts the length of reports by 25 to 50 percent without sacrificing content, reduces writing time by about the same percentages, and reduces time required for approval through channels.

I have found that I can skim a report organized and written in this way, reading only the first sentence of each paragraph, and then do a creditable job of summarizing the report. This means that technical information processing people could probably prepare an abstract of such a report quite readily, and searchers could do a better job of finding the desired information.

SUMMARY

In summary, use of summarizing topic sentences focuses the writer's and reader's attention on the immediate subject. It helps organize sentences into paragraphs and paragraphs into sections. Use of summarizing topic sentences to expand an outline results in better, shorter, and more useful reports that get approval quicker. Knowledge of types of paragraphs (introductory, transition, text, review, and summary) and types of movement (narrative, descriptive, etc.) will help a writer understand and use the tools of his trade to do a better job. Length is partly dictated by style (modern paragraphs are rather short), but it should be dictated by what the writer has to say about the immediate subject (topic).

REFERENCES

1. Porter G. Perrin, *Writer's Guide and Index to English*, pp. 110-156, 4th ed. (revised), Scott, Foresman and Co., New York, 1965.
2. Herman M. Weisman, *Basic Technical Writing*, pp. 315-317, Charles E. Merrill Books, Inc., Columbus, Ohio, 1962.
3. James R. Newman, *The World of Mathematics*, vol. 1, pp. 552-554, Simon Schuster, New York, 1956.
4. Melba W. Murray, *Engineered Report Writing*, pp. 35-36, rev. ed., The Petroleum Publishing Co., Tulsa, Oklahoma, 1969.

Transitional Devices for the Writer

KURT M. MARSHEK

Abstract—Transitional devices fasten together words, ideas, and thoughts; and they enable particularly the expository writer to develop a cogent and coherent article that his reader can easily follow. This paper describes and then presents general guidelines for the effective usage of six types of transitional devices.

I. INTRODUCTION

In this journal, Orth[1] states that an essential problem of abstracting "... revolves around the use and choice of transitional devices." This problem is not only associated with abstracting but is central to all technical writing.

To aid the technical writer, the use of transitional devices is examined here in detail; the objective is specifically to answer the following questions: What are the various types of transitional devices? Why are transitional devices important? What general rules can be given for usage?

II. DESCRIPTION AND IMPORTANCE

Transitional devices may be described as logical or mechanical connections between or among ideas or topics and may be words, phrases, clauses, sentences, or paragraphs. Typical transitional words are *and*, *also*, *yet*, *alternatively*, *thus*, *therefore*. Transitional phrases that are commonly used are *in addition*, *on the other hand*, *for example*, *thus far*, *as well as*. Transition may also be accomplished by repeating a main idea or an important term from the preceding paragraph in a clause or by asking and then answering a question.

Transitional devices are important because they show relationship, relative importance, sequence, and the completeness of ideas. They connect apparently unrelated elements giving order and motion, thereby alerting the reader to the progress of thought: which point is being discussed and which will be covered next.

Transitional devices are especially important between sentences and between paragraphs. They make each sentence part of a paragraph and each paragraph part of the whole rather than a separate element. This coherence pulls sentences and paragraphs together and makes reading easier and clearer.

III. TRANSITIONAL DEVICES

Six transitional devices useful for achieving clarity and variety are (Orth lists one through five):

1. insertion of transitional words and phrases
2. repetition of key words and phrases or summarizing a main idea
3. use of numbers or letters for enumeration or listing
4. use of pronouns and demonstrative adjectives to refer to key words
5. use of parallel grammatical structure for ideas of equal importance
6. use of sequences: time-place, cause-effect, mathematical, question-answer.

The six transitional schemes listed above are effective methods for connecting thoughts and ideas. These devices will now be discussed in detail.

1. *insertion of transitional words and phrases*

There are many transitional words and phrases. These can vary in meaning, temper, and tone. Some transitional expressions have been categorized and listed in Table 1. Use of this table gives the writer a large choice, aids him in selecting the correct expression, and gives alternatives to overused expressions. Reference to the table also gives a quick reminder of the possibilities available and serves to provoke thought. Items are categorized in terms of *what is done with ideas;* i.e., add, compare, detail, repeat, revise, and summarize, or with respect to *what the ideas show;* i.e., purpose, result, solidarity, time, and position.

Table 1: Transitional Expressions and Conjunctions

add, continue, or introduce ideas
Additionally, Also, And, Another way, A second method, As well as, Besides, Equally important, Finally, First, Further, Furthermore, If desired, In addition, Last, Lastly, Moreover, One further remark is appropriate here, Remember, Second, Subsequently, Third, To begin, With the above

Manuscript received November 10, 1975.

The Author's with The Mechanical Engineering Department, University of Houston, Houston, Texas 77004.

[1] M. F. Orth, "Abstracting for the Writer," *IEEE Trans. Professional Communication*, vol. PC-15, June 1972, p. 43.

Reprinted from *IEEE Trans. Prof. Commun.*, vol. PC-18, pp. 320-322, Dec. 1975 *compare ideas*

After all, Alternately, A similar analysis shows, But, However, In a similar manner, In comparison, In contrast, In many problems, In practice, Instead, In like manner, In the latter case, In this connection, Likewise, Nevertheless, Nonetheless, Notwithstanding, On the contrary, On the other hand, Otherwise, Similarly, Still, Whereas, Worse, Yet

detail ideas (intensification, exemplification, specification)

As an application, As an example, Basically, For example, For instance, Generally, In general, In particular, Namely, Often, Specifically, That is

repeat ideas

Again, As before, A second time, As has been stated, As I have said, In other words, In review, Once more, That is, To reiterate

revise ideas

Are now given by, Becomes, Can be written, From, Gives, Produces, Reduces to, Replaced by, This leads to, Yields

summarize ideas

In effect, In essence, In other words, In short, In summary, Let us briefly review the steps in the analysis, Several remarks need to be made at this point, Summarizing, The foregoing discussion illustrates, To sum up

show position

Adjacent to, Beyond, Here, Near, On the other side, Opposite to, The former

show purpose

For this purpose, For this reason, In order to, To this end, Toward this objective, With this goal

show result

Accordingly, A review of the literature reveals that, As a result, Consequently, Hence, In as much, In consequence, In spite of, In view of these considerations, So far, The foregoing discussion illustrates, Therefore, Thus, Thus far, To meet this additional difficulty, Wherefore

show solidarity, doubt

Admittedly, Certainly, Conclusively, Fortunately, Hopefully, Indeed, In any event, In fact, In reality, Obviously, Of course, Ordinarily, Perhaps, To be sure, Truly, Undeniably, Without any question

show time

After an hour, day, year, ..., Afterward, A little later, As will be seen, At length, At present, At this point, Earlier, Finally, Here, Heretofore, Immediately, In the meantime, In those days, In what follows, Meantime, Meanwhile, Presently, Recently, So far, Then, Today, Ultimately, Will now be developed, Will now be illustrated, Will now be invoked

2. *repetition of key words and phrases or summarizing a main idea*

The repetition of key words (word echo) and the use of synonyms are simple transitional devices. As remarked by Tichy,[2] "When main words are repeated, the central idea of a paragraph is stressed, and thus coherence and emphasis reinforce each other. Such repetition is often hardly noticeable because synonyms and synonymous phrases may be used as well as repetitions of a main word. Indeed, even antonyms provide connections, for example, hot with cold, dry with wet, economical with wasteful, busy with idle."

There are several reasons for using word and phrase repetition. These include: leading the reader to the next material, forcing him to remember a basic concept, and, if the reader is skimming material, he is bound to find the major points if they are repeated.

3. *use of numbers or letters for enumeration or listing*

A list of items or ideas can be mechanically but effectively connected by enumeration or tabulation.

4. *use of pronouns and demonstrative adjectives to refer to key words*

Pronoun linkage is another common transitional method, and this type is rarely disturbing, bothersome, irritating, or noticeable. Here a pronoun is used which refers to the persons, things, or ideas discussed in the preceding sentence or paragraph. For instance, the pronouns *he*, *she*, *they*, and *it*, might refer to nouns in preceding sentences. Demonstrative adjectives also may be used for linking ideas. This linkage is usually effected during revision, and thus loosely connected ideas may be joined. For demonstrative adjectives, *the* may be replaced with *this*, *that*, *these*, or *those*.

5. *use of parallel grammatical structure for ideas of equal importance*

Parallel grammatical construction is another means for knitting ideas together. This type of connection links thoughts firmly, makes comprehension easier, and in most cases makes the use of other transitional devices unnecessary.

6. *use of sequence: time-place, cause-effect, mathematical, question-answer*

The time-place sequence may often act as a transitional device, especially when a process is being described. The reader is given a time and place for the first action. Then a second action follows in both time and place. The reader progresses through events in time and place.

Cause-effect transition involves *either* a time *or* a place sequence. Cause always precedes effect in nature, but this does not have to be the case in writing.

The mathematical sequence determines the explanations given for mathematical processes. The presentation of the mathematical development forms the structure for the remaining discussion.

In the question-answer sequence, the author poses a pertinent question and then answers it.

In the next section, some general rules for the usage of the six transitional schemes just described for connecting thoughts and ideas are given.

IV. GENERAL RULES

Good usage of transitional devices results in smooth, inconspicuous movement from one idea to another. But most

[2] H. J. Tichy, *Effective Writing for Engineers, Managers, Scientists.* New York: Wiley, 1966, p. 276.

writers do not fail to use transitional expressions. They overuse and misuse them. This ineffective usage may burden, irritate, distract, annoy, and confuse the reader.

Some general guidelines for transition from one topic to the next are given below:

First, the sequence of topics must have a logical connection. If the parts do not fit together, it is impossible to join them, and even if they could be joined, the assembly would not be correct.

Second, the writer should be aware that placement of transitional words and phrases at the beginning of a sentence gives them more stress than if the word or phrase is used inside the sentence. And since many transitional devices in themselves have no meaning, inside placement is preferred.

Third, the writer should prefer the organic (word echo, sequence, parallelism) to the mechanical connection (transitional words and phrases). Organic transition is smoother, less conspicuous and, in most cases, does not distract from the main topic or idea.

Fourth, repetition is best accomplished for connecting sentences by repeating the most important word contained in the preceding sentence. To connect paragraphs, repeat in the opening sentence the most important word or main idea contained in the preceding paragraph.

Fifth, in long reports, summaries should be given at the beginning and end of each section or chapter. But for variety, different words and many different sentence forms and lengths should appear.

Sixth, pronouns and demonstrative adjectives referring to antecedents in the preceding sentence should be placed at the beginning of the sentence. This is a more efficient and a less obtrusive connection.

Seventh, transitional devices that imply a logical connection of ideas where none exists should not be used. Instead, a conjunction should be employed.

Eighth, whenever there exists a series of similar grammatical elements (subjects, objects, and modifiers) which grammatically perform the same function and grammatically have the same nature, the writer can mechanically achieve parallelism by merely placing the series elements in parallel with one another.

Ninth, a variety of transitional devices like those listed in section II should be used.

Finally, as Orth[3] states: "The choice [of transitional devices] must be one that contributes most to clarity, exactness, conciseness, and directness."

ACKNOWLEDGMENT

The author wishes to acknowledge the assistance of Mrs. Donna McLoughlin, Managing Editor of the *Threshold*, University of Connecticut, Storrs, Connecticut.

[3]Orth, op. cit., p. 44.

CREATIVE DESIGN AND RHETORICAL INQUIRY: REPORT WRITING STRATEGIES

Victoria M. Winkler
Visiting Associate Professor
Humanities Department
University of Michigan
Ann Arbor, MI 48109

Technical report writers, be they technical communicators or engineers and managers, face two related problems in consistently producing effective and instrumentally useful reports: (1) solving the organizational problem that prompted the report writing task and (2) designing and writing the report itself. Using creative problem-solving strategies in tandem with the rhetorical inquiry process can aid novice report writers in becoming better problem-solvers and report writers. These strategies also aid experienced writers in overcoming "gumption traps."

Experienced professional communicators devise their own methods for organizing and writing reports--methods that work for them most of the time. In some cases, they can even teach novice report writers their processes of composing. In most cases, however, the stages in the process are implicit, sometimes unconscious and lack distinct boundaries. Attempts to transmit an experienced writer's composing processes are often "product" rather than "process" oriented, that is, they concentrate on arrangement and editing. The fact that writers do develop strategies for composing--even if they cannot be readily transmitted--and the fact that these strategies work for their originators indicates that a systematic procedure for gathering data, arranging it to form concepts, and designing and writing reports is possible. No one procedure is guaranteed to work effectively for all writers all of the time, but having a tried and tested procedure to follow improves the writer's chances of success.

The usefulness of teaching problem-solving strategies in the composition classroom have already been discussed by Richard Young in "Problems and the Composing Process" and in his text *Rhetoric: Discovery and Change*. In this paper, we intend to suggest how and why these problem-solving strategies can be fruitfully applied by communication professionals in designing and writing technical reports. We have found that technically-oriented professionals, such as engineers and scientists, have already developed *implicit* problem-solving techniques. In the design courses that they take as undergraduates, some of these strategies are made *explicit*. They are actually taught. Since problem-solving procedures can be taught, such instruction helps students who have already internalized the process to become conscious of what they are doing so that when they get stalled--as they inevitably do--they can begin again to generate viable alternatives and ideas. For students who are not particularly astute problem-solvers, teaching these strategies provides the students with a provisional procedure which improves their general problem-solving skills.

Introducing a problem-solving orientation to technical and professional communicators provides the novice professional with five important tools:

1. A general procedure (a set of questions or operations) which is provisional, but which helps him to discover and recover relevant data and arrange it in a logical order.
2. A set of parameters determined by the problem itself which helps him limit or expand his investigation by identifying the "knowns" and "unknowns" of the problematic situation.
3. A set of criteria which tell him when he has completed the investigation or solved the problem.
4. A method for evaluating his solution against the restrictions and constraints imposed by the problem.
5. A method for evaluating the rhetorical situation (audience and purpose) and for designing an effective report to communicate his findings.

The purpose of this paper is to outline a two-part strategy for designing and writing reports. The first part is based on the design principles taught to engineers and the second is based on report writing strategies taught to technical and professional writers at the University of Michigan. These composing procedures are meant to be guides or ways of proceeding that increase the writer's chances of consistently producing effective reports. Rhetoricians call such procedures "heuristics," or methods of invention and discovery. Heuristic procedures differ from algorithms (or rule-governed procedures, like mathematical formulas) because they are not prescriptive techniques that will always guarantee success. Heuristics are provisional strategies which increase a writer's possibility of success. By combining problem-solving heuristics taught to engineers in their design courses with the rhetorical strategies developed by Mathes, Stevenson and Young, we can come closer to plotting the nebulous stages in the composing process from the pre-writing, information-gathering stages to the rewriting and editing stages. We are convinced that the combination of these procedures forms a powerful heuristic for designing and writing reports. We will begin our discussion by describing the mechanical design procedures and the process of rhetorical inquiry.

The Problem-Solving Orientation

All communicators, whether they are engineers, managers or technical writers, assume two major roles: (1) the role of investigator or problem-solver and (2) the role of communicator. Problems often occur in recognizing both of these roles as important and in learning to "change hats" from one to the other effectively. Both by temperament and training, engineers initially tend to be better investigators than communicators and technical writers tend to be better communicators than investigators. However, both of these roles operating in tandem are crucial for the technical and professional communicator's success in an organization. Combining technical and rhetorical problem-solving heuristics helps to integrate these dual roles.

Engineering educators have responded to the need for teaching engineers to become effective problem-solvers by teaching creative design principles (various problem-solving techniques, such as brain-storming) as in integral part of their design courses. The term "design 'principles'" underscores the importance relegated to these problem-solving strategies and the attitude often taken toward them. Design principles are presented both visually and pedagogically as processes. (See Figure A: Design Strategies.) The stages in the process can be neatly plotted in flow diagrams with inputs, outputs, decision boxes and feedback loops. These diagrams are almost too neat and take on the appearance of algorithms or rule-governed procedures. They are accorded a high degree of reliability for aiding engineers in arriving at solutions to open-ended problems (or problems that admit of more than one possible solution). Most engineers are familiar with the problem-solving methodologies common to their respective disciplines and they tend to equate these strategies with the scientific method and the rules of strong inference.

We recognize that these design principles are heuristic procedures, or ". . . explicit plans for analyzing and searching which focus attention, guide reason, stimulate memory and encourage intuition."[2] They are heuristic probes for directing the problem-solver's exploration and for making him a "better guesser." They are recursive plans which enable the problem-solver systematically to generate and to discover information, to provide a perspective on that information and to suggest ways of organizing it. In sum, heuristics, such as the problem-solving strategies we will discuss, present writers with a set of procedures for exploring problems and shifting perspectives which lead to ideation and concept-formation. These heuristic probes are not merely information retrieval techniques, although that is an important function that they serve. Their power derives from their inventional possibilities: they are methods of discovery and invention which can lead to the formation of new concepts and the creation of new knowledge.[3]

Professional communicators can develop more effective modes of composing by supplementing the implicit knowledge of how to structure reports that they already possess with the conscious use of the problem-solving heuristics we are suggesting. In the next section, we will discuss the epistemological assumptions underlying the use of problem-solving heuristics and provide a working model of how they can be used in designing technical discourse.

FIGURE A: DESIGN STRATEGIES

Vidosic, Elements of Design Engineering

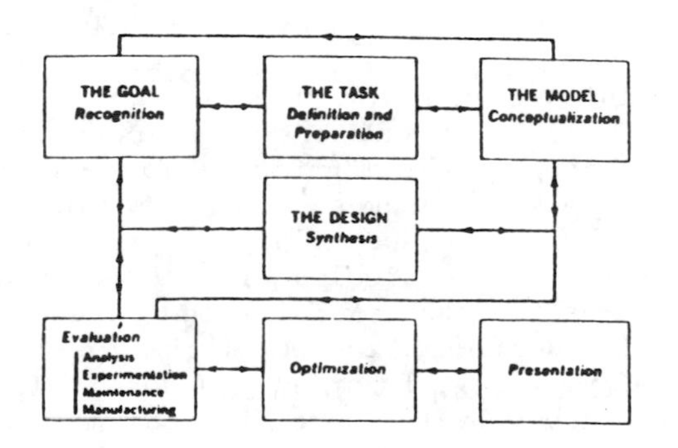

Contextual and Perspectival Nature of Problems

Technical discourse lends itself naturally to the problem-solving approach due to the nature of the subject matter. This statement is not meant to imply that modified versions of this approach would not work equally well in other kinds of discourse.[4] We believe that writing is not just a matter of putting words on paper, but it is a conceptualization process which expands our perceptions and our knowledge. It is with this assumption in mind that we advocate the use of problem-solving strategies as inventional and arrangement tools.

Problem-solving activities are highly perspectival and operate in a definite context. In other words, problems do not exist in a pure state out in the world. Problems are only "problems" for someone. Richard Young agrees that "problems arise from inconsistencies among elements of the individual's cognitive system. We do not find problems, we create them."[5] Problems take shape when we perceive inconsistencies in our image of the world, such as our desire to achieve the release of the American hostages in Iran without precipitating a military confrontation or alienating other Islamic states. Problems are special kinds of psychological events.[6] They derive from a felt difficulty, or in Bitzer's terms, an "exigence, an imperfection marked by urgency; a defect, an obstacle, something waiting to be done, a thing which is other than it should be.[7]

The felt difficulty or exigence creates tension in the individual which motivates him to find a solution and to resolve the difficulty. The problem-solver examines the problem in its context and takes into account any restrictions, criteria or constraints upon his actions, as well as what changes his solution would require either in himself, in his surroundings or in others.

The problem-solving process, therefore, begins with a felt need, a perceived difficulty, a psychological event--the recognition of a problem to be solved. In many cases, the communication professional does not discover the problem, but is made aware of it by an engineer, a manager or a customer. In such cases, the communicator must quickly assess the problematic situation, bring his own perspective to bear on it and come to as complete an understanding of the difficulty as possible. The problem then becomes "his" problem. This stage is highly perspectival and leads to the formulation of the problem statement and an analysis of the problem motivated by the individual's desire to remove or resolve the difficulty. It is important to note that due to the perspectival nature of problems, what might be viewed as a design problem by an engineer, could be viewed as a production problem by a manager and as a marketing problem by a technical writer. The context of the problem becomes paramount along with the various perspectives that different problem-solvers bring to bear on it. The problem-solver has not only to find a viable solution to his problem, but he also must discover a way of implementing it. The implementation of the solution involves a change in either the individual himself, the world around him or in other people. If the problem-solver has to affect change in the world or in others, he will first have to persuade them that change is desirable and advantageous. The problem now assumes a distinctly rhetorical dimension.

Technical and professional communication, like engineering, is an "applied" science. The professional communicator has to know not only what, but how. It is not sufficient to discover a solution to a problem, that solution has to be communicated and adopted to change the behavior of an organization. Professional communicators can aid their technical counterparts best by beginning where the technical investigation usually breaks down--designing and writing the report. The need for reciprocity and flexibility between the investigative and reporting roles and their respective perspectives on the problem emerges at this stage of the process. This is true whether an engineer and technical communicator are working cooperatively on a problem or whether the technical communicator is both the investigator and the report writer. To meet this need, the problem must be reevaluated in terms of its organizational context and rhetoric situation. Once problem-solvers have arrived at workable solutions to their technical problems, the process is not complete until the solutions have been communicated to an audience of decision-makers and the decision-makers have been persuaded to act on them.

Using a problem-solving approach to designing reports not only limits and defines the content of a report, but it provides systematic procedures for generating and organizing data without losing sight of the rhetorical situation--the audience and the constraints on action. Having discussed the value of learning problem-solving skills and applying them to designing technical reports, we will now present a working model of the problem-solving heuristic.

A Heuristic for Designing Technical Reports

The design principles devised by E.V. Krick[8] for use in mechanical engineering design courses and the process of rhetorical inquiry advocated by J.C. Mathes and D.W.

Stevenson[9] provide complementary problem-solving heuristics to aid professionals in solving both technical and rhetorical problems associated with designing effective technical discourse. An understanding of both design principles and rhetorical strategies proves beneficial to technical and communication professionals because the problem-solving strategies in both areas are roughly equivalent to the preparation, incubation, illumination and verification stages in the creative problem-solving process. The sole difference between the mechanical design process and the rhetorical inquiry process is one of focus: in the mechanical design process, the engineer conducts a technical investigation and generates a solution to an open-ended technical problem while in the rhetorical inquiry process, the communicator analyzes the rhetorical situation and generates an appropriate form of technical discourse to communicate the results of his investigation to the organization or to the public. Just as the technical investigation yields the content of technical discourse, the rhetorical investigation yields its form. In both processes, the professional must define a problem, analyze it in detail, generate alternative solutions, evaluate, compare and screen these solutions and document the results. Both processes are recursive and dynamic rather than linear and sequential. When used effectively, they can stimulate thought and help generate creative solutions.

There are a wide variety of problem-solving strategies resulting in concept formation as Thomas T. Woodson mentions in the Introduction to Mechanical Design.[10] A quick glance at the outlines of some of these strategies reveals the similarity between them. (See Figure B: The Problem-Solving Process: A Variety of Descriptions.) The steps in these processes are roughly equivalent. It is not surprising that all of the strategies that Woodson lists bear a generic resemblance to the scientific method. What surprised us was that most of these strategies appear to be variations of the process of rhetorical inquiry which has its roots in classical rhetoric.

Using what Aristotle calls the art of "invention" implies retrieving information from long-term memory to generate the necessary data to compose a discourse. The heuristic developed by Young, Becker and Pike, on the other hand, taps not only memory, but also the spontaneous and creative generation of new information. "Tagmemic invention"[11] and the rhetoric of inquiry that it fosters provides a dynamic process for generating new information and for formulating new concepts. The most compact representation of the tagmemic heuristic is the nine-celled particle-wave-field matrix reproduced from Rhetoric: Discovery and Change (See Figure C: The Heuristic Procedure). The nine-celled matrix guides the problem-solver's exploration of the problem as an isolated static unit (the particle or feature mode of perception), as a dynamic flowing process (the wave or manifestation mode) and as a network of complex relationships (the field or distribution mode.) Each of these three modes of perceiving the problem requires a change in perspective on the part of the problem-solver. The changes in perspective help the problem-sover to understand the complexity of the problem and to explore it in a systematic way. Moreover, the changes in perspective reenforce the perspectival nature of problems and they foreshadow the need for shifting perspectives from a technical investigator to a communicator to complete the problem-solving process. Besides the systematic shifting of perspectives, there are two other important features of this heuristic stressed by Mathes and Stevenson: the systematic audience analysis and the formulation of a problem statement.

FIGURE B: The Problem-solving Process:

PROCESS	Design Process (Asimov)	Thought Process (Wallas)	Professional Method (Ver Planck & Teare)	Engineering Method (Smith)	Problem-solving (Buhl)
STAGES	Analysis	Preparation	Define problem	Preliminary analysis	Recognition Definition Preparation Analysis
	Synthesis	Incubation Illumination	Plan & treatment	Statement of question	Synthesis
	Evaluation				Evaluation
	Decision			Solution	
	Optimization	Elaboration	Execute plan		
	Revision		Check as a whole	Check	

Once a problem has been recognized, the creative design principles can be employed to further define the problem and to search systematically for solutions. The process itself can be divided into two phases with five specific steps in each. Phase One is part of the investigative and technical problem-solving stage, while Phase Two is part of the report writing or rhetorical stage:

FIGURE C: The Heuristic Procedure

	Contrast	Variation	Distribution
PARTICLE	1) View the unit as an isolated, static entity. What are its contrastive features, i.e., the features that differentiate it from similar things and serve to identify it?	4) View the unit as a specific variant form of the concept, i.e., as one among a group of instances that illustrate the concept. What is the *range* of physical variation of the concept, i.e., how can instances vary without becoming something else?	7) View the unit as part of a larger context. How is it appropriately or typically classified? What is its typical position in a temporal sequence? In space, i.e., in a scene or geographical array. In a system of classes?
WAVE	2) View the unit as a dynamic object or event. What physical features distinguish it from similar objects or events? In particular, what is its nucleus?	5) View the unit as a dynamic process. How is it changing?	8) View the unit as a part of a larger, dynamic context. How does it interact with and merge into its environment? Are its borders clear-cut or indeterminate?
FIELD	3) View the unit as an abstract, multidimensional system. How are the components organized in relation to one another? More specifically, how are they related by class, in class systems, in temporal sequence, and in space?	6) View the unit as a multidimensional physical system. How do particular instances of the system vary?	9) View the unit as an abstract system within a larger system. What is its position in the larger system? What systemic features and components make it a part of the larger system?

Phase One: Creative Design Principles

1. Problem Formulation
2. Problem Analysis
3. Data Search
4. Decision and Evaluation
5. Complete Specification of the Solution

Phase Two: Rhetorical Inquiry Process

1. Problem Reformulation (Problem Statement)
2. Audience Analysis (to determine the rhetorical situation)
3. Arrangement: Basic Report Structure
4. Preparation of the Prototype Draft
5. Editing, Proofreading and Verifying the Basic Design (See Figure D: Creative Design and Rhetorical Inquiry.)

Although at first glance these stages in the process appear to be linear and sequential, they are actually recursive and dynamic. The problem-solver generally does not move from stage one to stage five without numerous repetitions and reversions to earlier stages of the process. To demonstrate the recursive nature of the process, we will compare the stages in the design process used in the technical investigation of a problem with those in the process of rhetorical inquiry for the remainder of this paper.

1. Problem Formulation and Reformulation

In the first stage of the problem-solving process, the felt difficulty or exigence is recognized and is viewed as a "given state" or a static unit such as "coal in the ground." The "given state" is juxtaposed with a "desired state" such as "energy to the people." The procedures used here (such as free association, recall and brainstorming) examine the problem from a unit and a dynamic perspective. Using brainstorming techniques and free association to explore the possibilities of the "given state" and the "desired state," the problem is considered without evaluating the merits of the suggestions or searching for comprehensive solutions. The problem is explored. This stage of problem formulation involves broadening the scope of the problem to explore many possible avenues and to foster conceptual blockbusting. These activities stimulate the investigator's thinking in new ways which invite creative solutions to problems.

Problem formulation is similar to the heuristic strategies advocated by Young, Becker and Pike. They propose a systematic inquiry of the problem and the unknowns necessary for solving the problem simply to stimulate the writer's thinking and to amass as much data as possible about the problem. The heuristic using the nine-celled particle-wave-field matrix serves as a strategy for exploring the problem in a purposive way while furnishing a data base which can later be evaluated and rearranged to discover (or uncover) a reasonable solution.

FIGURE D: Creative Design and Rhetorical Inquiry

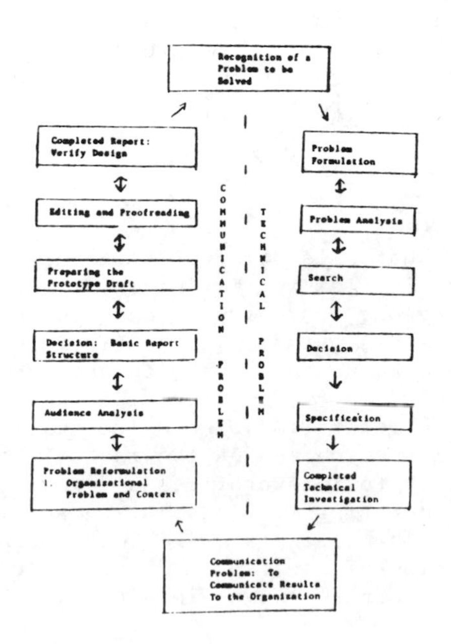

Before a problem-solver can complete his technical tasks and begin his report writing task, however, he must solve the communication problem engendered by his technical investigation: To communicate the results of his investigation persuasively to an audience of decision-makers and to win their assent. In his restatement of the problem to discover its rhetorical dimensions, the writer's goal is to determine the problematic context of the report in the organization. In other words, if the problem-solver is asked to redesign part of the ignition system of an automobile to meet certain specifications, he must understand why such a redesign is necessary. Is the present design perceived to be inefficient? Has it been subject to frequent failures? Have there been customer complaints? Will it meet recent federal emission standards? Is there a more economical way of producing the part? Until the communicator has asked such questions and becomes aware of the organizational problem that prompted his investigation, he cannot effectively design a report to meet the organization's needs. In his restatement of the problem, the writer must not only state the problem and the organizational context, but he must also specify the technical tasks or assignment that he was given and the rhetorical (or instrumental) purpose of the report.

The advantage of introducing the particle-wave-field heuristic at this stage of the composing process is two-fold: it serves as an aid to invention and arrangement and it complements the brainstorming that occurs in the mechanical design group discussions. The drawback with relying solely on brainstorming is that it works best only in groups. Engineers often work in teams, so brainstorming proves beneficial as a point of departure for exploring open-ended problems. However, the problem-solver should also be equipped with heuristic strategies to guide his problem exploration and analysis when he works independently. The particle-wave-field heuristic operates as a powerful probe that can be used individually as an explanatory, suggestive and predictive device both to retrieve information and to process it.

The entire nine-celled matrix need not be brought into play in every instance. It generally serves as an excellent thought-generator when communicators "run dry" or reach impasses. The built-in shift in perspectives in the matrix reaps additional benefits in aiding the problem-solver to change roles from a technical investigator to a professional communicator. Moreover, the fact that the nine-celled matrix builds in a systematic shift in perspective tends to reenforce the unconscious and implicit shift from a unit to a wave perspective that generally occurs during group brainstorming. It forces the third shift to a field or network perspective which places the problem within the organizational context. This last shift is essential for effective communication of the solution.

2. Analysis of the Problem and Its Context

As we have already noted, the group brainstorming and classification and division lead rather naturally into a unit and a dynamic perspective on the problem. The field or network perspective which includes the problematic context often emerges only after we examine the problem as an isolated instance and/or as a dynamic process. The use of the nine-celled matrix draws as naturally to this next cognitive step. In rhetorical terms, the field view proves valuable because it encompasses the rhetorical situation and leads to an analysis of audience and purpose.

In the problem analysis stage of the mechanical design process, the problem is defined in detail. It is analyzed into its component parts and the input and output specifications are considered. The real and imaginary restrictions are examined and the production limits, cost limits and performance criteria are established. This second stage is similar to the processes of division and classification in classical rhetoric. As the problems are analyzed and each of their parts is considered, the investigators begin to realize that the solutions may vary depending on how a certain set of data is classified or interpreted. They also discover that many initial restrictions turn out to be false restrictions depending on the outcome of an individual's analysis.

Having stated both the problem in terms of the organizational context and his tasks and the purpose of the report, the investigator switches hats from a technical investigator to a professional communicator. He shifts his perspective by examining the problematic context and his audience.

The systematic method for audience analysis advocated by Mathes and Stevenson consists of identifying possible audiences for a report by constructing an egocentric organization chart which targets specific individuals in their roles in the organization. It also determines their proximity to the writer and their knowledge of his daily activities and concerns. (See Figure E: Egocentric Organization Chart.) After identifying the audiences, Mathes and Stevenson suggest that they be characterized according to their operational, objective and personal characteristics. (See Figure F: Form for Characterizing Individual Report Readers.) Although novice writers have to force themselves to analyze audiences systematically, experienced writers do this automatically. By constructing an egocentric organization chart to identify audiences and to determine the route the report will travel in the organization and by characterizing the audiences to determine what questions they will want answered, the communicator is ready to classify his audiences. Audiences may be classified in the following way:

1. Primary audiences--decision-makers who act on the basis of the information the report contains.
2. Secondary audiences--audiences who are affected by the decisions and actions recommended by the report.
3. Immediate audiences--audiences who route the report or transmit the information it contains.[12]

Once the report writer has completed his analysis of the communication problem engendered by his investigation and has analyzed the audiences of his report, he is ready to slect the appropriate structure for the report itself.

FIGURE E: Egocentric Organization Chart

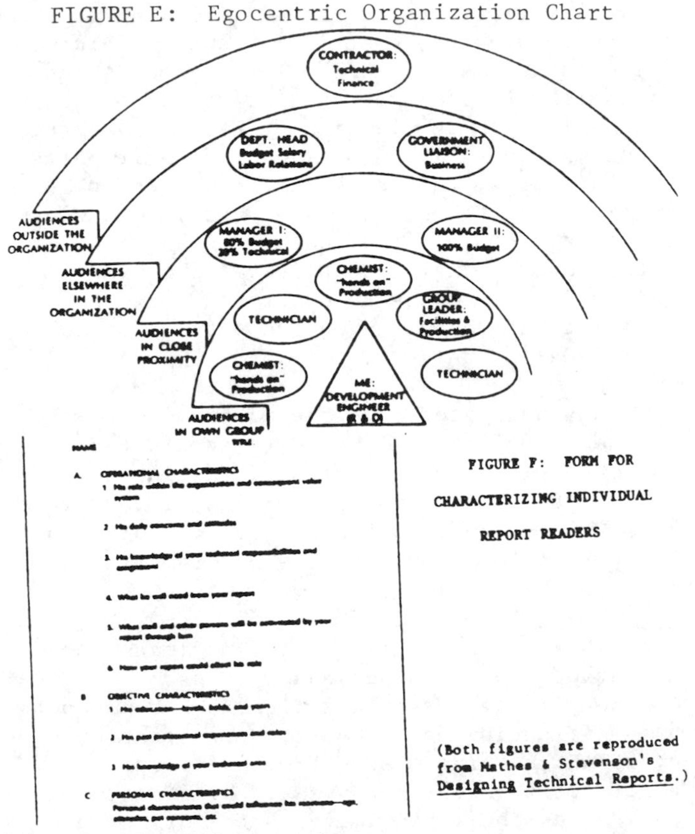

FIGURE F: FORM FOR CHARACTERIZING INDIVIDUAL REPORT READERS

(Both figures are reproduced from Mathes & Stevenson's Designing Technical Reports.)

3. Search and Arrangement

Using the problem breakdown from step two (problem analysis) of the design principles, alternative solutions are generated by the investigators using any of the tools of inquiry, invention or research that are available to them. The search for possible solutions precludes evaluation. The search phase is a recursive process, and the key to searching successfully for solutions is to keep an open mind until all of the evidence is in and all possible solutions--regardless of how impractical they might seem initially--are accumulated.

To help him to organize the mass of data that he has generated during his technical investigation, the report writer is asked to think in terms of a two component structure for his report moving from general to particular between components. After he has determined whether his primary purpose is to inform or to persuade, the communicator selects the basic structure for the discussion component of his report. (See Figure G: Structuring the Discussion Component.) The discussion component, or body of the report, presents the investigative procedures and the discussion of the results of the investigation complete with conclusions and recommendations. The appendices, which are even more technical and detailed in nature than the discussion, are completed and attached to the end of the report. The process of selection that determines what should be elaborated on in the discussion component and what should be relegated to the appendices depends on the purpose of the report and the audiences it addresses. Finally, the writer composes an opening component or "Executive Summary" which includes an overview of the problem, the technical tasks and the rhetorical purpose of the report and a summary of the recommendations and conclusions.[13]

4. Decision-making and Preparing the Prototype Draft

During the decision stage of the design heuristic, the investigator's critical judgment and evaluative skills are called upon. The alternative solutions are arranged in different configurations to produce various combinations of alternatives. All of these newly generated alternatives, as well as the previously generated alternatives, are evaluated against the criteria established in the problem analysis (state two). At this time decisions are made concerning the best possible solution to the problem.

Just as the investigators apply a problem-solving approach to arrive at their proposed designs, they reapply a similar strategy to design the report which will effectively communicate the results of their investigation. Having selected the basic report design, the writers verify the basic report structure and write the prototype draft. (Two suggested report designs are presented in Figure G: Structuring the Discussion Component.) Many of their earlier calculations and technical segments may be used in this draft, but they must be carefully edited to ascertain whether they are fulfilling the general purpose of the report and whether they are contributing to the basic design. It is at this time that the writers make decisions concerning layout, formating and the inclusion of visuals in the discussion and the appendices.

FIGURE G: Structuring The Discussion Component

1. **Basic Structure for a Persuasive Report:**

   I **The Problem Statement and Solution**

   II **Criteria to Evaluate Evidence**

   III **The Support (internally ordered either in descending order of importance or as positive and negative evidence)**

   IV **Restatement of Conclusions and Recommendations**

2. **Basic Structure for an Informative Report:**

   I **Introduction:**

   - A. **The Problem: An explanation of the organizational problem that gave rise to the report.**
   - B. **The Objective: A statement of the assignment.**
   - C. **The Method of This Report: A forecast of the structure.**

   II **Background:**

   - A. **Previous Work: An explanation of what already has been done about the problem.**
   - B. **Specifications: An explanation of detailed instructions or specifications that served as the basis for the present work.**

   III **Experimental Procedure:**

   - A. **Materials: A description of equipment used.**
   - B. **Methods: A step-by-step description of the procedure followed.**

   IV **Discussion of Results: An extended discussion and explanation of precisely what was learned.**

   V **Synthesis:**

   - A. **Problem Restated: A restatement of the report's objective and of the problem that gave rise to the report.**
   - B. **Summary: Reviewing the main points.**
   - C. **Recommendations: Explaining subsequent action or posing specific questions for investigation.**

(Both of these suggested structures for the discussion were adapted from Mathes & Stevenson's _Designing Technical Reports_.)

5. Specification, Editing and Proofreading

In the mechanical design process, the specifications must be written and drawings made after the design decisions are made. Finally, the report is written which completely specifies and communicates the proposed solution to the technical problem. It is at this stage that the writer's energies are focused solely on stylistic and arrangement problems--the kinds of concerns most frequently attributed to rhetoric. He must now review the communication problem engendered by his technical investigation in order to design a report structure which will clearly and persuasively present his proposed solution to an audience of decision-makers. The decision-makers will act on the basis of the report either to accept or to reject the writer's conclusions and recommendations. Since the report is all the decision-makers will see and since it will be almost the sole basis for their for their judgment, the report must not be a hastily compiled afterthought. Rather, the report _is_ the logical culmination of the investigation which communicates the results to those who must act upon them and to those who will be affected by the resulting changes. To bring about the desired change in an organization, the communicator must learn to communicate the results of his technical investigation clearly, effectively and persuasively. He must attack the rhetorical problem and analyze the rhetorical situation with as much care as he did the technical problem.

Conclusion

The complementary nature of technical and rhetorical heuristic procedures enable both investigators and novice communicators to see the problem-solving process whole--from the felt difficulty or exigence to the communication of their solutions to an audience of decision-makers. They recognize that their role as communicators complements and completes their role as investigators. They realize that even before they reach the end of their technical investigation, they must solve a correlative problem: the design of an effective communication tool. The measure of success that their reports attain depends on how well they conceptualize the audiences and purpose of their reports and how effectively they execute them. The end result of this cooperative venture in combining problem-solving heuristics is better conceptualization of the technical problem and better communication of the results of the investigation. The factors contributing to this conclusion include five basic tenets:

1. Novice communicators learn that composing is a process that involves both their technical skills as investigators and their critical skills as rhetoricians.

2. The composing process can be aided by more effective methods of gathering and organizing information, that is, by the use of heuristic problem-solving strategies.

3. The differences between shifting roles from investigator to communicator are minimized when the bulk of the technical investigator becomes part of the inventional and arrangement phases in the composing process.

4. The analysis of the rhetorical situation to determine audience and purpose is a primary contributing factor to deciding on the basic format of the report and ranks equal in importance to technical competency and ingenuity in investigative procedure.

5. The composing process, like the technical problem-solving process, is only complete when it is redefined as part of a larger process--the process of cognition leading to concept formation.

REFERENCES

[1]Richard E. Young, "Problems and the Composing Process," a paper delivered at the NIE Writing Conference, Los Angeles, CA (June 13-15, 1977) and Richard E. Young, Alton L. Becker and Kenneth L. Pike, Rhetoric: Discovery and Change (New York: Harcourt, Brace and World, Inc., 1970). See Chapters 3-10.

[2]Richard E. Young, "Invention: A Topographical Survey," in Teaching Composition: Ten Bibliographical Essays, ed. Gary Tate (Fort Worth: Texas Christian University Press, 1976), p. 1.

[3]Classical rhetoricians traditionally divided the process of formulating speeches into five basic arts: (1) Invention or gathering information and formulating a thesis; (2) Arrangement or deciding on the basic structure appropriate to the content and audience; (3) Memory; (4) Delivery; (5) Style. Arrangement and style have been stressed by composition teachers to the exclusion of invention for the past century. Inventional strategies--such as the heuristic procedures we are discussing--have been making an important comeback since the mid-1960's. Continuing research on invention and the composing process is currently being conducted in various universities, including Carnegie-Mellon and the University of Michigan. Most researchers are attempting to trace the steps in the composing process of actual writers to determine the relationship between composing and concept formation.

[4]For a comprehensive discussion of heuristics in teaching composition, see Richard E. Young's "Paradigms and Problems: Needed Research in Rhetorical Invention," in Research on Composing, eds. Charles R. Cooper and Lee Odell (Urbana, Ill.: NCTE, 1978), pp. 29-47. For a general discussion of heuristic procedures and their relationship to "tacit" knowing, refer to Michael Polyani's Personal Knowledge (Chicago: University of Chicago Press, 1958).

[5]Young, "Problems and the Composing Process," p.2.

[6]For additional discussion of problems as psychological events, see Young's "Problems and the Composing Process" and D. Gordon Rohman and Albert O. Wlecke, Pre-writing: The Construction and Application of Models for Concept Formation in Writing, Cooperative Research Project No. 2174 (East Lansing: Michigan State University, 1964).

[7]Lloyd F. Bitzer, "The Rhetorical Situation," a paper delivered at the Central States Speech Association (April 1967), p. 6.

[8]E. V. Krick, An Introduction to Engineering and Engineering Design (New York: John Wiley & Sons, Inc. 1969).

[9]J. C. Mathes and Dwight W. Stevenson, Designing Technical Reports: Writing for Audiences in Organizations (Indianapolis: Bobbs-Merrill Co., Inc., 1976). See particularly Chapters 2-6.

[10]Thomas T. Woodson, Introduction to Engineering Design (New York: McGraw Hill Book Co., 1966), p. 22.

[11]Tagmemic theory was originally developed as a linguistic theory by Kenneth and Evelyn Pike. It was later applied to the composing process by Young, Becker and Pike.

[12]For a complete explanation of audience analysis, see Mathes and Stevenson, Chapter 2: "Audience Analysis: The Problem and a Solution," pp. 9-23.

[13]Additional information on designing the opening component can be found in Mathes and Stevenson, Chapter 5: "Designing the Opening Component," pp. 59-82.

Writing Better Technical Papers

L. L. FARKAS

Summary—**This article discusses the writing of technical papers from inception to final editing. It considers first the need for writing a paper and also the audience for which it is intended. It then describes the steps of writing the paper: the summary, the outline, the illustrations, the first draft, the reviews and editing, and the final draft of the paper. Examples of these steps are given so that the writer can understand them fully and can readily adapt them to his own writing project. The article stresses establishing the need for the paper and then its development of thought and text logically so that the writer can make his point with the greatest possible impact.**

How do you go about writing a better technical paper for a magazine or a symposium? Where do you start? How do you organize it, write it? What does it need to be acceptable? To answer these questions let us discuss the details of writing such a paper from its inception to submittal.

Establish the Need

Before you put a single word on paper consider the subject of your article. Is it significant? Timely? New? Does it propose a different approach to an old problem? Are you sure that it is not just a restatement of an existing system or method? Is it really worth writing about? These are basic questions but they are important because they will help you determine, even before you start, whether or not you should take the time to plan and write the paper.

Audience

The next consideration is your audience. Who are you writing for? Are you trying to reach a specific, highly specialized group interested in higher mathematics and intricate technical details, or are you aiming your paper at the general public? What do you want your paper to do: educate, raise discussion, or simply impress your neighbors?

Honest answers to these questions are necessary. You have to establish the need for the paper, the type of persons it will interest, and the specific effect it should have upon them. Armed with this type of information you are then in a good position to start work on your paper.

The Summary

Now for the mechanics of writing. Set down in a few sentences exactly what you want to tell the reader.[1] Are you going to show him how to build a better mousetrap, or will you present a significant development in the state-of-the-art? State briefly the major points the article should convey. This is the first step to insure that your thoughts are going to move in a straight line.

If you have difficulties in summarizing what you want to tell the reader, you have a problem that must be resolved before going any further. Do not skip over this portion with the comment that you know perfectly well what you want to say and therefore you do not have to summarize it. Put it down now, even if it hurts to do so, and then examine what you are saying. It is revealing how often ideas that sound perfectly good when you think about them are not very clear or significant when expressed in black and white. Clarifying the thoughts at this point will make writing the complete paper easier for it will in essence define the boundaries within which the paper must be written.

The Outline

The next step in writing a technical paper is to make an outline. In your summary you have stated roughly what you want to tell the reader. Now you must decide how you plan to do this. What sequence of thoughts are you going to use? What points do you want to emphasize and what place in your paper are you going to do this? Generally an outline indicates the arguments or evidence you will use to prove to the reader that your new widget or idea is really good and that you know what you're talking about. This permits you to check quickly, before you start to write, whether the organization of your paper strengthens the major points you have written down in your summary.

There are, of course, many types of outlines. You can develop your paper chronologically, using time as the factor that determines your outline division. Or you may arrange your paper to evolve logically, starting from a premise followed by examples and/or arguments to prove the premise, finally leading to a conclusion that restates the premise. For instance, your outline may look like this:

1.0 Premise: System tests insure product quality.
2.0 Arguments:
- 2.1 Lack of system tests permits shipment of defective systems.
  - 2.1.1 Example 1
  - 2.1.2 Example 2
  - 2.1.3 Example 3
- 2.2 System tests reduce defects.
  - 2.2.1 Example 4
  - 2.2.2 Example 5

3.0 Conclusion: System tests insure product quality.

Manuscript received May 27, 1964; revised October 19, 1964.
The author is with the Martin Company, Orlando, Fla.
[1] Also refers to the audience at a symposium or convention.

Reprinted from *IEEE Trans. Engrg. Writing and Speech,* vol. EWS-7, pp. 31-34, Dec. 1964.

On the other hand, you may wish to approach your subject inductively, presenting examples from which you draw a conclusion. Or you can start with a concise statement of a problem or objective, show how the problem can be solved or the objective reached, and end up with a conclusion that shows the advantages of your procedure or the effects of reaching your objective. Whatever method you use, the thing to remember is that the outline sets up the basic blocks and from these you can readily determine if your structure is sound: that is, that you are covering your subject fully, effectively, and without diverging from your main points.

One further consideration: you should estimate the length of your article or presentation. Technical articles seldom go over the 3,000 word limit. This is ten to twelve standard type pages typed double-spaced. Technical presentations longer than 30 minutes must be good to hold the attention of the audience. Using a speech rate of two minutes per page, that means a fifteen page text. Try then to estimate the number of pages each division of your outline will contain. A little work at this time will prevent you from widely overshooting your word limit on the first draft and save you some painful cutting later on.

Review of Outline

Let us assume that you have made an outline containing four or five main parts. Now, keeping in mind what you stated in your summary, check these parts to see if they all support your original thought. Examine their sequence to make sure that each part is in the right place to produce this support. Sometimes one point will appear completely out of sequence. It is a good point, supplying important information, but left in its present position, it not only leads the reader away from what you want him to think, but it also confuses him. In that case you must ate or eliminate the part in which it appears. Also hat you have enough parts to cover fully the points ant to make. Conversely, make sure that you are swamping the reader with too many repetitive or unimportant facts. It is much better to present two or three important examples or to elaborate on a few representative ideas or principles than go into a mass of details. So choose your points carefully. Choose them for their significance and for their impact upon the reader.

Illustrations

Closely allied to an outline is a plan for illustrations. In a technical paper you will often need sketches, drawings, tables or photographs to illustrate your text. Unless you plan for these you may end up with your paper completed minus the necessary illustrations. Examine the main points in your outline and decide where an illustration will help clarify the thought. You may want a sketch of a test set-up, a drawing or a photographic slide of equipment. At another point you may want to present a chart to discuss data obtained in an experiment. A drawing may be required to show construction. You may even want to show how you derived a particular formula, although here a word of caution is in order. Unless your paper is aimed at a highly technical group vitally interested in your theoretical approach, leave out the complex formula. Remember, the size of a general audience decreases in direct proportion to the complexity of the mathematics. In any event, plan the illustrations for your audience. Make them large enough to be readable and, above all, make them add something to your text.

Once you have decided on the illustrations needed, you can make a separate illustration plan, or note the requirement beside the various points listed in your outline. The plan will act as a reminder that you must either do the work yourself or get someone else, like a presentation group, started so that you can have the illustrations ready with the paper. The notes in your outline remind you to refer to the illustrations as you write the paper. Some authors forget this. They add beautiful illustrations but they make absolutely no reference to them in their text. Thus, not only has the time spent producing illustrations been wasted but, worse, the help they could have provided to make the paper more interesting or convincing has been thrown away.

The First Draft

With your outline completed, you can start writing the first draft of your paper. At this time do not worry about language, grammar, or spelling. Just concentrate on developing the points you have indicated in your outline. Write down all the significant data you want to include under each point. If, as you write, you have some thoughts that do not specifically follow the outline but which you now think are pertinent, put them in. Essentially what you want to do is to fill in the blocks, adding the details, the examples that are needed to make your thoughts clear. This phase can often be done rapidly, writing at top speed until the complete draft of the article is finished. Of course some people may find that they work more slowly, completing only parts of the paper at one sitting. It does not matter. The goal is to get the whole draft written as quickly as possible, without deviating too much from the outline, so that for the first time you can see the complete article.

Here the question may arise, how long should my draft be? It should be long enough to cover all the points listed in your outline and, if possible, it should be close to your estimated length.

The draft is generally written or typed double or even triple-spaced to leave plenty of room for changes and corrections. And while it is not imperative that it be typed, the cleanly typed copy will make it easier to read. It also gives you a better feel for the impact the full article will have upon the reader.

Review the Draft

When the first draft has been completed it is worthwhile to read it as an entity. Reading it aloud helps or, if you have a tape recorder available, tape it and then listen to it. At this time all you should do is correct the most obvious errors in English and note where you need to add or delete information. After that lay the draft aside for a week or more. This is done to gain perspective, for as long as you are close to the writing you cannot truly evaluate it. The time delay helps make you forget the immediate details so that when you read the paper once more, you can edit it more objectively.

Editing

Editing is one of the most important phases of writing a paper. Don't be scared by the word "editing." It means nothing more than examining your paper slowly, critically, and making the necessary changes to smooth the language, insure that your thoughts are expressed clearly, and that you make your points well. This is the same type of review accorded the outline except that now you are checking the details of your writing to see how closely they follow your plan. Granted that this is hard work, but if you want to be effective and produce a good paper, it must be done carefully.

In editing, you assume that you are the reader and that you must be interested, entertained, taught, or convinced by what you read. Your paper should accomplish this, first, by making it easy for the reader to read and understand your thoughts and, second, by the weight of the logic and evidence presented to support your conclusions.

To check on the first point, read the article again for flow of language. Do the sentences flow smoothly into each other? Do the words sound right? Are they simple and clear? Sometimes the thought can be confused by too long a sentence. Chop the thought into easily digested doses by using short sentences, varied occasionally by a longer one where detailed explanations make more words necessary. If desired, tape the article again and listen to the playback. The awkward constructions and muddled sentences will stick out and can readily be corrected.

As you work to smooth the language also examine the text from the point of view of reader interest and understanding. Is your first sentence startling? Does it arouse curiosity? Will it make the reader want to continue reading? Then look at the way you express your thoughts. Are you really saying what you want to say in the shortest and most powerful way possible? This brings up one of the toughest editing tasks. If any portion of your article appears to drag, if there is any extraneous material, or if any part is out of proportion to its weight in supporting a particular point, don't hesitate to cut. I realize that the article is your creation, and that every part of it is sacred, but you have to be ruthless. A pet word, showing off your knowledge of technical terms, but which is vague or confusing must be yanked out and replaced by a simpler, better-known term. A good rule is to cut your draft by 20 per cent. You say that is impossible, but you would be surprised how much it will improve your article.

Also, in presenting your thoughts to the reader, do not ever assume that he knows as much about the subject as you do. Too often trade names, abbreviations, and functions are mentioned without explanation on the assumption that the reader is in the same business and therefore is totally familiar with the terms. At other times it is strictly an author oversight. Being completely at home with his subject, the author overlooks the fact that another person lacks the detailed knowledge, or even some of the fundamentals, which he so casually omits. Remember that it is better to err slightly on the side of details, particularly for complex technical subjects, than to leave the reader with a lot of gaps which he has neither the time nor the inclination to fill.

When your are reasonably sure that the reader will understand what you are saying, then you must examine how well your article can convince him. If you have done your homework well and considered this phase in your outline, the basic logic should be there. Now all you need to do is check that your development of each point is effective. For instance, if you want the reader to agree that system tests insure product quality, every statement you make, every example you give, must support this contention. The evidence you present may be positive, showing the advantages of running system tests; it may indicate the disadvantages of not performing them; it may indicate the effects of engineering tests, of field tests. Your argument may proceed from the known to the unknown; it may evolve logically or chronologically, but to be pertinent, every sentence written must proceed to stress, to restate, perhaps by examining different aspects of the subject, to develop and to extend the basic thoughts you indicated in your summary until, by the weight of evidence and logic, the reader is convinced and agrees with your conclusions.

In this task make sure that your examples are good, for a vivid example not only can clarify a statement and evoke reader interest but often can act as the clinching argument. For instance, if I say that lack of a system test causes failures, the statement by itself does not have much weight, but if I add: "for example, when flight testing missile Y in the desert, we omitted the system test to meet a scheduled commitment. But we forgot one thing, the one thing that nearly killed us . . . etc.", then I gain reader interest and can more easily make my point.

One important consideration: if you have any doubt about any portion of your article, even when it is finished and ready to be sent out, do not hesitate to rewrite. Often the writer will let the article go even though he feels that the language is fuzzy or the thought is not completely clear. With the normal reluctance to do extra work he will rationalize that the reader will understand his point. Not so! If you have any qualm about any part of your article you can be sure that the reader will stumble over it. So rewrite the portion to make it better. You may have to struggle to do so, but once it has been done you will feel that is was worth the effort.

Final Draft

Now you should be ready to write your final draft. This is your calling card to the editor so you have to make it good. The draft should be typed double-spaced with a good ribbon on 8½ by 11 inch white bond. Colored paper might look pretty but it does not impress the editors.

Start half-way down the first page with your title, name, and under it, your present affiliation. For example:

The Hybrid Life of the Electron
by
John C. Shufholder
Senior Engineer
XTA Corp.

The title should be fairly short and so conceived as to arouse interest or pose a question. It must cause the reader to pick your article from the others offered him.

In typing the text, leave a margin on each side, wider on the left, so that the editor has plenty of room to write his comments or instructions to a printer or typist. Also, do not forget to make a carbon copy. You might need it in case the original is lost. Now as you type you may want to make minor corrections. Go ahead. The aim is to make this version as good as possible. But be sure that the copy is clean and without typing errors. Remember that a sloppy, over-corrected, and dirty manuscript creates the impression that the author thinks and works the same way. This may not be true and your paper may be the work of genius, but do not handicap yourself by adversely conditioning your first reader.

At this time you should examine your final illustrations with the following points in mind:

1) Are the illustrations adequately tied in to the text?
2) Do they really add understanding or weight to the paper?
3) Are they too complex? Or is the printing or drawing too small to be effective?

Do not let someone else handle this phase on the excuse that he knows more about illustrations than you do. This is your paper and, as the author, the quality of everything that goes into it is your responsibility.

Before submitting your final draft it may be worthwhile to ask a qualified person to review and criticize it. At this time ignore congratulations on your writing ability or praises of your technical virtuosity. What you are looking for are answers to the following:

1) Am I defining my problem or my purpose well?
2) Is my development clear and understandable?
3) Do I make my point(s)?

If the answers to these questions are in the affirmative and you have generally complied with the previous suggestions, you should be on your way toward having your paper published or accepted for presentation.

WHAT TO DO ABOUT COBBLESTONE WRITING

George H. Douglas
University of Illinois

Some time ago a colleague of mine brought me a sheaf of his working papers and scholarly articles. He wanted me to read them, comment on them, edit them, in fact do anything to them that would improve their writing style. It seems that he had submitted some of these papers to academic journals but without much success; in most instances he had met unkind rejection. It might be, said this young assistant professor eagerly anticipating promotion, that the editors had disapproved of his subject matter. On the other hand, he had a feeling that some quality of his writing might have been held against him. After glancing briefly over what he had written, I had to agree that it probably had.

The writing in question was in no way extraordinary; it was really the kind of thing that one frequently encounters in academic journals, technical papers, scientific reports, and so on. It was not very good writing; yes, if one wanted to be uncharitable, one could say that it was just plain bad writing. But its badness was due not to sheerly technical defects of grammar or syntax nor was it due to imprecision of thought, at least not the kind of fuzzy thinking sometimes associated with the neophyte specialist. For these and perhaps some other reasons my author's departmental colleagues felt—although very uneasily—that "everything seems to be all right." Of course the writing was clearly not all right, but its weaknesses were in that gray area which has always been difficult for the writing teacher to get hold of.

Perhaps it would help to quote a sample paragraph; I think it is fairly typical, although there are places where the writer is more lucid than this and others where he is more unintelligible. I should add by way of introduction that he is a social scientist whose field is labor and industrial relations. The paragraph:

> Thus, although the factors which produce similarity relations among occupations are either unknown or unquantified, these factors will govern occupational mobility patterns. Grouping occupations on the basis of similarity in mobility patterns is, then, an alternative to delineation of the complex interplay of factors which tie occupations together and determination of the proper weight per factor, per occupation. Indeed, the types of occupations which cluster on the basis of similarity in mobility patterns will at least to some extent reflect the relative importance of underlying determinants (skill level, industry, subject matter, work field, material) of the cluster.

Now this is the kind of thing that writing teachers have to deal with all the time in colleges and universities. In some fields the end-product may be better, in others, worse. There's an unpleasant, but alas very common, disease of writing at work here. The symptoms are clear enough in general outline, but I must confess that in the past I've always taken the easy road when dealing with the problem. Like most of my colleagues I've satisfied myself by commenting: can't you write simpler; can't you write more briefly and concisely; can't you eliminate big words and professional jargon. The trouble with this advice is that however sound it may be, it doesn't really do much to clear up the things that really produce the bad writing. My writer will argue (quite rightly) that there really isn't much professional jargon in the passage you select for analysis. He will admit that shorter sentences might be a help, but after cutting down his sentence length you will see that the writing is really not at all improved.

The main weakness of technical writing, indeed of professional writing of all kinds, is usually not that it is long-winded, or that it is overstuffed with big words (even though these charges can also occasionally be made) but that it is compact; it suffers from hardness and density. This hardness and compactness can be endured if it is short in duration, if it goes on for a sentence or two; but it becomes unendurable when it goes on sentence after sentence, paragraph

Reprinted with permission from *Tech. Writing Teacher,* vol. 5, pp. 18–21, Fall 1977.

after paragraph. I call the kind of writing that I'm talking about "cobblestone" writing; it is a kind of writing that gives the impression of a trip down a cobblestone road in a one-hoss shay, an experience of bumping from one hard side to another.

But what do I mean by hard ideas? What is it that gives rise to the cobblestone effect? This is not as easy to explain to the student of writing as one might like. Very often the writing teacher puts it all down to a matter of abstraction. The author of the *Random House Handbook* advises his readers to "beware of unrelieved abstraction." Probably good advice, although cobblestone writing can come from any kind of "hard" or dense language or phraseology, not merely from abstract or general words.

Consider the writing of the engineer. Very seldom is his writing plagued by excessive abstraction; he may have to convey technical, specialized information, but it is usually quite concrete in nature. Consider the following passage against which no charge of abstraction can be made; still, it is technically dense. It is cobblestone writing by any standard of judgment.

> Doric automatic integrating digital microvoltmeters are specifically designed to measure loud-level dc voltages in the presence of severe electrical noise and environmental changes. Low-level signals must be handled by an instrument capable of extracting a few microvolts from a dc source buried in ac noise. Zero drifts are critical error sources at the microvolt level, and accordingly Doric engineers directed themselves to the problem of developing an instrument free of drift problems.

Most of the words here are concrete, specific, thing words. They are technical words to be sure. But it is not really technicality that is the source of the mischief in this kind of writing. It is overall compactness and density. The writer is essentially a cataloger of ideas, a listmaker; he has no notion of the narrative element of writing; he has no understanding that to be readable his ideas have to be interpreted, explained, pre digested. Cobblestone rhetoric as found in the writing of engineers and scientists usually takes the form of listing of hard concrete ideas, each of which has exactly the same weight and configuration as every other. There is no story line; there is no beginning, middle, and end; everything is cast in the same gray monotone; there are no transitions, no lights and darks, no ups and downs. Never does the writer say "Look here, this is what is important, that other idea is subordinate detail." Every idea is exactly as important as every other, as in a list of nuts and bolts—300, 1/8" capscrews; 250, ¼" capscrews; 240, 3/16" capscrews, etc. But writing is not a listing; writing is interpreting and explaining.

The difference between writing on the one hand and cataloging or cobblestone rhetoric on the other might be illustrated by the following little schematic table:

COBBLESTONE WRITING
(Cataloging or Nonwriting)
Idea I
Idea II
Idea III
Idea IV

WRITING
Introduction or Overview
Idea I
examples
explanations
comments
interpretations
transitional ideas
(when needed)
Idea II
———
———
———
Summaries as needed

In cobblestone writing ideas are shoved tightly up against one another; there is no breathing room; the reader has no chance to digest or understand one concept before he is hit hard with another. Every idea is an undigestible iron nugget that has to swallow in an unconverted form. The idea, the sentence, the concept may not necessarily be unintelligible, or difficult by itself; the trouble comes when the reader is subjected to a constant succession, a constant bombardment of uninterpreted detail.

For a moment let's go back to the passage of engineering writing that I've quoted above. With the introduction of a little more narrative element, with a little more effort at interpretation, it becomes more nearly like writing. It is not, of course, popular writing for the man on the street. Nor should it be, although this is sometimes the foolish and unrealistic goal frequently urged on engineers or scientists by English teachers. There is no requirement that technical writing should be popular writing, only that it be writing. So here is the same passage as above, with a little more attempt to provide interpretation and narrative flow:

> In the past six months Doric has been marketing its new automatic integrating digital microvoltmeters which are specifically designed to measure low-level dc voltages in the presence of severe electircal noise and other environmental disturbances. Low level signals—down to one microvolt per digit—present special problems which can only be solved by an instrument which is totally dedicated to the task of extracting a few microvolts from a dc source buried in ac noise. Zero drifts, which could be ignored at the 10V level, are critical sources of error at the microvolt level and must be eliminated. So Doric engineers started from scratch and developed an instrument absolutely free of zero drift problems—an instrument which outperforms meters costing thousands of dollars more.

This second passage is of course quite a bit longer than the first. A very significant point too when it is remembered that very often the advice given to technical writers is "be brief," or "say things the shortest way." This advice, given by English teachers with the best of intention, backfires and encourages the technical writer to do the very sort of thing he shouldn't do. The most characteristic fault of technical writing is not that it is long-winded, but that it is too compact, too stuffed full of hard and uninterpreted detail. It is a kind of wishful thinking on the part of some teachers that brevity is going to solve all the technical writer's problems. (The wishful thinking is probably due to a hidden desire for technical writing to come to a quick end and go away.) The technical writer is usually quite brief enough. What he doesn't know is how to get away from the density of cataloging or listing—from the mercilessly rough road of cobblestone rhetoric.

Let's go back for a moment to the learned social scientist whose paragraph I quoted earlier. Where do we start with him? How do we get away from cobblestone writing and back to writing (I say *back* to writing because I assume that before he became ensnared in the entanglements of his own professional terminology, he could probably write pretty well.) Before any progress can be expected from him it must be emphasized that his writing is bad, not because it would be totally unintelligible to the specialist (for this is not true—the co-specialist could probably struggle with the passage and eke some meaning out of it), but that the very same thing can be said in a way that does not force major acts of interpretation on the reader. The assumption quite true—that the co-specialist *can* make the interpretation, can muddle through, does not mean that he should have to. It is the writer's function to provide a smooth and easy ride rather than a series of headache-producing thumps.

The technical writer then has to be urged to politeness, to generosity. He has to be urged to put out the extra effort that is required to make dull cataloging into narrative flow. He has to make the technical easy (but not necessarily nontechnical) as a matter of courtesy. He has to see it as his job to make his ideas readily accessible. He has to tell a story with a beginning, a middle, and an end. He has to anticipate the kinds of places where the reader would have questions, where he would have to fill in the gaps himself. Making writing easy is hard work; it's not simply the process of chopping and cutting that many writing teachers too readily assume. But it is not the kind of hard work that requires complex skills.

The skills that are needed are part of the educational equipment most people possess if only they are urged to put them into practice. Sometimes the writing teacher can encourage writing by having the specialist shift outside his own specialty where he will appreciate the value of easily understandable exposition. Another approach would be to have the specialist write stories or how-to materials for children in his own field or perhaps for the ordinary general reader. The exact kind of practice is not as important as the recognition that the cobblestone ride is not really writing, that writing only comes into being with narrative flow and full interpretation of detail.

How to get dullness out of your writing

There are rules about writing and rules about writing, some of which work while others are simply rules. A well-known public relations man gives some practical suggestions.

BY LUCIEN R. GREIF
President
Greif Associates, Inc., Chappaqua, NY

WARNING: Do NOT give up after the first paragraph. Things get better. Read at least two paragraphs; then if you still feel you cannot pick up a few pointers, turn the page. No hard feelings. Coming up: "the first paragraph."

This is an article about writing. The object is to show the reader how to write well. The point is that good writing is not easy. The reason good writing is not easy is that many writers make some very basic mistakes. The most basic mistake they can make is to be repetitive. The other basic mistake they make is repeating everything. The next basic mistake is being redundant or being too long-winded, or continuing to write to a point when that point has already been made, and then explaining it just to be sure that the point just made is understood. The most unforgivable mistake is starting almost every sentence with "The," such as this paragraph, which was carefully constructed as a prime example of deadly dull writing. The reason this article is so dull is that . . .

Are you still with us? Bless you for heeding our warning. Without it, we might very well have lost you. In fact, that first paragraph is so smashingly horrid, we were ashamed to proof-read the galley! But your persistence shall be rewarded, because we now come to the pay-off: *Ten ways of checking your copy* to be sure it has at least a fighting chance for survival in the jangling jungle of articles clamoring for reader attention.

Count the number of sentences in your copy.

Next, count the number of sentences which start with the subject, or are generally of the subject-verb-object variety. If they constitute 70% or more of your copy, worry. (Count any plain or modified noun — or pronoun — as your subject.)

Look at it this way: out of all the sentences you've written, how many are not simple declarative statements? Don't you ever ask a question — even if only rhetorical? Every once in a while, come on with a bang! Or a quick phrase Just for a change. If your copy consists almost entirely of declarative statements, expect to lull most of your readers to sleep long before they reach your third paragraph.

Next, count the number of sentences which start with the deadliest of all letter combinations, "Th." Such as the, this, that, there, they, these, those, etc. More than 50%? Back to the drawing board.

Count the sentences starting with "It." You can take double-credit for this one. (A) It ranks with "Th" words for lack of power, and (B) it increases your "subject starter" score. Solution? Get rid of it! Er, them.

Look at the sentence length. Are all sentences about eight to ten words long? Should make for pretty snappy copy. But throw in a few long ones now and then, for change of pace. On the other hand, if after reviewing the material you so painfully prepared — and it truly was a labor of love, where every word, every inflection counts — if, then, you find that your average sentence is so constructed as to contain sometimes thirty, sometimes forty, or more words — on average — perhaps you should reconsider, lest you lose some of your less sophisticated readers. Remember, most of us have a 12th-grade reading mentality.

Count the number of commas, semi-colons, other punctuation marks in your typical sentence. More than two — not counting final periods — and you're on your way to writing complex copy.

Embarrassing Questions

Now that we're finished with the small pieces, let's go to the larger ones: paragraphs.

What is the average length of paragraph? One sentence? Not enough

A CHECKLIST FOR CLEAR COPY

1. **Count the number of sentences in your copy.**
2. **How many of them start with the subject?**
3. **How many of them start with 'th'?**
4. **How many start with 'it'?**
5. **Do your sentences vary in length?**
6. **How many sentences have more than two punctuation marks?**
7. **How many long paragraphs do you have?**
8. **Are your paragraphs too long or too confused?**
9. **Have you eliminated unneeded words?**

Reprinted with permission from *Advertising and Sales Promotion*, pp. 37-38, Aug. 1969.

— except for emphasis, in deliberate cases. A whole page? Much too long. When copy is double-spaced, you should generally find from two to four paragraphs on a page.

Here's a truly critical question: what are you *saying* in each paragraph? What is the subject matter, the key point you're trying to make? Write it down in the margin. When you're all done, look at your marginal comments for insight into your organizational abilities. If you're making the same point more than once, it's a clue to a wandering mind, a hint that your article probably needs some rearranging. Put apples and pears together. Make each point forcefully, but once you've made it, lay off.

While you're about it, ask yourself: for whom has this article been written? (How do you see the typical reader? What is his educational or business or general background? What are his interests?) What is the main point the entire story is trying to make? Most important of all, have you made it???

Next in line, screen your work for excessives and waste words that may have crept in. Have all unnecessary adjectives been eliminated? Have all triple-compound modifiers been deleted? Have awkward sentences been straightened out? And just when you're convinced that your copy has reached its irreducible minimum, go back and trim another 30%.

▶ Ultimately, you must reach the most embarrassing question of all: "Is this the best work I'm capable of doing?" Try and answer that one!

One last point. Someone will no doubt ask if it's possible to assign numerical scores to these various dullness-indicators. Not really. But we'll attempt it anyway.

a) Too many Simple Simon sentences (subject-verb-object), 50 points.
b) Too many sentences starting with "Th," 20 points.
c) Too many openings with "It," 10 points.
d) Too many long, involved sentences, or too much punctuation, 30 points.
e) Paragraphs much too long, 10 points.
f) Paragraphs repetitive or confused, 50 points.

As far as we're concerned, it only takes 100 points for writing to be deadly dull. Obviously, anyone can win!

Lucien R. Greif is president of Greif-Associates, Inc. (Chappaqua, NY 10514), a firm specializing in preparing feature technical articles.

How Does Your Writing Measure Up . . .

. . . at These Nine Vital Points?

J. R. GOULD
with commentary by
N. G. SHIDLE

NOTE—*If you're like most chemical engineers, you've probably noticed that a big chunk of your time is spent in writing. And, again, if you go along with the majority, you probably don't enjoy it.*

Well, if such is your predicament, this article is your meat.

Our author, Jay R. Gould, is the director of the Technical Writers' Institute of Rensselaer Polytechnic Institute and a well-known teacher of technical writing. His nine check points can mark the route to better writing for you. Norman G. Shidle, whom we've asked to do the commentary, is editor of the SAE Journal and the author of "Clear Writing for Easy Reading." His somewhat different viewpoint will aid you—especially when you write for publication.—EDITOR.

Can you . . .

√ **Force your writing to serve you?**

√ **Make your personality pay off?**

√ **Keep your story on the beam?**

√ **Gage your material requirements?**

√ **Get off on the right foot?**

√ **Buttress your text with the right example?**

√ **Keep the reader moving with you?**

√ **Brake to a stop at the right place?**

√ **Guarantee pleasant, profitable reading?**

ONE EVENING not long ago, I sat in the Engineers' Club in New York with a prominent young engineer, the vice-president of a large construction company. During our conversation he brought up the subject of the writing problems that confront the engineer.

"Every engineer has to write at some time or another," he said. "Of course, all of us in our college days had visions of passing the writing job over to our secretary, or even the office boy, but in reality it has turned out differently.

"Today the engineer is responsible for all kinds of communication jobs. Reports have to be turned in to government agencies, inter-company memoranda have to be written, and articles must be prepared for trade journals. Also, if the engineer wants to get ahead, he may find it necessary to deliver papers before professional societies. Yet we often find ourselves unprepared to do the writing job."

That engineers can't write is a myth that engineers themselves have nurtured, just as my friend at the Engineers' Club was doing. He's a victim of self-delusion.

This delusion is that writing is an art; that very few people can write; that it's hard work (1) and that it better be left in the hands of other people.

It Can Be a Science

Writing can be an art, but it can also be a science. And it is nonsense to say that very few people can write. Very few can write well enough to win the Pulitzer Prize, but most people, if they took the pains, (2) could write adequately and effectively.

---

(1) It is hard work. That's why so many people do it badly. And the hardest part of the work is the clear thinking which must precede.—NGS

---

(2) Right again. But the pains must give birth to well-organized thoughts. Just suffering isn't enough.—NGS

However, my friend was right when he said that every engineer has to write at sometime or another. It is the vast increase in writing assignments that has thrown many engineers into a tailspin. Since the war, it has been estimated that the volume of written work streaming from the average company has increased four hundred fold.

Can something be done to offer the engineer certain short-cuts to lick his writing problem? I think it can, and it's on this assumption that I am offering these nine check points for writing.

Writing That Has to Be Done

But before we arrive at our checklist, let us look at a few of the writing situations that involve the engineer.

Certainly he dictates letters, if he doesn't write them himself. Only a few weeks ago I had a letter from the publications manager of a utility company commenting on how hard it is to get people to be simple and direct when they are writing business letters.

This man went on to say, "When a man writes a personal letter, his tone is usually light, he sounds as if he enjoyed writing the letter, and what he says sounds like good normal conversation. 3 But the minute he begins a business letter, oh boy, bring on the jargon!"

But to go on with our list: The engineer may have a hand in preparing manuals of one kind or another; he may have to write promotional copy. And all the time, if he's an asset to his company, he is forever explaining that company and its acitvities to the general public through what he says and what he writes.

Analyzing the Problem

✓ So because your writing assignments are so varied, I submit this as Check Point 1 in your short-cut to writing. Size up the situation. Analyze the writing problem in at least three ways.

The first of these is to find the purpose 4 behind the writing—why you are setting it down on paper.

Are you briefing someone in your company on a new development connected with your product? Perhaps all your reader requires, then, is a plain answer to a plain question. Cut out the frills. Make it simple and terse. Imagine that every sentence you write is costing money. Get the information down and stop worrying about how it sounds.

But perhaps your reader wants more than plain information. He's not completely sold on your proposal. In that case, you have to be more persuasive. Deliberately select your vocabulary.

The writers of adjustment letters are told never to use the word *complaint* in writing to dissatisfied customers. That word is guaranteed to lose friends. There are many, many such words in business communication, some that are pleasant, some that leave a bad taste. Watch your words.

Who's Reading It

But besides considering the purpose behind our writing, we must pay attention, a lot of attention, to the man who is going to read it.

In some writing jobs he may be a single reader, the president of your company, another person in your office. And if he is someone you know, I'm sure that you keep a mental picture in front of you while you are writing. Will Mr. D. understand this, you say. Or, Fred is a crank on grammar and punctuation. I'd better check and recheck.

But do you use the same care when your readers are more complex, the members of a professional society or the stockholders in your company? You should. Every reader, whether an individual or a composite, has his peculiarities.

The Situation Sets the Style

The third division under our first check point is this. What kind of material are you using and to what extent will it condition your writing? Engineering material by its very nature often calls for a certain style. But this doesn't mean it has to be "jargon." It doesn't necessarily have to contain a statement such as this from a government agency: "This office maintains a chronological suspense upon matters to which a response is expected."

Material prepared by technical writers for technical readers, usually need not contain as much illustrative material as material written for nontechnical readers. And on the other hand, some material cannot be "popularized" no matter how much you try. What you gain in interest value you lose in accuracy.

So check Point 1. Analyze the writing job in its three aspects of what you want to accomplish, the kind

Shidle comments . . .

3 It's good to have our writing sound like normal conversation; but fatal for it to be like normal conversation. Conversation and writing are two different forms of communication; require different techniques. Conversation is received through the ear. The speaker sets the pace. Repetitions and emphasis necessary and possible in conversation are loose and wordy when copied in writing. Get stenotype notes of any high-grade conversation you've been in, if you doubt this.

4 Define the purpose as well as find it. Force yourself to write down in one sentence the main point you want to convey. Do this before you start to write. Then what you write becomes simply an amplification, explanation, or proof of that main idea.

It's the easiest way to insure good organization adapted to the special purpose of the particular piece of writing.

of reader you have, and the special conditions that the material itself demands.

Where YOU Come In

At this point a statement by a colleague of mine at Rensselaer comes to mind. Of engineering writing, Prof. Douglas Washburn has this to say: "Certainly a degree of objectivity is necessary, but frequently technical writing seems to strain so far in this direction that it gives the impression of having been written by a machine rather than a human being."

√ Let your personality show through your writing. This will be Check Point 2. Show that the writing has been done by a man, not a machine. **5**

What would you say if you were invited by your employer to talk over a deal with him, only to find that you were to be on one side of a screen, and he on the other? The channel of communication would be poor to say the least. Yet this analogy, ridiculous as it is, is minor compared to what often happens to the engineer when he sits down to write. He sets up his own screen.

He has been conditioned to such an extent by something called "technical style" that his native way of saying things has been throttled, and, as Prof. Washburn remarked, he has become a machine.

How to Stick to the Subject

It has been said that the idea back of every good piece of writing can be expressed in a single sentence. **6** Teachers of writing call it the thesis sentence.

√ Check Point 3: Decide what the subject of your piece is going to be—then stick to it. **7** Every writer as he is working, just as I am doing now, thinks up ideas, with incidents to enliven his writing. Sometimes they belong; more often they don't.

So stick to your topic even if it means pitching a pet idea into the waste-basket. One way to achieve this singleness of purpose is to draw a box on a piece of paper, and in it write your thesis sentence. **8** And from time to time, especially as you pass from one point to another, glance at the box and see if your new material will go into the box. If it won't, discard it. Be particularly ruthless in revision.

How Much Material Do You Need?

√ Check Point 4 is for you to assemble all the material that you possibly can, and if some of it seems beyond your needs at the time don't worry. Material for your piece of writing will come from many sources: research, things you've seen, experiences you've had. Especially don't overlook that last source. Your experiences are part and parcel of your ideas.

So gather as much material as you can. Nothing gives you a happier feeling than to know that you have reserves to draw upon. It will prevent a thin piece of writing.

How to Get Going

Many writers feel that they must start with something called an introduction. My personal reaction is this. If you think you need a warm-up period, by all means write an introduction. But later on, when you are revising, see if you can't cut it off. **9**

√ Check Point 5, then, is to take particular care with your beginnings. Elizabeth Ogg, a professional author specializing in interpreting technical material to the layman, says this: "Beginnings are strategic. They may make a difference in whether your writings get read or passed by."

The initial paragraph of your writing, then, not only presents the material to the reader; it induces him to read it in the first place. **10**

The beginning, however, must be compatible with the rest of the piece and with the reader. A controversial statement, a quotation, something lifted from today's newspaper, all these will dangle the bait in front of the reader. **11** But each in its own place. A report on an important process for a man who must use the process will demand little in the way of a fresh beginning. But your colleagues may be more wily. Their reaction may be: "Oh, are we going to hear that again?"

An article before me, written by an engineer for engineers and about a Canadian steel man starts out this way: "When Adam Smith was a boy, he wanted more than anything else to grow up to be a mechanic." A name strikes our attention, a story is being told.

A bulletin on photography, aimed at serious photographers, begins less visually, but it has a positive

Shidle comments . . .

5 Very desirable . . . but a touch dangerous for the unskilled amateur. Trying to put "personality" into his writing, he often ends up merely verbose and trite. Too often he turns cliches. Might be safer if this were moved back to Number 9 instead of being Number 2.

6 In my opinion, this is by far the most important point of all. The man who actually does this before writing has a hard time doing a really bad piece of writing —if he sticks to his thesis.

7 Amen!

8 Amen again. And why not work this into your lead sentence? Everything else will flow naturally and effectively from it.

9 Cut it off! Don't just "see if you can't cut it off." Do it.

10 And here is the reason for cutting off the introductory "warm-up".

11 Be Careful! Like the advice to put in "personality," this sort of "human interest" lead can come out badly more often than well for the everyday writer. I believe that the writer's job is to interest the reader in his thesis—and the sooner he starts doing it the better.

approach and a deliberate goal in mind. It says, "The aim of most photographic effort is to make a print of satisfactory quality." Not sensational, but useful to the purpose and the reader.

Don't Forget to Show 'Em

√ Check Point 6: Less interpretation and more illustration. 12

A writing assignment of any length can be broken up into a number of segments or main points. But if you want your main points to do more than go in one ear and out the other, you must illustrate each of them.

There is a difference between illustration and interpretation. When a writer says *in other words*, you can be sure that he is going to interpret for you; unfortunately, the interpretation is sometimes more cloudy, more confusing than the first thing he said.

Get into the habit of providing a pungent example for each main point. Give the reader something to visualize, a good concrete down-to-earth example. Draw on your personal experiences. Every reader likes to read stories. Take him on that job you did. Let him see you in action. Keep a file of items which you think might be useful—then use them in your writing.

How to Hang On to the Reader

Not only must we provide examples for our main points, we must provide the means for the reader to get from one main point to another. Not being as familiar with the content as you are, he often loses his way. He says, "A moment ago you were talking about Korea. Now we seem to be in ancient Rome. How did we get here?"

√ Check Point 7: Give the reader signs and guides to show him the direction of your thoughts. Our language is lavish with such transitional devices: such phrases as *nevertheless, although, consequently, however, in addition to.*

When the reader sees these, he knows that you are shifting your direction, and he is being told to be on the alert. There are hundreds of such directional words. Sprinkle them liberally throughout your writing. 13

How to Put On the Brakes

√ The next Check Point, 8, pertains to the ending. While the beginning attracts attention, the ending of your piece should leave the reader with the feeling that he has accomplished something by having read it. 14 This feeling can be arrived at in a number of ways.

First of all, it is a good idea to make some demands on the reader, to ask him to do something. Don't let your writing "run down." And a way to avoid this is to give it a lift instead of sending it into a decline. Put some pressure on the reader. If nothing else, ask him to think about the ideas you have presented.

Or provide him with a summary of the main issues. Especially if the piece is of some length, this device is good not only for the reader, but also for the writer. It forces him to be logical in arrangement.

A third principle connected with the ending is always to save something for the ending. Make the circle complete. If you began with a story, end with a story. Take a tip from the skillful public speaker. Observe how he finishes with an anecdote or a punchline; when he doesn't have any more to say he stops cold.

Is It Readable?

√ And now before we come to an ending for this article, consider for a moment Check Point 9, the over-all principle of readability. Here I would like to cite the case of John Newsome, a third-year student in a special writing course I conduct.

John, a fledgling engineer, would like to have a side career as a writer of science fiction.

He reacted strongly when I criticized his first effort. It was full of purple passages and roundabout ways of saying things. John couldn't say, "The man went on reading the book." Instead he said, "His lips continued to read." And one very mixed metaphor bore the sentence "His body bore my scars on its handsome face."

His writing was unnatural as well as inaccurate.

Until the John Newsomes come to their senses, not much can be done about their writing. John somehow had the idea that when you turned to fiction you aban-

Shidle comments . . .

12 Very important—and infrequently done. Examples are much harder to dig up and express than generalities and opinions.

13 This is good advice. Used for exactly this purpose, the suggested words are excellent. . . . Trouble is the average everyday writer tends to toss in such words too frequently when they aren't needed. . . . Usually they are "empty" or "say-nothing" words. I'd suggest adding the advice: "When in doubt, leave them out."

14 Here's the only point on which the professor and the editor really differ. To me, the ending is of very minor importance as compared with the beginning. Reader surveys prove that scores will read the first paragraph of an article or report for one who gets to the finish. Most engineering writing that crosses my desk is most easily improved by putting its last paragraph first. Either article or report is most interesting when the conclusions come first and the article or report proves them. . . . Besides, the main points will thus have been registered with a maximum—not a minimum—number of readers.

(For eight years now, at least one engineering society publication has been applying exactly this technique to the rewriting of all papers. Only thanks and congratulations for the technique have come from readers; scarcely a single kick. And less than one half of one percent of some 1,600 authors have objected.)

doned normal ways of saying things and indulged in something called "pretty writing".

I'm afraid that something of the same sort prevails among engineer-writers—only in reverse; that when you write up technical material, you abandon good easy communicative writing and adopt something called "literary" style.

How to Boost Reading Ease

One piece seems easier to read because the writer has used comparatively short paragraphs and sentences. He has made direct statements, not cluttered them with modifications and amendments. The page looks easy because he has left plenty of white space.

If he has written a report or a research article, he has used crisp informative headings. He has used a vocabulary that has good visual words in it, words you can see, and feel, and hear. He has used names, names of people and places. The reader's eye is caught by the name of a famous scientist; he recognizes a product by its trade name.

In any analysis of writing, many more factors enter. A complete mastery of them belongs to the really professional writer. But most of us look upon writing as an extension of other work, as a vehicle for expressing what we do.

Where to Check Yourself

In any writing assignment, then, after you have completed your first draft, check it at these nine vital points. They are: 1. analyze the situation and see if you are doing what you had intended; 2. allow your own personality to show through your writing; 3. stick to the topic under discussion; 4. consider your material, and make sure that you have more than enough; 5. pay particular attention to the beginning; 6. check the main points for pertinent and interesting examples; 7. give the reader plenty of transitional devices; 8. provide a sharp and decisive ending; and 9. make the piece of writing as readable as possible.

Check these nine points in your writing, and help destroy the myth that engineers can't write.

Frequent and effective publication in engineering magazines have helped many engineers to fame and fortune. Here are a few suggestions to increase the chances of acceptance and the readership of your article.

KEYS TO GOOD ARTICLE WRITING

By JOHN M. CARROLL*

THE ENGINEERING magazine addresses itself to an audience that is neither as broad as the readership of a daily newspaper nor as specialized as the readership of an engineering report or technical manual.

The newspaper attempts to relate a new engineering development to the experience of the layman and may presuppose only a general public school background on the part of the reader. The manual attempts to give extremely detailed design or maintenance information and may often presuppose considerable specialized knowledge or experience on the part of the reader.

The engineering magazine attempts to relate new developments to the practice of electronic engineering over-all. It may presuppose adequate engineering training on the university level on the part of its readers, but should not presuppose an intimate acquaintance with each part of the field of electronics.

How Articles Are Selected

Engineering editors judge articles on the basis of timeliness, significance and technical accuracy. A development must be new, for it is not the place of an engineering magazine to repeat information already available in textbooks or existing articles.

A development should likewise be of broad significance to the field served by the publication. The large circulation of modern engineering magazines requires that each article represent an advancement of the art, not just a minor modification of someone's modification.

The technical accuracy of an article is, of course, of paramount importance. It is usually ascertained from the reputation of the author and his affiliation and by the experience and judgement of the publication's technical staff.

Articles are published to be read. A publishing company spends a great deal of money to find out who reads engineering publications and how they are read.

We know, for example, that few readers ever complete an article. A high percentage of readers drop out every time they are required to turn a page. This is a telling argument for brevity. Fig. 1 summarizes results of readership studies showing mean percentage readership decline for each two-page spread of a six-page article. Six articles were studied.

Readers are not a captive audience. They must be induced to read an article—and, an article must get down to business fast. Don't save the main point of the article for a masterful conclusion. Nobody may be left to read it. For this reason, an oral presentation seldom makes a good article without extensive rewriting.

*Managing Editor, *Electronics*, McGraw-Hill Publishing Co., Inc., New York, N. Y.

Reprinted from *IRE Trans. Engrg. Writing and Speech*, vol. EWS-2, pp. 78-82, Dec. 1959.

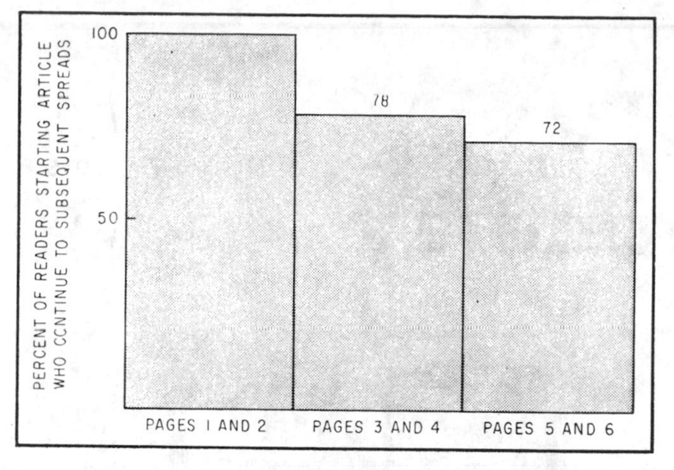

Fig. 1—Mean percentage of readers continuing a six-page engineering article from first spread to subsequent spreads.

A good engineering article must first arrest the reader's attention, then induce the reader to read it, and finally convey the maximum of unambiguous information with the minimum of effort on the reader's part.

The main features of an engineering article are: headline, opening photograph, bankhead or deck, lead paragraphs and the subheadlines. Fig. 2 is a flow chart illustrating these features of a typical engineering article. Opposite each item is a summary of what it should do for the reader.

Display Elements

The headline and opening photograph can do the most to attract the reader. The headline should be lively and should appeal to the largest group of readers possible. A headline should also be concise. Headline writing is a fine art that takes years to learn. Do not be discouraged if an editor decides to change your headline.

A good opening photo should be of high photographic quality, a glossy print of course, for best halftone reproduction. It should be well composed with a single center of interest. Furthermore, it should have a person in it and should set the mood of the article.

An article on radar front-end design, for example, might show the author tuning up the local oscillator. Such a photo projects the reader into the story, establishes a bond between reader and author. A good opening photo greatly increases the readership research on percentage readership of illustrated versus unillustrated articles. Twenty-nine articles were studied: 12 illustrated, 17 unillustrated.

Many publications use a deck or bankhead between the headline and the text of the story. Such an element functions to get the reader into the story. It should not be merely a summary of the article and certainly not a set of quantitative specifications. Think, rather, of how you would go about talking a busy colleague into reading your article, stating your point in 50 words or less. Specifications can best be presented in a prominent box or table.

The first 200 words or so of an article should work closely with the bankhead. Such a lead or introduction should avoid specialized technical jargon. It should be readily intelligible to any worker in the field. The lead should state the problem as it existed, how it was solved, and the most important technical implications of this solution. Avoid long historical essays. Prior work can be credited most efficiently in superscript references.

The main text should proceed to reason deductively, from the general to the specific. It should be outlined into subsections of 200 words or so, each set off with an appropriate short subheadline.

In modern engineering articles the text, display elements and illustrations work closely together. In moving from the general to the specific, the illustrations proceed from a system diagram to more specific mechanical drawings or electrical schematics. Fig. 4 illustrates how construction of a piece of electronic equipment is illustrated.

Outlining An Article

For example, consider an article about a new piece of equipment. Of course, all good articles are not centered about equipment development. Today perhaps the most significant articles deal with radically new components, new theoretical bases for design, new production techniques and new systems applications.

The article should begin with a systems description based on a functional block diagram. In such a diagram, a block may represent a piece of equipment or a single stage within a piece of equipment. The blocks are linked with single lines indicating signal or information flow. Text in a block diagram should be contained within the blocks. Schematic symbols should not be used on block diagrams.

The article should then consider each circuit or important subunit of equipment individually, proceeding in the usual direction of signal flow—from input to output.

Each unique circuit should be presented schematically and its novel features pointed out in text. To make a

Using Tv in Astronomy

HEADLINE—ATTRACTS LARGEST POSSIBLE NUMBER OF READERS —"INTEREST"

OPENING PHOTO – PROJECTS READER EMOTIONALLY INTO ARTICLE—"CURIOSITY"

Modern electronic techniques aid astronomers in locating variable stars. Unique flying-spot-type closed-circuit television system compares two photographs and displays any difference between them.

BANKHEAD-TELLS READER WHY HE SHOULD READ ARTICLE—"MOTIVATION"

OF THE VAST NUMBER of stars visible on a clear night, certain types are known as variable stars. This means the amount of light that they radiate is variable. Some of these stars show irregular variation, but most are of a certain periodic type that have periods of a day or more. This periodic type of star has become an important tool in astronomy as the mean energy radiated in a certain band of wavelengths can be derived from the period of light variation. The apparent mean brightness, measured by the observer on earth, will give the distance to the star.

LEAD PARAGRAPHS—EXPLAIN PROBLEM, ITS SOLUTION AND SIGNIFICANCE-"INTRODUCTION"

Detecting Variable Stars

SUBHEADLINES-SUSTAIN INTEREST THROUGHOUT ARTICLE —"CONTINUITY"

Fig. 2—Main elements of an engineering article and what each element does for the reader.

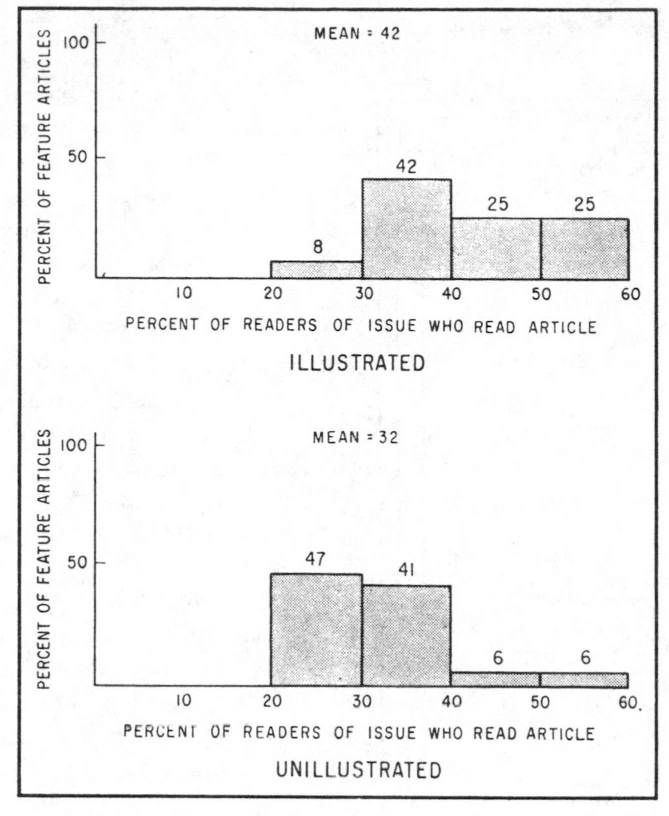

Fig. 3—Relative mean readership of articles with and without a prominent introductory photograph.

circuit schematic more meaningful, component values should be given. Circuits should be simplified by omitting, where practicable, power supplies, heater circuits, reference designations, terminal boards, test points, fittings and connectors. Remember, circuits presented should be unique. There is no merit in publishing circuits for the sake of circuits. Symbols should be communications preferred types from ASA-Y32.2 from which the familiar Mil 15A derives.

After circuit descriptions can come experimental test results with applicable charts and graphs. On graphs, write the ordinate legend with units at the left-from bottom to top. Write the abcissa legend with units at the bottom from left to right.

Use as few grid values as possible. Use stubs instead of full grid lines. Omit plotting points and avoid legends on the face of a graph. Avoid unusual line construction and do not attempt to convey information by varying the line weight.

An equipment article may finish up with a physical description of the equipment. Mechanical drawings and top and bottom close-up chassis photos may be useful. Avoid callouts, or other inked lettering on photographs. The halftone screens used by most magazines are too coarse to reproduce them.

Material that must be included, but which adds little to the reader's information, should go to the end: kudos to co-workers, praise to the boss and credit to armed forces contract numbers belong here. Of course this material should be stated as briefly as possible.

References and a bibliography complete the article. References are works specifically cited by superscript numerals in text. A bibliography is merely a list of works consulted in the course of the project. For electronics publications, both should cite works using the style adapted by the publications of the IRE.

As to writing style, it is well to keep sentences short and use the most familiar word that will convey meaning. There are several excellent books on expository writing. See, for example, Gunning, "The Technique of Clear Writing." A word about captions: keep them short and avoid the obvious. It is superfluous to say: "block diagram" under a perfectly obvious block diagram.

"Do's" of Article Writing

Here are some pointers that will help you in getting your articles into engineering magazines.

1) Carefully match your article to the magazine to which you submit it. Consider whether the readership of the magazine is the group you want to reach with your article.
2) Before preparing an article, contact the editor. Tell him of your plans, but don't expect him to buy a cat in a bag. An experienced editor can tell a lot

from a 200-word summary, a three-level outline of the text and a list of the illustrations you propose to use. If your article is to be based on work already described in internal engineering reports, send along a copy.

3) Be brief with your article. Engineering magazines can seldom use articles over 3,000 words in length. Estimate 250 words a double-spaced typewritten manuscript page and about two manuscript pages a printed page.

4) See that your article is adequately but not over illustrated. Estimate two and one-half illustrations per printed page. Neat pencil sketches are perfectly acceptable, since established magazines maintain drafting rooms to prepare their illustrations for the engraver. Be careful of copies made by a process duplicating equipment. Sometimes they are not completely legible.

5) Offer your article exclusively to one publication. Don't play the field. Legitimate publications pay for articles, usually upon acceptance. However, be sure to indicate that your contribution is an original, exclusive article and not a broadcast handout. Otherwise, it may get short circuited to the wastebasket.

6) Be patient with the editor. A top-flight magazine may have an article reviewed by several editors, perhaps even by outside experts before deciding to buy. Don't expect an offhand or precipitate decision to accept or reject.

7) Cooperate with the editor in furnishing additional details, photos, component values or explanations. Your cooperation will help hasten publication.

8) Return galley proofs promptly and make only essential technical corrections. Do not argue matters of style with the editor. You know your subject, but the editor knows his audience.

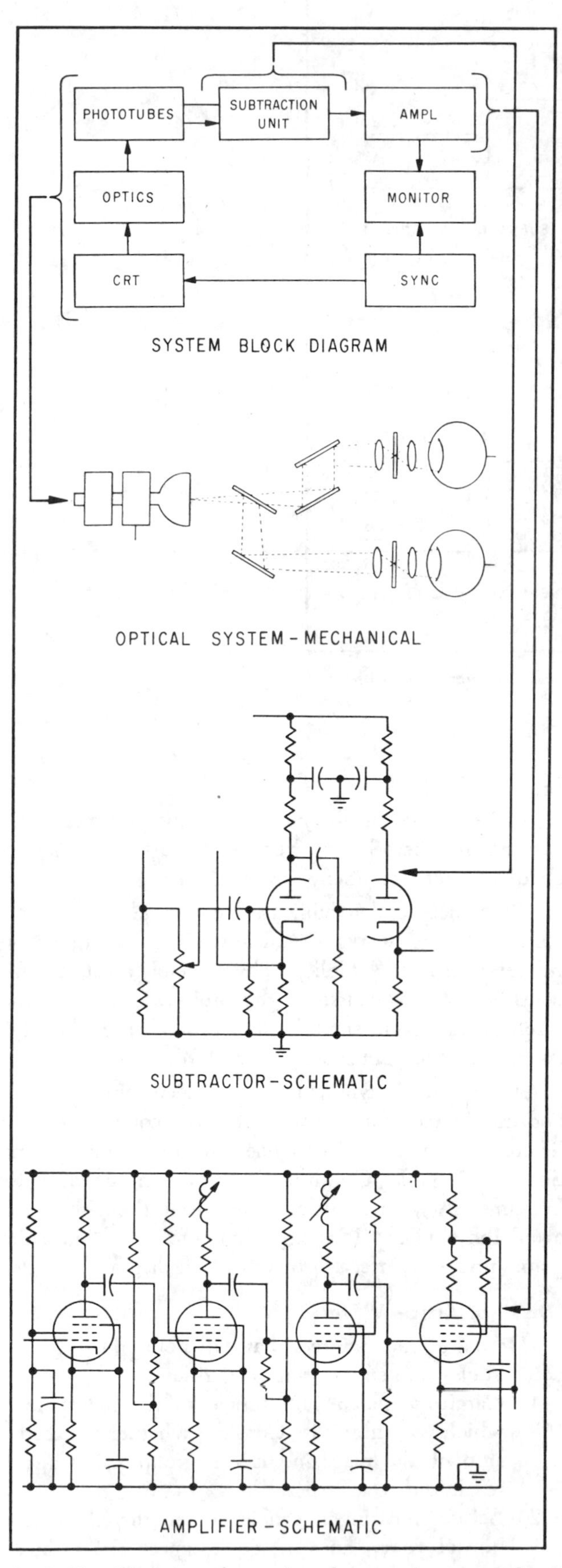

Fig. 4—How illustrations show points of engineering interest for the reader.

"Don'ts" of Article Writing

Here are one or two don'ts.

1) Don't attempt to withdraw an article for security reasons after acceptance. Publications assume you act in good faith, submitting only copy previously approved for publication.

2) Don't pay fees for placement, publication or for cuts. Legitimate publications seek out news. A good article will sell itself.

3) Don't discuss advertising with editors. The editorial and advertising staffs of legitimate publications operate completely independently.

Remember good articles are the lifeblood of an engineering publication and a valuable contribution to the profession. The editors are ready and eager to work with you in every way possible.

Part IV
Some Research Results

IF WRITING PAPERS and articles is part art and part science, so is the analysis of this pursuit. The section on rhetoric handles the "art" analysis; this one is designed to show what has been learned about technical communication using the methods of "science."

As Andersson tells us in the first article, we need useful research results because "measurement is the key to the future in communication." He suggests several vital parameters requiring measurement, but he is quick to point out that we must cautiously analyze the results and methodology of experiments that attempt to measure these parameters. With this caution in mind, we are better prepared to judge the "usefulness" of the research results presented in this section.

Davis reports on three experiments involving sentence length, errors in agreement, shifts in viewpoint, and misspellings. His results indicate that errors in expression reduce the effectiveness of communication; in addition, they cause the reader to question the writer's competency. Bram considers other factors contributing to readability. He differs from the originators of popular readability formulas and from Davis when he concludes that sentence length is not one of the predominant factors in readability; rather, he suggests that sentence structure has the strongest influence. Klare provides a broader look at readability research and results. He offers twelve guidelines for increasing readability and reminds us that "avoiding a mismatch between the reader's ability and the difficulty of writing is what readable writing is all about." Adding to our knowledge of how to avoid these mismatches, Lauer does a content analysis of various texts. His findings are that "writing is technical in proportion to its use of a specialized vocabulary," and that we must, therefore, adjust an article's language to the abilities of the intended audience.

Other sources of "noise" are then examined. Indeed, it's truly a sophisticated author who chooses the publication for an article on the basis of typography; yet Carte's research indicates that this does influence an article's readability. Accordingly, he provides guidelines on this often overlooked aspect of the publication process. Davis reports the results of an experiment that complements those he reported on earlier in this section. He explains that poor typing, poorly made corrections, and sloppy reproduction have an effect upon how a paper or article is received.

The last two articles in this part report on research and experiments designed to test several well-known and accepted "rules" of technical writing. McKee's research challenges one almost universally touted practice: "the value of the sentence outline as a necessary step in report writing is mainly a myth of high school and college teachers of English composition." Similarly, Bloom's and Goldstein's investigation of other accepted procedures produces some expected and some not-so-expected results.

All told, a potpourri of useful information for papers and articles has been generated by the world of research. And, as Andersson notes, the future of technical communication will be bright if there is a continuing search for those effective and useful measurement techniques that will result in technical writing that is more clear, more precise, and more readable.

Methods of Measuring Communication Results

ULF-L ANDERSSON

Communication Education Center AB
Stockholm, Sweden

ABSTRACT

Measurement is the key to the future in communication. To be able to develop better methods of communicating difficult material in shorter time, we need methods to measure the result. This article discusses what measuring methods can be used and which measuring instruments are needed.

Some of us have had a long and varied experience with communication work. But this is not enough. Experience and standard solutions on communication problems can be wrong. And besides that, experience is a very weak argument against the boss when he thinks he knows the answer. We need methods to measure the function of the information both as a whole and in details. There are basically two types of measurements we can use:

1. simple and inexpensive routine measurements and
2. special measurements for developing new forms of information or for refining old methods.

It is often said that it is impossible to get 100 per cent true measurements when we deal with human communication. This may be true—but we are not primarily interested in "true measurements," that is, in exact figures, but in useful results. We need results as precise as possible so that we can build our decisions on them. Even if a measurement is not right, it is not dangerous as long as we know the magnitude of the fault, that is, as long as we can base our decisions on the result with confidence.

What Shall We Measure?

Measurements are nothing in themselves. Since measurements guide us in decision making, we must be careful to measure only those things that are important for the function of the information.

I don't think any of my readers will have any objections up to this point. But when in our daily communication work we use measurements, we are likely to forget all these self-evident bases.

Let us look at an example. When we in Sweden were going to change from left-hand traffic to right-hand traffic a special governmental office was created with the initials HTK. Among the responsibilities this office had was the information about the changing operation.

Six months before the changing day, the HTK office published a press release saying that "the Swedish people were not yet prepared." They drew that conclusion from a questionnaire the office had sent out the results of which showed that it was obvious that very few people in Sweden knew the meaning of the letters HTK! The "researchers" who did this "measurement" applied what they had learned in their previous occupation in advertising. They did indeed make a measurement—but they were not aware of what it was that they measured. When you put out a product on a market, it should be of interest to find out how many people know the product or company name. But in this case the problem was to determine if people were prepared for the right-hand traffic, not if they were able to decode the letters HTK.

Another example which may not be as exotic as left-hand traffic follows. Today we have many formulas, such as the Fog Index, devised to measure clearness in communication. What answers do we get when we use these formulas? We count the number of words, the word length, the number of affixes, how many personal references there are, and so on. And then we put these figures in a formula and get a result, which is said to be a measurement of how difficult the text is.

But still, we don't know anything about the difficulty of that text because difficulty is not primarily dependent on the number of words per sentence, and so on, but is dependent on:

1. the idea-chain in a text,
2. the structure of the text (paragraphs and sentences), and
3. the concepts we use for explaining the idea.

All of these factors are very difficult to measure; we cannot just depend on word count.

Reprinted with permission from *J. Tech. Writing and Commun.*, vol. 1, pp. 109-113, Apr. 1971.

The results we get from yardstick formulas are not only inexact but so inexact that they are not useful. We measure totally irrelevant features in the text. We could as well count birds flying by and from that figure, by special calculations, predict a space trip to the moon.

Vital Decisions

When we construct information, we have many decisions to make. Some of them we are not aware of—others we feel as a heavy burden. To be able to make the decision we ourselves need information. Let us say we are giving an explanation of a "device." What decisions are we then forced to make?

We must ask these questions: Is the explanation good enough for its purpose? Will the reader be able to do what he needs to do after he has read our explanation? Can we measure this? My answer is yes. But not yet, in most cases, for a reasonable amount of money.

For one type of information, the instruction, in many cases it is rather easy to measure the function. The reader can show by his action that he has understood. It is much more difficult when we deal with some form of descriptive information, that is, background material which the reader needs to be able to think out for himself.

Noise From the Test Situation

Testing such information is, however, still very much a hazard. We don't know how much noise we will get from the test situation itself and the reader will likely behave differently when he finds himself observed than when he is alone.

On rare occasions it is possible to lower this noise. For example, in Sweden some tests have been done on tape-recorded instructions used for training industrial workers. To the instruction-tape recorders have been connected chart recorders which record all stops and reverse plays. The user of the tape instruction will not notice this second recording, but the researcher can afterwards analyze the charts from several users of the taped instruction and find out if the difficulties (indicated by the stops and reverse plays) are on the same places in the instruction. Different versions of the instruction can thus be tested and the one with the best overall function can be selected. Special difficult passages can then be rewritten and tested again.

An example of a test method which is not so sophisticated is the "unprepared-read-it-aloud" test which everyone can very easily use. It is based on the observation that a person reading a text for the first time, will stop and reread a difficult word. The problems arise not from the difficult words but from the fact that the idea being expressed is difficult. These interruptions indicate the amount of work being done by the brain. In some cases, however, they may arise from outside disturbances. This test can at least give some hints on what may be wrong in a text.

The Need for Calibrated Instruments

I have given a few examples of test and measurement methods. There are others—but not very many. What we badly need is many more measuring methods, more "instruments" to use, and more calibration work where we can compare the instrument readings with the actual turn out. And, too, we need more knowledge to be able to interpret the figures.

To get more than vague hints on what is wrong with a piece of information, we need measurements for vital parameters. We need to measure:

1. Complexity (chapter structure, paragraph structure, sentence structure, concept complexity, picture complexity)
2. Abstractness (subject abstractness, concept abstractness)
3. New concepts (how many, how difficult, how important for the needed effect)
4. Need for background knowledge (being able to read, being able to calculate the transfer function for a servo system)
5. Work load during the reading (energy consumption)
6. Time used (even that is difficult to measure!)
7. Sensitivity to outside noise (old habits)
8. Longtime memory content

Objective Measurements

In all areas of human activity the possibility of objective measurements has contributed very much to the evolution. What types of objective measurements can we communicators use?

We need to notice people's reactions to information. We can use physical measurements of sweat, blood pressure, heart beat, breath, eye movement, and so on. We can try to record the behavior of people consuming our information. We can ask questions and try to draw conclusions from the answers. But don't forget that in most cases the goal for the information is that people shall be able to do things other than answer questions. It can be misleading to depend too much on questions and answers.

Even if the only instruments for observing results are ourselves, this is better than nothing. But we must put our observations down in figures to make them easier to evaluate—so they can be trusted. Putting observations into figures is as if instead of saying to your wife "you are almost everything for me," you say, "you are 0.96 for me."

But instruments must be calibrated. Even we ourselves must be calibrated if our observations are to be of any use. And this is another big problem, I don't think that NBS has any standards we can use.

What is calibration? It is a method to refer the measured figures back to a single standard; either it is 36 barleycorns taken from the middle of the ear, or it is 1,650,763.73 wavelengths in vacuum corresponding to the transition between the levels $2p_{10}$ and $5d_5$ of the Krypton-86 atom.

For us as communicators, it is very difficult to find the standards. But it is a problem we must solve. Our standards should be primary, not secondary standards on a very low level. We cannot put up one instruction book as a standard and then try to get all other books to measure up to it. And it is important that the measurement methods don't limit our freedom to find new ways to communicate.

In the bright future of communication we will have test laboratories at all information departments. In these test laboratories more complicated measurements on information will be done. Here also the individual communicator's instruments can be calibrated, for example, is Bob really 87% when he thinks he is. We can all help each other by searching for effective and usable measurement methods.

Does Expression Make a Difference?

Richard M. Davis

THREE EXPERIMENTS *tested the effect of departure from standard in expression (sentence length, errors in agreement, shifts in viewpoint, and misspellings) upon the effectiveness of a written technical communication. In the first experiment, the departures had no apparent effect. In the second, test materials were modified to allow subjects additional working time and significant effects favoring the unaltered forms of the variables were found. In the third experiment, subjects were told that they would not be held accountable for performance and the judgments made, and even stronger effects favoring the unaltered form of the variables were found.*

We all have opinions about just what makes an effective technical communication, and on many matters concerning style and expression people often take some pretty extreme positions. Some technical people maintain that they judge an author only by the facts and ideas that he presents; as long as these come through, they are the only reasonable basis for judgment. I will not comment on the obvious paradox here, but I might note that bright young PhD's in science and engineering sometimes take this position. Others believe that precise expression with conventional grammar, syntax, and mechanics are essential to an effective written communication. The form of the expression and everything about the communication should follow local and general conventions to avoid detracting from its effectiveness.

I suspect that most technical communicators lean generally toward the latter school of thought—especially inasmuch as they justify their existence, in some measure at least, by their ability to follow the general conventions of written technical English. Just which of these positions (if either) is right, and how right it is, of course, is pretty much a matter of opinion. There isn't much direct experimental evidence to go by; not on matters of style, expression, grammar, and mechanics anyway.

In this article I describe the results of three experiments intended to measure the effects of certain departures from good practice (or what most of us would probably call good practice) in expression upon the effectiveness of a written technical message and to suggest something of what they might mean. They are part of a continuing series in which the effects of variations in content, organization, and presentation are being tested. Because they involved the testing of something over 1850 subjects in 15 different audiences, it is not possible to present full detail on all points. This is available in references 1 to 3.

EXPERIMENTAL APPROACH

As has been indicated in previous articles in this journal,[4] the approach is quite straightforward. In brief, a communication is written to meet the demands of an assumed situation. Controlled alterations are then made in the communication to produce several versions of the same message. A test is developed to measure the degree to which the purposes of the communication are attained. Each member of a homogeneous audience is assigned (on a random basis) one version of the communication. The test subjects (readers) read their assigned versions of the communication, and then all are given the same test. Appropriate statistical analyses of the scores are made, and differences in performance on the test are assumed to be caused by the variables tested—the controlled variations in the message being the only difference in the test materials and administration. To the degree that other situations are similar to the one assumed, messages organized and presented in the manner found to be most effective in the experiment should be more effective in attaining their purposes than those organized and presented in the other forms tested. Comparison of the results of the tests of definably different groups of test subjects should indicate the most effective forms for differing groups of potential readers.

TEST MATERIALS

The reading passage and criterion test used in these experiments were those used in previous experiments in this series. Since they appeared to make the intended discriminations in the earlier experiments, their continued use permitted direct comparison of the results obtained in different experiments.

The reading passage is a 3619-word description of a simple machine. The machine has not been produced commercially and nothing has been written about it, so the test subjects could have no prior knowledge of it.

The criterion is composed of four measures: comprehension, reading time, judgment of the author's competence as a writer, and judgment of the author's knowledge of his subject matter. Gross failure of a communication in any one of these areas might wholly offset its effectiveness in others and nullify the effectiveness of the whole. Certainly other measures could have been taken and might well be used in other experiments with other sets of assumed conditions.

EXPERIMENT 4

In the first of the three experiments concerned, three variables in the expression were tested and seven test audiences were used.

Variables. The probable size of available test audiences suggested that three test variables be used with two conditions of each. They should be elements in the expression of the reading passage about which the readers could reasonably be assumed to have expectancies whose violation would be evident to them. Further, the variables should lend themselves to objective measurement. Those chosen for this experiment were sentence length, agreement of subject and verb, and shifts in point of view.

1. *Sentence Length.* Sentence length is commonly accounted one of the prime measurable factors affecting reading ease. It is an element in most readability formulas and is usually a matter of concern in courses in technical writing. The base reading passage contains 184 sentences (19.67 words per sentence). The alternate form was made by running sentences together to form compound sentences and by incorporating some sentences as modifiers or subordinate clauses in others. The information contained was not otherwise rearranged and no information was added or deleted. The alternate form contains 98 sentences (36.9 words per sentence). Thus the information contained in the two versions is the same, but sentence length in one is almost twice that of the other.

2. *Subject-Verb Agreement.* Errors in agreement between subject and verb are certainly among the most readily apparent departures from standard in expression. Sometimes it seems that readers who can recognize no other violation will recognize these. Twenty errors in agreement were introduced into the text as the alternate treatment of this variable. They were intended to be the sort of errors most commonly found in poor writing—logical or believable errors if there can be said to be any such thing.

3. *Shifts in Point of View.* Books on technical writing and course instructors often recommend consistency in tense, mood, and person. Unnecessary shifts in

Based on a paper presented by the author at the STC 22nd International Technical Communication Conference, Anaheim, Calif., May 16, 1975.

point of view are felt to have a jarring effect on the reader and require that he reorient himself to the viewpoint of the author. In this variable, 44 shifts in person, tense, and mood were introduced into the passage in no particular pattern and with no apparent logic.

Experimental Design. As three variables were tested and there were two conditions on each variable, eight versions of the test passage were used in a 2 × 2 × 2 factorial experiment. One (the base passage) contained the unaltered form of all three variables, one contained the altered form of all three, and each of the remaining six versions contained one of the possible combinations of altered and unaltered forms of the three variables.

Audiences. Seven audiences (approximately 750 subjects) drawn from the student bodies of six institutions of higher learning were tested. The participating institutions were the Air Force Institute of Technology School of Engineering and School of Logistics, Colorado State University, University of Cincinnati, University of Western Australia, and Wright State University.

In the three experiments discussed in this article, all subjects were native speakers of English, they were unfamiliar with the subject matter of the reading passage before it was given them, and none knew anything about the author. Audience contrasts in this experiment were made on the basis of sex, age, intelligence, technical inclination, experience with written technical communication, instruction in technical writing, degree of technical training, and variety of English spoken. Details of the contrasts made are presented in the report on the experiment.

Results. In this experiment, only one effect was found at the 0.01 level of probability and seven were found at the 0.05 level of probability. The directions of the differences found were varied—in some cases the base form of the variable appeared to be the more effective, and in some the altered form appeared the more effective. Because of the relatively small number of statistically significant differences and their apparent weakness, they may be viewed as arising by chance. So, in effect, it appears that by the measures taken and by the criterion used, the variables did not affect the effectiveness of the message in the seven audiences tested. This came as a considerable surprise.

The obvious first reaction might well be that the experimental method is not valid—that the methods used will not detect real differences in performance or the judgments. But in the three previous experiments in which some 2300 subjects were tested, the same procedures appeared to make the appropriate discriminations. Several other possible reasons for the absence of effect by the variables suggest themselves, two being of particular interest here.

First, the subjects had to work quickly to get through the reading material and complete the test in the time available. In fact, they had to work more quickly than did the subjects in the earlier experiments and they had little time to review what they had read. It seemed possible that they moved so quickly and that their attention was so directed at the content of the technical message that they simply did not react to the expression as they might have otherwise. A similar experiment with shorter versions of the message would help to determine whether this was the case.

Second, because the subjects were not told the purpose of the test, differences in performance and judgments may have been minimized. Some of the less sophisticated subjects may have been reluctant to rate the author's knowledge of his subject or his competence as a writer as low as they felt it to be. The spread of ratings within cells was considerable and even in those cells with comparatively low means, there were usually several high ratings. These may have been sufficient to cause a leveling effect and to obscure what would otherwise have been significant effects in the statistical analysis. To determine whether this was the case, an experiment might be run in which the subjects were told that they were participating in an experiment and that they would not be held accountable for their performance. This might reduce any inhibition on the part of the subjects to grade the author as low as they felt he deserved and, perhaps, produce more consistent judgments and additional statistically significant effects.

EXPERIMENT 6

The first of these possibilities was tested in the sixth experiment in this series. In it, the base reading passage was reduced from 3619 words to 2341 words—a reduction of 35 percent. Thus the time the subjects spent reading it would be reduced and time available to consider it and answer the test questions increased (total testing time in each experiment was 50 minutes). Apart from this, the test materials and the procedures used were exactly the same as those used in Experiment 4.

Test Variables. The probable size of available test audiences suggested that four variables in the expression of the message might be used. Three of them—sentence length, agreement of subject and verb, and shifts in point of view—were the variables used in Experiment 4 and were presented exactly as they had been in that experiment. The fourth (added) variable was spelling.

Twenty-two errors in spelling were introduced into the reading passage as the alternate form of this variable. They were chosen as common errors which should be recognized easily, and included such departures as *handel*, *incorect*, *apear*, *aproximately*, *leaver*, and *responce*. It seemed that readers who might be uncertain of their judgment on some of the other variables could be certain that these spellings were incorrect.

Audiences. The four audiences (approximately 825 subjects) tested were drawn from the student bodies of the Air Force Academy, the Air Force Institute of Technology, Southern Alberta Institute of Technology, and the University of New South Wales.

Audience contrasts were made on the basis of sex, age, intelligence, technical inclination, experience with written communications, technical training, and variety of English spoken.

Results. In this experiment, three effects were found at the 0.01 level of probability and nine were found at the 0.05 level of probability. In eight of the effects, the unaltered form of the variables concerned appeared to be more effective than the altered form, three effects were interactions which could not be interpreted clearly, and only one effect indicated that the altered form of a variable was more effective than the unaltered form (this would be expected by pure chance). All variables were involved in one or more significant effects, effects were found in all audiences tested and by each of the four criteria (comprehension, reading time, and the two judgments). The effects found were more numerous and generally stronger than those found in Experiment 4, and they supported the unaltered forms of the variables as more effective than the altered forms. Evidently the shortening of the reading passage and the increase in time available to react to it produced—to some extent at least—the intended effect (reading time in directly comparable audiences was reduced by 5.21 minutes). Despite this, the effects were neither as numerous nor as clear-cut as might have been hoped.

EXPERIMENT 9

In the ninth experiment in this series, a primary intention was to test the second possibility just postulated—the possibility that subjects' performance and judgments may have been inhibited because they had been told nothing about the purpose of the test administration.

Variables. Two of the variables used are of concern here: one in the expression of the message and the other in the instructions given the subjects.

The variable in the message was simply a combination of the altered forms of the four variables used in Experiment 6. The base passage contained the unaltered form of the variables, and the contrasted form contained the longer sentences, errors of agreement, shifts in point of view, and misspellings.

The variable in the instructions given the subjects is of particular interest. Two of the four audiences tested were told nothing about the purpose of the test administration. Like the subjects in the previous experiments, some may have assumed that their scores would somehow affect their standing in the courses in which they were tested or their classification at the institutions in which they were enrolled, some may have guessed that they were being used as subjects in an experiment (but they could not be certain), and some may have guessed any of a dozen other possible reasons for the test administration. These were referred to as the *uninformed audiences;* they had been told nothing about the purpose of the test administration.

The other two audiences were given exactly the same test instructions except for the addition of a statement that their performance on the test would not affect their course grades or their classification at their school. They were told that data was simply being gathered about them as a group. Thus there should have been a more relaxed general attitude on the part of these subjects and less concern about individual performance. The intention, of course, was to determine what effect this might have upon performance and the judgments made. These were referred to as the *informed audiences*.

Audiences. The four audiences (approximately 275 subjects) tested were drawn from the student bodies at the Air Force Institute of Technology School of Engineering and from Central State University. The groups tested at the two schools were contrasted on the basis of age, sex, intelligence, technical inclination, technical training, and experience with written technical material; but the contrast of primary interest is between the two audiences tested at each school—the *informed* versus the *uninformed* audience.

Results. Two effects caused by the variable in the expression were found at the 0.01 level of probability and two were found at the 0.05 level of probability. As only one variable in the expression was used, many fewer effects were possible than in the other two experiments described. The proportion of significant effects found was higher than in the other experiments, the effects found were stronger, and each of them indicated that the unaltered form of the expression was more effective than the altered form. Further, three of the four effects were found in the *informed* audiences, and both of the effects at the 0.01 level of probability were in the *informed* audiences. Even where the difference in scores, times, or judgments was not sufficient to be considered statistically significant, mean differences almost uniformly favored the unaltered form of the variable. Evidently the statement in the test instructions did what it was intended to do as more and stronger effects were found in the *informed* audiences than in the exactly comparable *uninformed* audiences—and they were all in the direction that we might have expected. **Bad expression reduced the effectiveness of the communication.**

CONCLUSION

What does all of this mean for us? What use is it? Didn't we already know that departure from standard in expression could reduce the effectiveness of a written communication? Well, maybe. Or, anyway, most of us have a feeling that they do and have advocated its validity in discussions with bosses, deans, and potential customers as partial justification of the value of our services. But as writers, editors, and teachers, we have an obvious vested interest in that view and our objectivity may reasonably be questioned.

So here is some experimental evidence—tentative and limited though it may be—developed through controlled and objective experimentation in which results were analyzed by accepted statistical measures. It indicates that departures in expression may indeed impair the effectiveness of a written communication. They may reduce comprehension, increase reading time, and reduce the reader's judgment of the author's competence as a writer and his credibility. This last effect, the reduction of credibility, should be a very telling argument with even the staunchest or the most obtuse believer that only the facts presented—not the expression—determine the effectiveness of his written technical message. Even with objective scientific and technical readers like himself, departures from standard in expression may well reduce the effectiveness of the communication.

The experiments reported here are, of course, very limited; and the variables tested are only a few of the many concerned in the expression of a written communication. But at least they are a start. Obviously, many other variables in the expression of this message and others in the content, organization, and presentation might be tested, as might variations in other kinds of messages produced under other sets of conditions. I intend to continue the experiments. If any of you want to participate—fine. I can use all of the help I can get. And if you develop new approaches or more effective experimental methods, if you carry out your own experiments and derive better results—fine. I'm all for you and will be glad to help out in any way I can.

REFERENCES

1. *Effective Technical Communications, Expression—Experiment I*, TR 69-5, Air Force Institute of Technology, Wright-Patterson AFB, Ohio, 1969. (AD 691210)
2. *Effective Technical Communications, Expression—Experiment II*, TR 72-3, Air Force Institute of Technology, Wright-Patterson AFB, Ohio, 1972. (AD 741753)
3. *Effective Technical Communications, Expression—Copy Preparation—Motivation*, TR 74-7, Air Force Institute of Technology, Wright-Patterson AFB, Ohio, 1974. (AD 784226)
4. Richard M. Davis, "Effective Mechanical Description: Experiment 3," *Technical Communication*, 14(3):28 (1967).

Sentence Construction in Scientific and Engineering Texts

V. A. BRAM

Abstract—Readability of professional technical material was assessed by scientists and engineers. Less than 33 percent proved "easy" to read and understand. Further investigation showed that sentence structure had the greatest effect on readability.

IT HAS been aptly said that

> an author with unique and indispensable information has his readers at his mercy,

and that his work is often

> like a quarry rich in precious ore but hard to work. 'Let those who want the ore,' the author seems to say, 'dig for it' [1].

This was, in fact, what readers had to do when they assessed the readability of scientific and engineering texts recently. The deeper they had to dig, the more difficult they found the text. Less than 33 percent of the material tested proved easy to read, which is a severe indictment of communication skills among professional people today.

In order to establish what it was that made individual texts easy or difficult, readability scores were correlated with scores for a number of other factors: familiarity with the subject matter, understanding of content, vocabulary, and sentence structure. It was the last of these that had the greatest influence on assessments of readability, and so proved worthy of further consideration.

Sentence Length

Authors who produce books on technical writing agree that sentence length is an important influence on readability, but they disagree about the optimum length. A. E. Derbyshire says,

> Very easily read sentences are not more than 10 to 15 words long [2].

Robert Gunning states,

> Nearly all reading material that wins a large audience has an average sentence length of less than 20 words [3].

While Graves and Hodge stress,

> A sentence may be as long as the writer pleases provided that he confines it to a single connected range of ideas [4].

Reprinted with permission from *The Communicator of Scientific and Technical Information*, no. 34, p. 3, Jan. 1978; copyright 1978 by the Institute of Scientific and Technical Communicators, Ltd., Hatfield, Herts., England.

The author is with the Department of Liberal Studies and Communication, South Glamorgan Institute of Higher Education, Cardiff, Wales, United Kingdom.

In order to test these ideas, the average length of sentences in easy, average, and difficult texts was found. The results are shown in Table I; the scale ranges from 0, easiest, to 10, hardest.

Only four texts had an average sentence length of 20 or fewer words, which was Gunning's target for easy reading. All four were easy to read, but so too were a number of others with longer sentences. However, Gunning himself concedes that this figure should be only a rough guide.

Although it was generally true that easy texts had shorter sentences than difficult ones, there were exceptions to this rule. For example, passage 1 had readability scores of 1.9 and 2.6, indicating easy reading, yet its average sentence length was 32 words. The following extract from this paragraph is even longer (59 words):

> Yet it is only in the past ten years or so that oceanographers have begun to make a systematic study of the surface organisms, and it is possible at present to give no more than a description of some of the animals found at the surface, as nothing is known of their physiology and little of their ecological relationships.

On the other hand, passage 19 had readability scores of 8.2 and 6.0, indicating difficult reading, and an average sentence length of 23 words:

> Similarly, a specific inhibition can provide a clue to a potentially rate limiting step in a metabolic pathway. As always, when studying dose-dependent kinetics, one relies on absolute accuracy of the assay results which probably range through several orders of magnitude in those studies.

Clearly, sentence length alone is not a reliable guide to the readability of a text. So the structure of the sentence was then examined.

Sentence Construction

In the following sections, only sentences from easy and difficult texts are examined. Those from average texts are omitted. Although this oversimplifies the picture, it provides a clearer distinction between what is and what is not an effective sentence. It in no way invalidates the discussion or subsequent conclusions.

Easy Texts

Texts with readability scores between 0 and 4, which were classed as easy reading, came from a cross section of scientific and engineering prose. They were extracts from an undergraduate (UG) textbook, popular journals, third-year undergraduate reports, and industrial reports. Since they all dealt with completely different topics, and were intended for different audiences, they had little in common so far as content or aim

Reprinted from *IEEE Trans. Prof. Commun.*, vol. PC-21, pp. 162-164, Dec. 1978. (Reprinted with permission from *The Communicator of Scientific and Technical Information*, no. 34, p. 3, Jan. 1978; Copyright © 1978 by the Institute of Scientific and Technical Communicators, Ltd., Hatfield, Herts, England.)

TABLE I
READABILITY OF SCIENTIFIC AND ENGINEERING MATERIAL [a]

Passage Number		Source	Average Sentence Length	Readability Score[b] Specialists	Nonspecialists
1	Easy	Popular journal	32	1.9	2.6
2		Popular journal	19	0.65	0.72
7		UG textbook	25	2.3	3.4
16		UG project	22	2.1	2.2
22		Industrial report	20	2.6	3.8
23		Industrial report	19	2.3	3.9
24		Industrial report	15	1.6	3.6
4	Average	Popular journal	31	4.7	4.1
5		UG textbook	27	4.6	3.7
6		UG textbook	28	4.6	6.7
11		Research journal	29	5.6	4.9
12		Research journal	27	4.2	6.6
14		M.S. thesis	26	5.6	5.5
21		Industrial report	26	6.5	4.4
9	Difficult	UG textbook	32	8.0	7.3
15		Ph.D. thesis	43	6.3	7.5
19		Industrial report	23	8.2	6.0
20		Industrial report	33	6.2	6.6

[a]There was no correlation between readability and sentence structure in six of the 24 texts.
[b]Ranges from 0, easiest, to 10, hardest.

was concerned that would account for their uniformly low scores.

Generally speaking, there were two distinct types of sentences in easy texts. First, those consisting of *a single statement*, and possibly additional information that qualified that statement. Second, those that contained *two or more statements*, of which either one or both could be qualified, too.

The majority of sentences in easy texts contained a single statement with no further information of any kind. For example,

> From place to place on the body the skin is a study in contrasts,

and

> This layer can be removed by the action of ions.

There were longer examples of this type of construction too:

> In effect the foundry has had to carry out a closely controlled chemical reaction in very adverse conditions of humidity and temperature.

There is only one verb in each of these sentences, so there can be only one statement. Most of the remaining sentences contained an additional clause, which gave extra information about the main idea (all additional clauses are italicized):

> Man's most obvious sign of age is the dry, wrinkled, flaccid skin *that marks his late years*,

and

> H.C.L. off-gases from the production plant contain entrained material *which is deposited in the recovery and neutralization systems.*

Both clauses give information about the word directly preceding them, hence enlarging the basic statement.

In this type of construction, readers are confronted with a single idea, and one qualification of it, and are able to grasp small amounts of information quickly and easily. However, an author who produces this type of prose is always open to the claim that short, abrupt sentences become tedious, and far from aiding a reader may actually irritate him, so that he loses his concentration. The redeeming factor in easy texts was that these sentences did not make up a whole passage. They were combined with longer, slightly more complicated sentences. Their importance was that they formed the majority of the text and placed no strain on the reader.

The rest of the sentences in easy texts contained more than one statement. In the following examples there are two or more which are joined by linking or joining words that are italicized. Where such signals are missing the symbol // indicates the division in the sentence. For example,

> The ionized particles in the bunsen flame are very reactive *and* easily burn off the layer of grease,

and

> This product has been named Botheric P.O. *and* is now available in commercial quantities.

Of course, there were longer examples of this type of construction, too:

> Chemco has been closely associated with the growth of this industry since the war, // marketing ethyl silicate under the name Botheric P.O.

It is clear that all these examples contain two statements of equal importance, for if the subject of the first two sentences were repeated instead of the word *and*, then two completely separate sentences would be created. In the third, both the subject and the verb have been isolated in the first part of the sentence, although they belong to the second part, too:

> *Chemco has been* closely associated with the growth of this industry since the war. *Chemco has been* marketing ethyl silicate under the name Botheric P.O.

There were also examples of sentences containing two statements, with extra information about one or both of those statements:

> Solids *which are not subject to external forces* maintain their shape indefinitely, *and* from this fact it is assumed *that on the whole, constituent molecules and atoms move about some mean position.*

In this case both halves of the sentence contain extra information. In the first, the subject *solids* is defined more explicitly, while in the second the italicized clause explains what is *assumed.*

But even these sentences with more than one statement and with additional information were neither long nor difficult to read. They did not put any strain on the reader at all.

Difficult Texts

Texts with readability scores between 6 and 10 were classed as difficult. They were extracts from a Ph.D. thesis, a research journal, and industrial reports.

None of these texts contained sentences consisting only of a single statement. Where there was one statement there was *always* additional information with it. In the following example there are four extra pieces of information which expand on the main idea:

> A compound *used/to test this view* was 5β-cholest-6-ene (partial structure 1) *in which the hydrogen at C-5 proved inert to photosensitized oxygenation conditions/that readily effected abstraction of the C-5 hydrogen at 5-cholest-6-ene (2).*

The main statement in this sentence is "A compound was 5β-cholest-6-ene (partial structure 1)." Everything else qualifies this.

There are two clauses between the subject and main verb. The first would normally read, "a compound *which was used*," while the second specifies why it was used, "*to test this view.*" After the main verb *was*, the third clause gives more information about partial structure 1, "*in which the hydrogen proved inert to photosensitized oxygenation conditions*," and the last clause provides details about these 'oxygenation conditions.'

Clearly, if this type of construction is one of the simpler kinds in difficult texts, there can be little respite for the reader when he faces the more difficult sentences, too.

Sentences in difficult texts containing more than one statement were the most complicated of all. They were all long and contained a great deal of information compacted and pressurized to fit the shape of the sentence. For example, in the following example at least three distinct sentences are trying to break out:

> The consideration of the system's impulse response sensitivity to transfer function coefficient variations, in particular the points *at which the first and perhaps higher derivations are zero*, has been shown *to provide a means of rapidly determining the existence of and correcting for coefficient variations from the nominal specification value.*

This particular example also illustrates the obsession scientists and engineers have with stringing nouns and adjectives together. This is what makes the first part of this sentence so very difficult to digest in a single reading. It might have been easier for the reader if the first two lines had read,

> The sensitivity of the system's impulse response can be considered in relation to variations in the transfer function coefficient.

It is at this point that the author includes an aside which separates the subject and main verb of the sentence, *has been shown.* The reader now has to grasp what has been shown in relation to the information he was given in the first part of the sentence. This sentence is far too long and too tightly packed with information to be read comfortably. Had the author divided it into three, then the remaining two sentences could have read,

> This is particularly true at points where the first and perhaps higher derivations are zero. It provides a means of determining the existence of coefficient variations and a way of correcting them using their specified values.

These three sentences would undoubtedly be more readable than the first mammoth offering. Further improvement would mean altering the basic structure even more, but even these slight alterations help the reader.

These sentences are characteristic of the long and complicated sentences found in difficult texts. They are unwieldly, unintelligible, and unnecessary. Only an author indifferent to the needs of his reader, or ignorant of the art of writing, produces such unpardonable prose as this.

Conclusion

Professional scientists and engineers who assessed the 24 texts used in these experiments did so spontaneously, and were probably not conscious of the syntactical patterns in the texts they were reading. Yet analysis of the grammatical characteristics of easy, average, and difficult texts substantiated their instinctive responses to each of these three groups.

Sentences in easy texts were shorter than those in difficult ones, and their internal structure was simpler. This meant that readers were presented with one or two statements per sentence, with no additional qualifying or explanatory information that might complicate the structure and make the sentence difficult to understand.

In difficult texts sentences were extremely long and complicated. In these texts authors expected their reader to grasp two or more important statements per sentence, plus additional information that often interrupted the main statement and made it difficult to follow.

But the results of this survey have far-reaching implications. All texts were produced by highly educated, professional scientists and engineers, yet less than 33 percent proved easy to read. It is not valid to suggest that these easy texts were less specialized than the others and that their subject matter made them easy to read, for texts drawn from the same type of document and of the same level of specialization are scattered throughout easy, average, and difficult groups of texts (see Table I). It is the skill of individual authors in manipulating sentence patterns that makes some more successful than others as communicators. This is one area of communication where training is possible and would produce a good return on any time and money invested in producing effective communicators.

References

[1] D. Bryson, *Effective Communication*, 2nd ed., Prentice-Hall, Inc., Englewood Cliffs, NJ, 1962, pp. 142–143.

[2] A. E. Derbyshire, *Report Writing: The Form and Style of Efficient Communication*, Edward Arnold, Ltd., 1971, p. 66.

[3] R. Gunning, *The Technique of Clear Writing*, McGraw-Hill Book Co., Inc., New York, NY, 1952.

[4] R. Graves and A. Hodge, *The Reader Over Your Shoulder*, Jonathan Cape, 1955, p. 129; Macmillan Publ. Co., Inc., New York, NY, 1961.

Readable Technical Writing: Some Observations

George R. Klare

ALL TECHNICAL WRITERS SHOULD BE CONCERNED *about making their writing more readable. But readability is a very subjective quality to most of us. In this article, Dr. Klare summarizes the findings of decades of research into readability and presents some concrete, objectively derived suggestions to help us make our writing easier to read, easier to understand, and easier to accept.*

Technical writing, more than most other kinds of written communication, should be readable. This judgment is not mine alone. A number of sources seem to agree, from the Armed Services to industry to books on technical writing:

- The Army recently issued a specification that sets desired and acceptable levels of readability for its technical manuals. The Air Force has proposed a similar specification, and the Navy has been considering readability requirements for some time.
- The Bell System has set a desired readability level for its new Task Oriented Practices (TOP) documents, and the Westinghouse Electric Corporation has included a section on readability in a new issue for writers.
- According to one recent study, ten books on technical writing stress readability or clarity.

Emphasis on readability in technical writing appears to be increasing, so perhaps some observations about it can be timely and helpful.

DEFINITIONS

What do we mean when we say that writing is "readable"? Briefly, we mean that the intended readers are able to

- read it quickly
- understand it clearly, and
- accept it readily (i.e., persevere in reading it).

When it is defined that way, all technical writers would probably agree that technical writing *should* be readable. Such a definition, however, says little that can be of value in the actual writing process.

Understandably, simple answers cannot be given. Writing and reading are much too complex for that; most of us spend most of our lives learning and improving upon these skills. Nevertheless, research is helping to provide useful information.

Concerns about readability can be pretty well summarized in two major questions:

1. How can one tell when he or she has written a readable document?
2. How can one write readably?

Though they may not seem to be, these are really two quite different questions, which can be referred to as "prediction" and "production" of readable writing. Furthermore, there is a good reason for what seems to be the inverted order of presenting them above.

PREDICTION OF READABLE WRITING

Readable writing is usually described in terms of the language characteristics a writer has used. Hundreds of such characteristics, or language variables, have been studied. These range from such simple ones as average word length in letters or average sentence length in words to such complex ones as the number of morphemes per 100 words or the sum of "Yngve word depths" per sentence.

Large numbers of passages that vary in such characteristics have been related statistically to the way large numbers of readers have read them. Readers' scores on rate of reading, comprehension, and preferences are typically of special interest. When a language characteristic varies with (is closely associated with) such scores, it can be called a characteristic of readable writing.

Typically, two such characteristics are combined statistically, one of them a word difficulty variable and one of them a sentence difficulty variable. A readability formula and scale are thus developed. The process is, of course, more refined than this description suggests, and results must be checked before use (e.g., for reliability and validity), but that is the basic procedure.

To illustrate the outcome, here is probably the most widely used readability formula, Rudolf Flesch's "Reading Ease yardstick."*

RE (Reading Ease) = 206.835 − 0.846wl − 1.015sl

wl = number of syllables per 100 words; and

sl = average number of words per sentence.

To predict the readability of a piece of writing, one must make the counts necessary to get *wl* and *sl* values (for a long piece of writing, 100-word samples are systematically chosen to save time). The formula score can then be computed, or one of several methods can be used in place of the computation. For example, a direct-reading chart is presented by Dr. Flesch in his book, *How to Test Readability* (New York: Harper, 1951). And a table is presented by James Farr and James Jenkins in "Tables for Use with the Flesch Readability Formulas" (*Journal of Applied Psychology*, June, 1949, *33*, 275–278).

Scores typically range along a scale from 0, which is very difficult to read, through 60–70, which is standard, to 90–100, which is very easy for any literate person. Typical magazines at these three levels are scientific journals, digests, and comics. Scores can be found, however, which are either above or below this range. For example, I have a passage of government regulations which scores −75!

A new piece of writing can be evaluated by counting the number of syllables per 100 words and number of words per sentence, and checking against the readability scale. This provides an answer to the first of the two questions: when is writing readable? Actually, a better way to put it is to say that it provides a prediction of how readable the piece of writing is likely to be for the intended readers.

A readability scale frequently has a set of bench marks that is especially helpful for this sort of prediction: reading grade levels. Comics are typically about fourth- or fifth-grade level in difficulty, for example, while digests are about eighth- or ninth-grade level and scientific journals about college-graduate level. These levels of difficulty correspond roughly to the last school grades typical readers have completed.

Ideally, writers should consider their intended readers' level of education, aiming their writing at *about* the same level of difficulty. Writing that is at a somewhat lower level seldom disturbs readers if it is skillfully done (unless, of course, one gets down near primer level when writing for adults). Almost everyone reads material below his or her reading level at times, especially in leisure reading. On the other hand, research shows that writing that is at a higher level of difficulty can cause problems. When the gap between a reader's skill level and the difficulty level of the writing exceeds two grades or so, reading becomes inefficient. That is, readers go through such materials rather slowly, with reduced comprehension or retention, and with lowered acceptance. Research also shows very practical consequences in that a reading ability-readability mismatch of this sort may cause unnecessary

- errors in following technical directives
- failure to use available technical manuals, and
- abandoning of instructional materials.

---

* Rudolf F. Flesch, "A New Readability Yardstick," *Journal of Applied Psychology*, June 1948, *32*, 221-233.

Reading is, of course, too complex for any formula to predict readability with perfect accuracy. Someone who is very highly motivated can read very difficult materials, where the mismatch between reading ability and readability is considerable. An example is income tax instructions; readers know what will happen if they make mistakes, and consequently many must read very slowly, re-read, and/or ask others for help. And even those who *can* read such material easily prefer a lower level of reading difficulty if they have a choice. A number of experimental studies (over three dozen to date) have compared comprehension scores on *more* versus *less* readable versions of the same topic. Most showed that the more readable versions produced significantly higher comprehension, but not all did (a major reason being that motivation tends to be quite high during experimental testing, causing readers to compensate through extra effort). But even where comprehension was not increased, those studies that used a preference measure almost invariably showed that readers preferred the more readable version.

Readability is important for effective technical writing, and formulas can help writers to predict how readable their writing is likely to be. Many hundreds of such formulas exist, and a number of them can be used with technical writing. Most of these formulas use simple "index" variables such as word length in syllables or sentence length in words in preference to the more complex variables. The simpler ones can be equally predictive, and are easier to count and use.

Unfortunately, some writers feel they can take the next step and adjust these index variables, or even "write to formula," in producing readable writing. This is not the case. Producing readable writing is a different and more difficult process than predicting readable writing. Index variables can be used in prediction because all they need to do is reflect the level of difficulty. The simpler they are (given equal predictiveness), the better. To produce readable writing, however, "causal" variables must be changed.

An analogy is that of changing the level of heat in a room. A thermometer will provide a good index of the temperature. But to alter the heat level, holding a match under the thermometer does little good.

Somewhat the same kind of approach must be taken if one wishes to change the level of readability of a piece of writing. Exact answers are not possible, but research has pointed to some useful suggestions. The following section takes up these suggestions as responses to the second question, raised earlier, of how one can write more readably, with examples from technical writing.

PRODUCTION OF READABLE WRITING

The suggestions presented here for producing more readable writing come from recent findings in psycholinguistics and readability, both of which are active research areas. They are therefore suggestions, *not* rules. Though the research supporting them is quite extensive, future research can always be expected to modify them somewhat. They must be applied with discretion; mechanical application can be more harmful than helpful.* In addition, they apply principally to general informative writing; application to technical writing requires further attention. For example, certain writing specifications must be followed, and many technical terms must be left unchanged because of their special meanings. But these suggestions can still apply. Technical writing seldom contains more than about 15 percent to 20 percent technical terms; so one still has plenty of room for changes in the other words. To make these comments clearer, I will use examples chosen from technical contexts where possible. (I am indebted to Doug Kniffin of Westinghouse Electric Corporation for modifying many of my general examples to fit technical terms.)

Once again, however, a word of qualification is needed. The suggestions given here are not the only important components of writing, even of writing designed to be readable. When one rewrites material to make it more readable, other changes may be needed—in organization, emphasis, inclusion of examples, etc.—in addition to those described in this section. The number of changes I suggest is limited because I have included only those variables having a research basis, and the relevant research is limited. Those presented consist of word and sentence variables only. Attempts to isolate other meaningful variables of this sort, e.g., for paragraph construction and usage, have not progressed very far. Nor have studies of organization or emphasis gotten very far as yet.

Despite these qualifications, what is already known about readability has clearly demonstrated value for helping readers to understand, to read efficiently, and to persevere in their reading. Studies show that these gains apply to technical writing as well as to other informative communications. The suggestions from the research will be given in the following paragraphs in terms of word changes and sentence changes. The term "changes" is a convenience meant to apply both when a writer changes his original writing and when he edits someone else's.

Word Changes. In order to discuss word changes, a distinction must be made between "content" words and "function" (sometimes also called "structure") words. The former refers to categories of words which deal primarily with content, that is, nouns, verbs, adjectives, and adverbs. The latter refers to categories of words which deal primarily with relationships, such as prepositions, conjunctions, articles, and interjections. An easy way to recognize function words is to look for those words which are customarily not capitalized in titles. The distinction between content and function words is important because the word changes described here usually apply only to content words. In fact, many apply only, or most specifically, to nouns and adjectives. Less research has been done, unfortunately, on verbs and adverbs.

Six word changes, or variables, are listed with examples. They tend to be related to each other (i.e., frequently used words are usually shorter than those infrequently used), but they are not close to being perfectly related; so one can often be applied when another cannot be, which is why each is discussed separately.

1. **Frequency or familiarity.** Words of high (versus low) frequency and/or familiarity contribute to more readable writing. This was found for nouns, particularly, but also for other content words. Compare *name* with *designation* and *part* with *component*, or *use* with *utilize* and *make* with *fabricate*.

2. **Brevity.** Shorter words (versus longer) tend to make reading easier and faster; generally found for content words, but sometimes also for function words. Examples are: *fail* versus *malfunction*; *turn* versus *rotation*, and *help* versus *facilitate*; or *too* versus *additionally*.

3. **Association value.** Words which call up other words quickly and easily also add to readability, especially when the other words (or their meanings) appear later in the text. Most of the studies have been made with nouns, but several with adjectives. *High* brings *low* to mind more quickly than *up*, and *light* suggests *bulb* more quickly than *lamp*; *tool* calls up associations more easily than *implement*, and *amount* more quickly than *quantity*.

4. **Concreteness versus abstractness.** Concrete words easily arouse an image in one's mind; they contribute more to readable writing than abstract words, which do not easily bring images to mind. The studies to date have been made only with nouns. Compare *fire* and *combustion*, *transistor* (for those in the electronics field) and *device*, or *oscilloscope* and *equipment*. The difference is easy to miss: *time* is abstract, while *timepiece* (*clock*) is concrete. Adjectives have been scaled for the related quality of vividness. *Filthy* is more vivid than *soiled*, and *brilliant* more than *gleaming*.

5. **Active verb versus nominalized form.** Nominalizations are words (usually verbs) made into noun form; the active verb form is usually more readable than the noun form. *Consider* may thus become less readable as *the consideration of*, *oppose* become *the opposition to*, or *suppose* become *the supposition that*.

6. **Reference of pronouns and other anaphora.** Anaphora are words or phrases which refer back to a previous word or unit of text. Clarifying what anaphora refer to and limiting their use (even if, occasionally, directly repeating what they refer to) help to make writing more readable. Examples of anaphora, besides pronouns such as *he* and *she* or *that* and *which*, are phrases such as *the above*, or *defined earlier*, or *in the third paragraph*.

Space prevents my describing further how these word variables can be put to work in making writing more readable. A possible approach is described in my book, *A Manual for Readable Writing*. Whatever approach you might find most convenient and appropriate for your purpose, a few words of qualification are essential. To begin, not all potentially hard words should be changed; some cannot be without an undesirable change of meaning. Then too, a good or better substitute word (or short phrase) cannot always be found or cannot always be inserted without undesirable sentence changes. Good editorial opinion

* My little book, *A Manual for Readable Writing* (Glen Burnie, Maryland: REM Company, 119A Roesler Road, 1975), presents research summaries, as well as a more complete description than this article of the suggestions themselves and their application.

is essential, and the final arbiter.

Sentence Changes. Before suggesting how to make sentence changes for improved readability, I should offer a few preliminary words of caution. Many of the changes which I shall describe come from studies using transformational grammar as an approach. The results are, as for almost all research, likely to be modified somewhat as newer data become available. Also, the studies have sometimes been done only with single sentences, since experimental work with passage-length materials is difficult. Consequently, the suggested changes are far from rules and should not be made mechanically.

1. **Brevity.** Shortened sentences and clauses contribute to more readable writing. However, not all long sentences are equally difficult to understand, so consider potential changes thoughtfully before making them.

   a. Conjunctions such as *but*, *for*, and *because* create more difficulty than *and*.

   b. Clause length is important apart from sentence length.

   c. Long sentences usually contain complex structures, but complexity may not be reflected in length.

   This suggestion does not mean "the shorter, the better," of course, at least insofar as writing at a child's level is concerned.

2. **Active versus passive voice.** The active form of a statement leads to easier recall and verification than the passive form. Active voice makes the subject of a sentence the actor; passive voice leaves the subject to be acted upon. Compare *The transistor amplifies the signal* . . . with *The signal is amplified by the transistor* . . . or *The current limiter protects the power supply* . . . with *The power supply is protected by the current limiter* . . . . Sometimes, however, emphasis resides less in the actor than the acted-upon, and this suggestion must then be disregarded.

3. **Affirmative versus negative construction.** Positive statements are likely to be verified more quickly and with fewer errors than negative statements (i.e., those with *not* or other negative words). For example, *The output of the nand gate is high* . . . is generally preferable to *The output of the nand gate is not low* . . . . With greater sentence complexity this becomes even more important. *This is the latest version*, or even *Is this the latest version*? are both preferable in most cases to *Is this not the latest version*? Times arise, of course, when emphasis demands a negative construction.

4. **Statement versus question form.** Statements, according to several studies, tend to produce better recall than questions. For example, *Could the other driver have been blinded by the sun*? could be changed to *The other driver could have been blinded by the sun*. Questions clearly have a place in providing emphasis and variety; further research is needed on when they hinder recall and when they help.

5. **Depth of words in sentences.** Word depth refers to the number of "commitments" a reader must store while reading; greater depth makes writing less readable. Here is a sentence with high word depth: *The vacuum tube, which was pictured on a preceding page, and which started an electronic revolution, is now almost obsolete for our purpose*. This sentence can easily be changed to a more readable form. For example: *The vacuum tube was pictured on the preceding page. Though it started an electronic revolution, it is now obsolete for our purpose*. Still other improvements might be made in this sentence, of course, for some readers.

6. **Embedding of words in sentences.** Self-embedded sentences have repeated parts within parallel elements; they are perfectly grammatical, but much harder to understand than non-embedded sentences. An example of a self-embedded sentence is *The transistor which the signal that the diode shorted drove failed*. Fortunately, sentences containing more than two self-embedded clauses are very rare in English, since they are almost incomprehensible even to skilled readers. If this level of complexity is desirable for some reason, a "right-branching" form with relative clauses is better. For example, *This diode shorted the signal that drove the transistor that failed*. This sentence could be simplified even further to help certain readers.

A few words of qualification are necessary when you are considering sentence changes. First, do not change all potentially difficult sentences to the simplest possible form; some variety is desirable. Second, try to combine sentence changes with word changes; studies show that changing only one or the other reduces the desired effect. In general, however, readers usually find simple, active, affirmative, declarative (or SAAD) sentences the most readable.

COMBINING PREDICTION AND PRODUCTION

To sum up, a readability formula can provide a quick and useful prediction of the reading difficulty of text—but *no more than that* should be asked. Once a writer has such an estimate, the formula should be set aside so that there is less temptation to "write to formula." Tinkering with index variables such as word length or sentence length alone can yield better formula scores; but this provides no assurance that a reader will be able to read more quickly, understand more clearly, or find material more acceptable.

Instead, more basic changes must be made in producing readable writing. Such modifications may take many forms, of course, and are not even completely understood as yet. In the use of words and sentences, however, some useful information can be provided. Before considering changes in words, three related probing questions might be asked: (1) Is the word longer than necessary? (2) Will intended readers know the word? (3) Is that particular word necessary? Research has suggested six qualities of words that can be helpful when changes are needed:

- frequency or familiarity
- brevity
- association value
- concreteness
- active verb form and
- clear reference.

Concerning sentences, three probing questions might again be asked: (1) Is the sentence longer and more complex than necessary? (2) Will intended readers understand the sentences? (3) Is the variety needed, or should I simplify the sentence? Six research-based suggestions can be helpful when sentence changes are needed. These are to use

- brevity
- active voice
- affirmative construction
- statement form
- lower word depth, and
- less embedding.

Combining word and sentence changes is more effective, generally, than changing either alone.

Once the text has been changed, the readability formula can be applied again. This process of writing, application, rewriting, and re-application takes time. Most writers like rewriting much less than original writing; even apart from that, most technical writers' time for rewriting is very limited. This process therefore seems questionable. Two observations should help to justify it, however,

1. The time one writer spends in making writing more readable can save time and effort for many readers, and reduce the errors they might otherwise make.
2. The process, because of the quick feedback provided by a formula, gets much easier with time; more accurate prediction and production of readability can become almost second nature.

Avoiding a mismatch between the reader's ability and the difficulty of writing is what readable writing is all about. And technical writing, more than most other kinds of written communication, should be readable.

How Technical Is Your Writing? —A Statistical Linguistics Question

Henri Lauer

WHAT MAKES A DOCUMENT *technical is not so much its topic as its writing style. A document becomes briefer and more precise, and it uses relatively more nouns and fewer qualifiers as its style becomes more technical. Reasons for this are given, and the conclusions are supported by an analysis of sample texts.*

"How is the weather?" I asked Oliver, as he stepped into my room.

"Fine," said he.

To me, this may have meant that the weather was sunny and warm; to you, perhaps, that it was bright and brisk; but to Oliver, it meant that we were having a downpour, for we had just gone through a drought and he had been praying for rain. What a wonderful word this was. Thus used alone, it could stand for anything from sunshine to rain! But because of this, it communicated nothing about the topic at hand: clear to none but to Oliver, it described only his personal reaction to the weather, his feeling and emotion. It was not a weather report. I told him this.

To my same question a week later, he answered, in his sometimes lyrical mood, "Across the deep, blue sky, a few big, white, wooly clouds are drifting, pushed along by a gentle breeze. The air is pleasantly cool and dry." This statement was a more informative weather report than his reply a few days earlier; and in future references, I shall call it "report 1."

There are, of course, other ways to describe this weather condition. One could say, for example, "There are cumulus formations; the wind blows at 8 knots; the temperature is 59 degrees; and the humidity is 62 percent." I shall call this "report 2" for further reference.

REPORT 1

Even though these two reports describe the *same* weather situation (in fact, they describe the very same elements of this situation), the reports are *different*. As we shall see later, report 1 is less technical than report 2. One of these reports does not make the topic—the weather situation—any different or any more or less technical. The way the topic is presented and the writing style of the report do, however. The style of the report depends on the words, the sentences, and their arrangement.

To be effective, a document should be as clear, brief, precise, and complete as possible. Although the writer provides the document with these features, it is the reader who evaluates them. We can compare our two reports to find out whether, how, and to what extent they contain these features and to determine what style characteristics make a piece of writing technical.

Report 1 is written in everyday language with sentences and words that are clear to anyone. In contrast to the ambiguous word *fine*, report 1 can be understood by as wide an audience as may be interested in the topic. Because the words used are familiar to a broad public, they also have broad meanings. Therefore for the report to be precise these meanings must be narrowed or pinpointed by adding qualifying adjectives and adverbs to the nouns.

Thus the clouds are said to be big, white, and wooly; the breeze is called gentle; and the air is cool and pleasantly dry. Like the word *fine*, these qualifiers still reflect the writer's (or Oliver's) personal feelings. The report is clear not only to Oliver but also to most of its readers. As the number of qualifiers for each noun is increased, the more precisely defined is the object which this noun describes. Each qualifier describes a particular feature of the object so that the group of noun and associated qualifiers forms a descriptive specification of the object.

In addition to being clear and precise, the report is brief and complete. It is brief because most of its words are necessary to obtain the desired precision. It is complete because it covers all the information items by which anyone deprived of special education and measuring instruments would be able to understand the weather situation.

COMPARISON WITH REPORT 2

Now look at report 2. It is written in the special (technical) language of meteorology. For example, the two technical words *cumulus formations* in report 2 take the place of the four nontechnical words (one noun and three qualifiers) *big, white, wooly clouds* used in report 1. The two technical words are used by people habitually or professionally concerned with the weather to designate big, white, wooly clouds occurring at an altitude of 5,000 to 15,000 feet. Thus the technical expression is briefer and handier to use than the group of everyday words it replaces. Brief, precise, and handy as it is, a technical word is clear to (or understood by) a smaller audience than the longer group of nontechnical everyday words.

Note also that qualifiers such as *gentle, pleasantly cool,* and *dry,* which give precision to report 1, reflect Oliver's personal impressions of certain weather elements. Just as for the word *fine,* these qualifiers may not convey

Table 1—Analysis of Various Levels of Writing

Grammatical category	Percentage of Words							
	Text A	Text B	Text C	Text D	Text E	Text F	Text G	Text H
Relational words (Articles, conjunctions, prepositions)	26.4	33.4	31.7	33.6	31.3	34.4	34.5	29.7
Action words (verbs)	17.7	14.2	16.5	17.6	19.0	16.4	14.2	17.6
Total relational and action words	44.1	47.6	48.2	51.2	50.3	50.8	48.7	47.3
Designators (nouns, pronouns)	28.4	27.0	30.3	30.7	32.4	32.5	35.8	37.4
Qualifiers (adjectives, adverbs)	27.5	25.4	21.5	18.1	17.3	16.7	15.5	15.3

Text A—Popular science: cosmology.
Text B—Popular science: evolution.
Text C—Scientific/philosophical theory: quantum physics.
Text D—Theory of an electronic-optical device.
Text E—Description of an electronic circuit.
Text F—Operating procedures: instructions for cooking fish and timbales.
Text G—Maintenance procedure: radio transmitter.
Text H—Mechanical assembly procedure: do-it-yourself kit.

identical impressions to others. Therefore in place of these, report 2 gives quantitative information, which is objective. The interpretation of this objective data is left to the report reader. This, again, limits the audience. To increase the size of his audience, the writer could have said, for example, "miles per hour" instead of "knots."

LEVEL OF LANGUAGE

The writer's choice of words determines the level of the language of a report. However, this word choice also affects the *texture* of the written material. We saw that a single technical word may replace a nontechnical noun and its qualifiers. This leads to two consequences:

1. A technical text tends to be briefer than a less technical text containing the same information items (the 22 words of report 2 present the same information as the 26 words of report 1).
2. A technical text tends to have relatively more nouns and fewer qualifiers than a less technical text. Fifteen percent of the words in report 1 are nouns and 50 percent are qualifiers. Report 2, on the other hand, contains 37 percent nouns and 18 percent qualifiers.

The two reports we have discussed are, of course, too short to indicate even a trend. To check how definite such a trend may be in actual writing situations, we have made a word count, by grammatical categories, of several samples which have been written for different audiences. The results are shown in Table 1.

The samples analyzed were from about 200 to 600 words each (average 425 words). They were taken from published literature produced by competent (some well-known) writers. As we go from left to right in the table, they become increasingly technical; and the percentage of the designators used goes from 28.4 to 37.4 percent, while that of the qualifiers goes from 27.5 to 15.3 percent. The percentage of words in other grammatical categories—the relational and action words—averages approximately 48.5 percent.

CONCLUSION

Writing is technical in proportion to its use of a special vocabulary current with those concerned with the topic it presents. Because this vocabulary is special, it is also precise without as great a need of qualifiers as everyday vocabulary. This makes technical writing brief. The more special the vocabulary, the more technical is the writing, and the more limited the reading audience to whom such writing is clear. Ω

Filtering Typographic Noise from Technical Literature

James Allen Carte

BY FAILING TO USE *various typography guidelines, technical communicators often introduce distracting "noise" into their documents. Research in typography can be applied by technical writers, editors, and illustrators to help filter out much of this undesirable interference.*

Most technical writers, editors, and illustrators in the electronics field are familiar with such "noise" sources as static in communications sets, own-ship screw beats that play havoc with aural and visual sonar reception, or "snow" on television and radar screens. Perhaps few have considered "noise" as something to be dealt with in the written word. This article defines noise as it applies to writing, mentions briefly numerous sources of the phenomenon, and elaborates on typographic noise.

When applied to writing, noise is defined as the degree to which the meanings of the writer and the reader are less than absolutely identical.[1] Stated another way, it is the *difference* between what the writer means and what the reader interprets the writing to mean.

After dwelling on noise in this context, the experienced technical writer and editor probably will think of several obvious sources, e.g., poor punctuation, poor sentence and paragraph construction, poor organization, poor selection of words, and failure to define the reader properly. A less obvious source, however, is the poor selection of typography.

Typography research done both recently and in the past can provide technical communicators with certain guidelines for making their documents more readable. Common factors to consider when deriving these guidelines are:

1. Line length, type size, and leading.
2. All-capital versus lower-case letters.
3. Italic versus bold-face type.
4. Hyphens at the end of lines in unjustified composition.
5. Justified versus unjustified columns.
6. Ink and paper color combinations.

Before discussing these typographic noise factors, I want to make it clear that I am not proposing that every document going out the door include all of the recommendations in this article. I am well aware of certain limitations put on some technical literature by such factors as customer specifications, production expediency, and layout goals. What I *am* suggesting is that you follow the guidelines when *practicable* to provide the reader with a more legible document.

LINE LENGTH, TYPE SIZE, AND LEADING

In 1940, D. G. Paterson and Miles Tinker, two psychologists at the University of Minnesota, conducted a study on the legibility of various combinations of line length, type size, and leading (space) between lines of type.[2] Their findings, substantiated in later studies by Tinker and others,[3] indicate that 10-point type with the line length ranging from 14 to 25 picas is easiest to read. Furthermore, they report that an ideal combination is 10-point type, 19-pica line length, and 1 or 2 points of leading between lines.

They add, however, that the following combinations of type size versus line length are acceptable: 8-point type, 17-pica line; 12-point type, 23-pica line; and 14-point type, 27-pica line. Table 1 provides a summary of these optimum combinations both in picas for direct application to cold-type equipment such as the IBM Composer and the Compugraphic Phototypesetter and in approximate inches for use with typewriters. The inches are rounded off to the nearest sixteenth of an inch on standard rulers.

Table 1–Optimum Combinations of Type Size and Line Length

Equipment	Type Size			
	8 points	10 points	12 points	14 points
Cold-type composers	17 picas	19 picas	23 picas	27 picas
Typewriters	2 13/16 in.	3 1/8 in.	3 13/16 in.	4 1/2 in.

CAPITAL VERSUS LOWER CASE

According to Edwin H. Stuart and Grace Stuart Gardner, careful optometric tests have determined that all-capital letters slow down reading speed from 20 to 50 percent, depending on such factors as the length and complexity of the word.[4] Thus with this sort of noise being introduced, it is difficult, it seems to me, to justify all-capital letters even on illustrations and viewgraphs, two common areas where they often are used by technical-publications specialists.

Obviously, all-capital lettering must be used when customer specifications so dictate; however, a good general rule to follow is–use them sparingly.

ITALIC VERSUS BOLD FACE

Stuart and Gardner also point out that italic type is 10 to 50 percent less readable than straight Roman type. They state flatly that solid paragraphs should *never* be set in italics. Indeed, at least one typography researcher, Rolf Rehe of Indiana University, recommends that bold-face type be used in place of italic even to show emphasis.[5]

In any event, if you do use italics in your technical documents, follow the same rule just mentioned for all-capital letters–use them sparingly. One final warning is never to use all-capital italic words. As Stuart and Gardner say, "They look like a picket fence in a windstorm."

HYPHENS IN UNJUSTIFIED COMPOSITION

According to Richard Hopkins, a professor of typography at West Virginia University, "Hyphenated words at the end of lines of type should *always* be avoided when striving for maximum readability in unjustified composition."

The reason for Hopkins' statement becomes quite clear when one analyzes the mental process the reader follows when he reads part of a word at the end of one line, scans back across the column to the remainder of the word at the beginning of the next line, and finally "marries" the two parts into a whole from which he must derive meaning.

Although I doubt that it takes the reader as long to perform this mental task as it has taken me to tell about it, end-of-the-line hyphenated words *do* introduce much unnecessary noise into the written-communication process. As a result, our old friend the "use them sparingly" rule is just as pertinent to hyphens at the end of lines of unjustified composition as it is to all-capital letters and italics.

JUSTIFIED VERSUS UNJUSTIFIED

Traditionally, we have considered justified composition (i.e., flush right) as the ultimate in appearance for published documents. Recently, however, many top advertising agencies have used unjustified composition even for body type, and Robert Townsend, who in three years took Avis Rent-a-Car from 13 years of operating in the red to earnings of $9 million, had his best-selling book *Up the Organization*[6] published with a ragged right margin. As a related sidelight, less than 25 words in the main-body text of Townsend's 220-page book are broken with hyphens at the end of lines.

This modern trend probably was influenced by research conducted by Stanley Powers for his master's thesis at the University of Florida in 1962.[7] Powers' study indicates that reading material presented in unjustified form can be read slightly faster than justified copy. Furthermore, he reports that readers seem to have little objection to the use of ragged right margins.

The implications of Powers' research are far-reaching in the publishing business. Is it possible that justified copy is not all we have thought it to be through the years? In a dollar-and-cents perspective, perhaps all published documents, including technical literature, can be made even more readable by using less-expensive, stand-alone, proportional-spacing composer or typewriter units in place of expensive, computer-run, justified-type systems.

COLOR COMBINATIONS

Useful guidelines for selecting optimum ink and paper color combinations also are suggested by Stuart and Gardner. Table 2 lists these combinations in the order of relative *visibility* (another term for *readability* and *legibility*).

CONCLUSIONS

By taking advantage of typography research and putting the applicable findings into practice, technical communicators can filter out much of the typographic noise that creeps into technical literature.

Table 2–Relative Visibility of Colors of Ink on Paper

Rank	Ink	Paper
1	Black	Yellow*
2	Green	White
3	Red	White
4	Blue	White
5	White	Blue
6	Black	White
7	Yellow	White
8	White	Red
9	White	Green
10	White	Black
11	Red	Yellow*
12	Green	Red
13	Red	Green
14	Blue	Red

* Also includes cream and ivory

Excessively long lines and too much space between lines are major contributors to typographic noise in technical documents, especially in proposals and reports done on typewriters. From a readability standpoint, copy typed on 10-point typewriters should be typed single spaced with a column width of approximately 3 inches. This suggests a two-column vertical format as opposed to the single-column format that is commonly used. Another format that could be used is three-column, horizontal (broadside), and single spaced. Although the latter suggestion is far from traditional, the layout flexibility would be quite significant. Table 1 summarizes type-size versus line-length combinations that provide maximum readability.

A single rule applies to three sources of typographic noise. When dealing with all-capital letters, italics, and hyphens at the end of lines in unjustified composition–*use them sparingly*.

A fifth possible source of typographic noise is justified copy. If we discard traditional prejudices, perhaps this source could be a blessing in disguise. The savings in money, time, and aggravation that can result from using ragged right margins would indeed be quite significant.

A final typographic noisemaker is incompatible ink and paper color combinations that generate interference that distracts the reader. Table 2 ranks the relative visibility of various color combinations of ink and paper.

REFERENCES

1. Melvin L. DeFleur, *Theories of Mass Communication,* David McKay Company, Inc., New York, 1970.
2. D. G. Paterson and Miles Tinker, *How to Make Type Readable,* Harpers, 1940, cited by J. K. Hvistendahl, *The Effect of Typographic Variants on Reader Estimates of Attractiveness and Reading Speed of Magazine Pages,* a communications research report published at South Dakota State University, March 1965.
3. Miles Tinker, *Legibility of Print,* Iowa State University Press, 1963, cited by J. K. Hvistendahl.
4. Edwin H. Stuart and Grace Stuart Gardner, *Typography, Layout and Advertising Production,* Edwin H. Stuart, Inc., Pittsburgh, May 1955.
5. Rolf F. Rehe, *Proposals for a Functional Typography,* master's thesis, Indiana University, August 1972.
6. Robert Townsend, *Up the Organization,* Fawcett Publications, Inc., Greenwich, Conn., February 1971.
7. Stanley Powers, *The Effect of Three Typesetting Styles on the Speed of Reading Newspaper Content,* unpublished master's thesis, University of Florida, 1962, cited by J. K. Hvistendahl.

Sloppy Typing And Reproduction In A Written Technical Message—An Experiment *

RICHARD M. DAVIS
Air Force Institute of Technology
Wright-Patterson Air Force Base, Ohio

ABSTRACT

An experiment was performed to determine the effect (if any) of sloppy typing and reproduction upon the effectiveness of a written technical message. The variables tested were the margins, the way in which corrections were made, and the reproduction of the message. Approximately seven hundred subjects in five definably different audiences were tested. Measures were taken of comprehension, reading time, judgment of the author's credibility, and judgment of the author's competence as a writer. Five main effects and five interactions were found at the 0.05 level of probability. Each variable, each measure of the effectiveness of the message, and each audience was involved in one or more of these effects. In each main effect and each interaction subject to easy interpretation, the unaltered form (good typing and good reproduction) of the variable(s) concerned appeared to be the more effective.

Introduction

We have all seen technical communications in which the typing was poor, corrections were badly made, and the reproduction was sloppy. Even before reading the text, we may sometimes have had an unfavorable impression of the author and his competence to report on the matter concerned—simply because his written report, whatever the subject, was presented in such a slovenly manner.

Because we see clean copy and reproduction in most of what we read daily, the author of the report in question has violated our expectancies. The farther we get into the report, the more we are bothered by the violations. The distraction is sometimes compounded to the extent that it becomes a real effort to follow the text. The author may then have an uphill battle to convince us that he can add two and two correctly.

Occasionally, one of my more practical colleagues will tell me that the physical presentation of the message is of no importance—that to a competent technical man (like himself) the only thing that really matters is the matter itself. But strangely, the same colleague is the one who rages the loudest when the reproduction of his own material is bad. He has been known to send an entire report back for reproduction of a new copy if one or two pages have noticeable background color. And woe betide the secretary who presents him with imperfectly typed copy. This, indeed, presents something of a puzzle. Just what effect do bad copy preparation and reproduction have—if any—upon the effectiveness of a written technical message? And is it possible to demonstrate any such effect by objective means?

In this article, I will describe a recent experiment intended to be a first tentative step in the determination and measurement of these effects. The experiment was the fifth in a series in which the effects of variations in the organization, content, expression, and presentation of a technical message upon definably different audiences are measured. Thus far, six experiments have been completed, and two others are in progress. Something over five thousand subjects in twenty-four audiences have been tested.

Communication Situation

The experimental approach is based upon a simplified view of the situation in which a written communication might be produced. The essential elements are the same as in other communication situations. For whatever reason, someone wants to tell another person something about a particular subject. Assuming that the medium by which the message will be

* The experiment described in this article is reported in greater detail in AFIT TR 71-1, *Effective Technical Communications: Copy Preparation and Reproduction: Experiment I* (AD 722054). It is available from the Defense Documentation Center and the National Technical Information Service.

Reprinted with permission from *J. Tech. Writing and Commun.*, vol. 2, pp. 43-55, Jan. 1972.

transmitted is written English, there must be an author or source, a purpose, a subject, and an intended audience. The possible conditions of each of these elements defy specification, and the possible relationships among them are infinite. In no two situations are the conditions of these elements nor the relationships between them exactly the same.

In practice, the author appraises the immediate situation and develops his message to fit his analysis of the case in hand. He draws upon his education, his experience, and whatever ability he may have in the particular variety of authorship required. He considers the subject, his purpose, his intended reader, and whatever local or universal forms and traditions may seem to apply. Then he writes and presents his message as seems to him most effective for the attainment of his purposes.

Experimental Approach

When the communication situation is viewed in this way, objective experimentation seems reasonable. It should be possible to determine experimentally the most effective content, organization, and presentation for given kinds of messages to given kinds of audiences for particular purposes. At least, this should be true to the extent that the elements in these situations can be defined, categorized, and reproduced under experimental conditions, and to the degree that it is possible to measure the attainment of a given purpose.

The experimental approach is quite straightforward. In brief, a situation is assumed, and a communication is written to meet its demands. Controlled alterations are then made in the communication so that several versions of the same message are produced. A test is developed to measure the degree to which the purposes of the communication are attained. Each member of a homogeneous audience is assigned (on a random basis) one version of the communication. The test subjects (readers) read their assigned versions of the communication, and then all are given the same test. Appropriate statistical analyses of the scores are made, and differences in performance on the test are assumed to be caused by differences in the effectiveness of the various forms of the message. To the degree that other situations are similar to the one assumed, messages organized and presented in the manner found to be most effective in the experiment should be more effective in attaining the assumed purposes than those organized and presented in the other forms tested. Comparison of the results of tests of definably different groups of test subjects should indicate the most effective forms for differing groups of potential readers.

Situation Assumed

With the exception of the audiences tested, the communication situation assumed is as in previous experiments.

Author

The passage is written by a man who is thoroughly familiar with the subject and who is experienced in technical writing.

Subject

The passage is a description of a simple mechanical device. Certainly this is one of the most common subjects in technical communications. It may be necessary to describe such a device to someone who is to design it, build it, modify it, sell it, transport it, demonstrate it, operate it, install it, or who may need greater or lesser degrees of understanding of the device for any of a dozen other reasons.

Purpose

The author wants to describe the physical structure and operation of the device to his readers so that they will have the understanding necessary to make the kinds of determinations or judgments about it that might be required in an actual working situation.

Audiences

The five audiences tested in this experiment were composed of approximately 700 students enrolled in three institutions of higher learning; the University of Dayton, the Air Force Academy, and the Air Force Institute of Technology. All subjects were native speakers of English, and none knew anything about the author of the reading passage (it was unsigned). As the device described in the passage had not been produced commercially, the subjects knew nothing about it until they read their test passages. For the purposes of the experiment, the audiences were defined as follows:*

Audience 14. Bright young women without known technical inclinations. They had not had extensive technical training, were not experienced in the use of written technical communications, and had not been instructed in technical writing. This audience consisted of 127 young women tested at the University of Dayton—those in essentially nontechnical curricula such as Fine Arts, Interior Decorating, Language, and Physical Education.

Audience 15. Bright young men without known technical inclinations. They had not had extensive technical training, were not experienced in the use of written technical communications, and had not been instructed in technical writing. This audience consisted of 178 young men tested at the University of Dayton—those in essentially nontechnical curricula such as Fine Arts, Physical Education, History, and Education. The audience is contrasted with Audience 14 on the basis of sex.

Audience 16. Bright young men with known technical inclinations. They had not had extensive technical training, were not experienced in the use of written technical communications, and had not been instructed in technical writing. This audience was comprised of 132 subjects tested at the Air Force Academy. It is contrasted with Audience 15 on the basis of technical inclination. Cadets at the Air Force Academy were assumed, as a group, to be technically inclined, while the young men in nontechnical curricula at the University of Dayton were not assumed to have known technical inclinations.

Audience 17. Very bright men with known technical inclinations. They had had extensive technical training and were experienced in the use of written technical communications, but they had not been instructed in technical writing. The audience was comprised of 83 subjects tested at the Air Force Institute of Technology, School of Engineering. It is contrasted with Audience 16 in that the subjects are defined as being more intelligent, older, and more experienced in the use of technical communications. Further, they had undergone extensive technical training.

Audience 18. Very bright men with known technical inclinations. They had had extensive technical training, were experienced in the use of written technical communications, and had been instructed in technical writing. The audience comprised 70 of the subjects tested at the Air Force Institute of Technology, School of Engineering. It is contrasted with Audience 17 only in that the subjects had recently had a course in technical writing. (See Table 1.)

Test Materials

The reading passage and criterion test used in this experiment were those used in the previous experiments in this series. They appeared to make the intended discriminations in the earlier experiments, and their continued use will permit direct comparison of the results obtained in different experiments.

Reading Passage

The reading passage is a description of a simple mechanical device—certainly one of the topics most often encountered in technical writing. The device is a "testing machine" on which the original patent was granted on February 3, 1959. The machine has not been produced commercially, and nothing has been published about it (beyond the original notice of patent). It is different in concept from commercially available "teaching machines" and no encompassing analogy could easily be drawn by test audiences.

The content of the passage was chosen as that which might be necessary or useful in describing the device's physical structure and operation. This included the function, size, shape, material, and weight of each of the parts and of the machine as a whole; the relative positions of the parts and the connections between them; and what the operator does to operate the device, what happens as a result, and how it happens.

The passage is 3619 words long and is composed of seven headed sections: an introduction; five sections, each of which describes one of the five functional units composing the device;

*** For ease in reference, the audiences are generally referred to by number throughout this experiment. Except in the first experiment in this series, audiences are assigned consecutive numbers in a single series.**

Table 1. Comparison of the Audiences Used In This Experiment

Audience	*Sex*	*Age group*	*Mean age*[a]	*Intelligence group*[b]	*Known technical inclination*[c]	*Experienced with written tech comm*[d]	*Tech training*[e]	*Instructed in tech writing*
14	Female	young women	(18 yrs 9 mo)	bright	no	no	no	no
15	Male	young men	(18 yrs 11 mo)	bright	no	no	no	no
16	Male	young men	(18 yrs 11 mo)	bright	yes	no	no	no
17	Male	men	(28 yrs 9 mo)	very bright	yes	yes	yes	no
18	Male	men	(30 yrs 1 mo)	very bright	yes	yes	yes	yes

[a] "Young people" (age around 20) are distinguished from "men" (age around 30). An author might be writing to an audience in one or another age group.

[b] "Bright" groups are those in early undergraduate years. Those defined as "very bright" were composed largely of students in graduate curricula. Most of the remaining subjects in this group had completed one degree and were in the final year of completing a second undergraduate program in a technical field. It is assumed that the mean intelligence of the "very bright" groups is higher as the subjects had completed one degree and had chosen to pursue another.

[c] Students in nontechnical curricula (music, literature, physical education) are distinguished from those in technical curricula (physics, chemistry, engineering). It is assumed that subjects in the latter group have technical inclinations or they would not be in technical curricula. Many in the "nontechnical" group may have technical inclinations, but this is not known.

[d] The two groups (Audiences 17 and 18) defined as being experienced in the use of technical communications were composed of officers who held technical degrees or were completing them. The few who had not yet completed a technical degree were experienced officers (most were rated pilots, navigators, or missile men) with six to eight years' experience with technical Air Force materials. Those who already held technical degrees were generally as experienced, though a few had moved directly from undergraduate degrees in civilian institutions to graduate programs at AFIT.

[e] Most of the subjects in Audiences 17 and 18 had had extensive technical training as aircrew officers, missile men, or for other technical specialties; most already had a degree in a technical field.

and a description of the device's operation. The passage is illustrated by ten line drawings. One, showing the unit as a whole, is placed at the beginning of the introduction. Five drawings, each illustrating one of the five major components, are placed at the beginnings of the respective component descriptions. The remaining four drawings, illustrating critical relationships in the operation of the device, are placed with the descriptions of the relationships they illustrate.

Criterion Test

Under the conditions assumed in this experiment, the author intends to communicate his subject matter to the reader in the most effective manner. When the content of the message has been determined, it must be communicated in the form most readily comprehensible to the reader and the least demanding of his time. It should be so written and presented that the reader will accept its accuracy and that it does not detract from his impression of the author's competence. A measure of the effectiveness of the message must, then, consider comprehension, reading time, the impression given of the author's knowledge of the subject matter, and the impression given of the author's competence in presenting it. The means by which these four measures are taken are described briefly below:

1. *Comprehension.* Measured by a 20-question multiple-choice test with four answer choices for each question. All questions required judgments, deductions, and inferences about the machine and its operation.
2. *Reading time.* Recorded by each subject as he finished reading his test passage.
3. *Author competence.* Impression recorded by each subject on a nine-point scale.
4. *Author knowledge.* Impression recorded by each subject on a nine-point scale.

Gross failure of a communication in any one of these areas might wholly offset its effectiveness in the others and nullify the effectiveness of the whole. Certainly other measures could have been taken and might well be used in other experiments with other sets of assumed conditions.

Test Variables

The probable size of available test audiences suggested that three test variables be used with two conditions on each variable. The variables should be elements in the appearance of the reading passage about which readers could reasonably be assumed to have expectancies and whose violation would be evident to them. The three variables chosen for this experiment were the left- and right-hand margins on the typed copy, the way in which typing errors were corrected in the typed copy, and the quality of the reproduction.

Margins

In typed copy we are accustomed to a uniform left-hand margin and a generally uniform right-hand margin—or one that is as nearly uniform as the typist can make it. Substantial variance from this convention should be immediately discernible to any reader. In the unaltered form of this variable, the copy was typed with the uniform margins that are normally expected. In the altered form there was occasional unevenness in the left-hand margin (as sometimes happens when the stops are not properly set on a typewriter or when the typist does not return the carriage fully after typing a line) and raggedness in the right-hand margin (as might be expected from an inexperienced or sloppy typist).

Corrections

Corrections are not usually evident on copy prepared by a good typist. When material is badly typed, however, corrections may become quite noticeable. In the unaltered form of this variable, any corrections made by the typist were not detectable in the reproduced copy. In the altered form, several typographical errors were deliberately made on each page of text, and corrections were made by "X-ing out" errors or by running lines through them. Occasional words were omitted from the normal typed line and placed above the line with caret insertions.

Reproduction

In the unaltered form of this variable, reproduction was clean.

In the altered form, the pages of copy were intentionally dirtied and mats were not cleaned before the pages were reproduced. In the fourteen pages of the altered forms, the following were included: thumbprints on two pages; careless lines across two pages; one page run with text noticeably crooked on the page, and four pages with several ink spots in otherwise clean areas.

Experimental Design

Because three variables were used and there were two conditions on each variable, eight versions of the test passage were used in a 2 X 2 X 2 factorial experiment. One (the base passage) contained the unaltered form of all three variables; one contained the altered form of all three variables; each of the remaining six versions contained one of the six possible combinations of altered and unaltered forms of the three variables. The eight versions of the reading passage may be represented as shown in Table 2. The numbers used on the table are for illustrative purposes only and have no significance beyond identifying the versions of the reading passage. The letters in each column indicate the conditions on the variables in one treatment of the passage. "*A*" indicates the unaltered condition of the variable, and "*B*" indicates the altered condition.

Table 2

	Treatment							
Variable	*1*	*2*	*3*	*4*	*5*	*6*	*7*	*8*
Margin	*A*	*A*	*A*	*A*	*B*	*B*	*B*	*B*
Typing	*A*	*A*	*B*	*B*	*A*	*A*	*B*	*B*
Reproduction	*A*	*B*	*A*	*B*	*A*	*B*	*A*	*B*

Test Administration

In each audience, reading passages were assigned to the subjects on a random basis. Subjects at the Air Force Academy and at the University of Dayton were tested by their class instructors in regularly scheduled classes. Because class sections met at various hours during the day, any variation in performance attributable to the hour of testing was normalized by the randomization in assignment of reading passages. Subjects tested at the Air Force Institute of Technology were all tested during the same hour in regular classrooms. The testing was done during Commander's Call, an hour set aside each week (Thursday at 11:00 a.m.) for any meetings, special training, or announcements involving all or a major portion of the student body. The test was administered by members of the faculty assigned as Faculty Advisors to the groups tested.

A measure of verbal ability (Test 10 of the California Mental Maturity Test, Level 5) was administered to each test group, and then approximately five minutes' instruction was given. Questions about procedure were answered and the subjects were told to begin reading their passages. When each subject completed the reading, he recorded his reading time, answered the test questions, and made the two judgments about the author. Three time warnings were given during the administration, and any who finished early were permitted to leave when they were done.

Statistical Measures

Because reading passages were assigned to the subjects on a random basis, readers of one version of the passage in a given audience might have been better readers than those assigned another version. If this were not taken into account in the statistical analysis, results would be open to serious question. As in previous experiments, preliminary analysis indicated a positive relationship between verbal ability scores and comprehension scores in each audience. No such relationship was found between verbal ability scores and the other measures taken of the effectiveness of the message. Therefore, to normalize the effects of any differences in verbal ability between cells (the readers of the various versions of the reading passage within an audience), analysis of covariance was used to analyze comprehension scores. Scores on the other three measures were analyzed by analysis of variance. Comparisons of total comprehension between audiences and groups of audiences were made by the Scheffé method.

Conclusion and Discussion

Effects of the Variables

Only one significant effect was found at the 0.01 level of probability. This was an uninterpretable second-order interaction in the Judgment of Author Knowledge in Audience 14. Five main effects and five first-order interactions were found at the 0.05 level of probability. Each variable tested, each audience, and each measure of the effectiveness of the message (comprehension, reading time, and the two judgments) was involved in one or more of these effects. In each main effect and in each interaction subject to easy interpretation, the more effective form appeared to be the unaltered (good typing and clean reproduction) form of the variable(s) concerned.

Somewhat fewer significant effects were found than might have been expected, and those found were at a lower level of probability than that accepted in this experimental series. But the direction of all main effects and all readily interpretable interactions was that which might have been expected. They suggested that the bad typing and bad reproduction impaired the effectiveness of the message.

Two possible reasons for the weakness of the effects suggest themselves. First, the alteration of the variables may simply have been insufficient to cause significant effects by the analyses applied. The variables themselves may cause effects but the altered forms used simply may not have been sloppy enough to produce effects strong enough for measurement by the criterion and the procedures applied.

The second possible reason lies in the time involved in the test administration itself. The subjects had to work quickly to get through the reading material and complete the test in the time available. In fact, they had to work more quickly than did the subjects in earlier experiments, and they had little time to review what they had read. It is possible that they moved so quickly and that their attention was so directed at the content of the technical message that they simply did not react to the copy preparation and reproduction as they might have otherwise.

A similar experiment with shorter versions of the message and greater variation in the copy preparation and reproduction should shed some light on these possibilities. Materials for such an experiment have been prepared, and the first audience (approximately 200 subjects) was tested in May at the General Motors Institute.

Comparisons of Total Audience Comprehension

All comparisons of total comprehension by individual audiences and groups of audiences are in agreement with findings in the four previous experiments. They support the categorization of audiences used in this experimental series and the hypotheses that 1) more intelligent subjects understand this material better than do those at a lower level of intelligence, and 2) at a given level of intelligence, subjects with known technical inclinations understand the material better than do comparable subjects without known technical inclinations.

None of the findings from the seven contrasts made was at variance with the results of contrasts made in the previous experiments. The only mild surprise was in the size of the difference between scores made by bright young men with known technical inclinations (Air Force Academy: Audience 16; mean score 9.74) and bright young people without known technical inclinations (University of Dayton: Audiences 14 and 15; mean scores 5.94 and 5.74, respectively). The scores in Audience 16 were about as expected from the testing of similar groups in previous experiments. The scores of Audiences 14 and 15 were somewhat lower than might have been anticipated.

The low comprehension scores in these audiences might be attributable to a difference in motivation. In all other audiences, subjects were told nothing about the purpose of the test. Many may have guessed that it was a part of an experiment, but they could not be certain. In Audiences 14 and 15 (University of Dayton) the subjects were told that the testing was a part of an experiment and that they would not be held accountable for their scores or for the judgments made.

To determine more exactly the validity of this assumption (fore-knowledge of purpose caused lower scores), a second audience at the General Motors Institute has been tested with exactly the same materials used in testing the first (referred to above). The two audiences were composed of comparable groups of students in the Freshman Composition course. The first audience was told nothing about the reason for the test administration, and the second was told that it was a part of an experiment. Comparison of the scores and the judgments should indicate the effect of fore-knowledge of purpose upon subjects tested in this and similar experiments.

Types of Outlines Used by Technical Writers

Blaine K. McKee
Colorado State University

The value of the sentence outline as a necessary step in report writing is mainly a myth of high school and college teachers of English composition. Though long revered by many of them, the sentence outline finds little acceptance among professional report writers.

That was the main conclusion of a survey of professional report writers. Those surveyed were a random sample of members of the Society of Technical Writers and Publishers, most of whom regularly write reports. Only five per cent of those surveyed use a formal sentence outline as a step in writing a report.

As a student in high school and college English courses I had been taught the importance of outlining in the preparation of a research paper. And the sentence outline was presented as the optimum. I found it difficult, however, to do the sentence outline until after I had finished the paper.

When I became a teacher, first of English composition and later of technical writing, the texts that I used advocated outline, especially the sentence outline, as a necessary step in the preparation of any good paper of more than a few pages.

My faith in sentence outlining grew less when students asked if they could turn in their sentence outlines after they had finished their long reports. I pointed out to them that the sentence outline was to be a help to them in the writing of their papers, not something to be written from their final work. They said that they found it impossible to do the sentence outline until they had finished the paper. Once the sentence outline was finished, it was easy to write the paper, as it was necessary to write the paper first to be able to do the outline.

Then I started asking students in all my courses how many had been required to write sentence outlines in high school or college English courses. The answer here was nearly unanimous. When asked how many had written the outlines before the papers, all but one or two hands in each class came down.

Having found that most of my students found the sentence outline of no help in doing research papers, I gave the classes the choice of any type outline. Nearly all of them chose a short, informal topic outline. But not all of the students found the sentence outline of no use. An occasional student found that he could use it and that it was very helpful to him.

INFORMAL SURVEY

Before I did the formal survey I checked with some friends in the business world to see what type of outlines they used.

Richard W. Dodge, Director of Technical Information, Westinghouse Electric Corporation, Pittsburgh, rarely outlines. Occasionally he jots down a few key words just to remind him of his sequence of thought, but for the most part he does his organization in his head. He checked with six of his Technical Information men who write articles and found that they used an approach similar to his.

Al Krieg, former Chicago District Director of Public Relations for U.S. Steel Corporation, does not use any written outline. He uses what he calls the "faucet" approach: he just starts writing and continues until the "faucet" runs dry.

THE FORMAL SURVEY

A random sample of 180 members of the Society of Technical Writers and Publishers were surveyed, and answers were received from 80, or 44.4 per cent. They were asked if they would use an outline for a paper of 3,000 words or so; if yes, what type of outline they would use; whether they would prepare the outline before doing any writing or after a rough draft; what advantages they found in their method; and what method they would advise an inexperienced writer to use.

The group that replied had an average of 5.5 years in their present jobs and had been in technical writing for an average of 12.5 years, ranging from two month's experience to 43 years. While nearly all did some writing as part of their jobs, some spent most of their time supervising the writing of others, while some were teachers. Some spent 100 per cent of their time writing while others spent as little as 5 per cent. On the average they spent 54 per cent of their time writing or supervising writers and, 19 per cent of their time editing.

The answers to the survey not only gave their methods of outlining but also told much about their technique of writing.

Only four of the 80, or 5 per cent, use a formal outline. At the other extreme, another four, or 5 per cent use no outline. The other 90 per cent used a topic outline or a combination of a topic outline and a sentence outline. Table I gives the types of outlines used and their percentages.

Table I – Types of Outline Used

Type		Percentage
Topic Outline		60.00
(Word)	(8.75 per cent)	
(Word and Phrase)	(51.25 per cent)	
Combination of Word, Phrase, and Sentence		30.00
Sentence Outline		5.00
No Outline		5.00
Total		100.00

VALUE OF OUTLINING

Most agreed that outlining is helpful, but again and again it was stressed that each person should experiment and find out what approach works best for him. H. Swift, Associate Professor of Communications at General Motors Institute, Flint, Michigan writes: "I suggest to my students that they should use a technique which is most compatible with their psychological make up. That is, what works for one person may not for another, and therefore each student has to discover for himself what is best for him."

Most in the survey, 95 per cent, use some type of outlining and find it a helpful tool in the writing of a report. Many point out, however, that research, note taking and "meditation" must precede any outlining. Some do a rough draft before they do the outline.

Dr. Francis H. Archard, Consultant for Technical Communications, Newton Centre, Massachusetts, advises, that "by all means a writer should–nay must–prepare an outline for every document, whether it be a one-page letter or a book."

Having an outline is a good way of getting started for many writers. J.R. Cathey, Senior Engineering Writer, Magnavox Company, Champaign, Illinois, finds it easy to "put the flesh on after the "bones" of the outline is prepared." Thomas G. Foster, Engineering Writer, Westinghouse Electric Corp. finds a blank sheet of paper "awesome and frightening" and to overcome this he tries to get something on paper as soon as possible, usually a loosely structured phrase outline.

For A. J. Kelly, Engineering Writer, with GD/Convair, San Diego, Cal., "the outline insures that no subject is forgotten once a writer becomes engrossed in research and writing of complex data."

NO OUTLINE (5 per cent)

Four use no written outlines as they find it unnecessary and in some cases a hindrance to creativity. William Wekoun, self employed writer from Aberdeen, Maryland, organizes his thoughts in his mind. Of forcing students to write outlines he says, "I am very much against requiring students to outline, turning in their outlines like good kindergarteners to prove that they have done them. I was forced to play that game as a student, and usually wrote the themes first, then the outlines afterwards."

Hal Shemia, Technical Writer Specialist for Aerojet-General Corporation, Azusa, Cal., compares the outline with pre-selection of colors by a painter. An article might suffer from the rigid restraint of an outline the same way a painting would suffer were an artist to restrict his use of colors in advance.

AN INDIVIDUAL CHOICE

Most are confident that their system of outlining probably would work for a beginner, but they stress that each should find his own way. Carl H. Harris, Writer/Editor, General Dynamic Boat Division, New London, Connecticut, says that "the advantage in my method of outlining is only that it suits me as an individual. I would suggest that the beginning writer try various systems until he finds one that suits him best."

Dewey E. Olsen, Senior Information Representative, IBM, Mohopac, New York, advises beginning writers: "If you can use an outline comfortably, do, because it helps structure your work. If you find it difficult to use one, do the outlining in your head and plough right into the writing."

UNCOMMON APPROACHES

As an aid to the beginner in writing a report, Mrs. Arlene D. Schaller, Technical Writer and Editor for Rohm and Haas, Philadelphia, suggests that they first write an abstract or summary of the report and then write the outline from it.

The Chairman of Humanities at Kalamazoo Valley Community College in Michigan, Peter D. Rush, tells beginners to first brainstorm to get all their thoughts down on paper, and then to write the outline.

The tape recorder is found useful by several in writing their papers. S. A. Miles, Vice-President, Miles-Samuelson, Inc., of New York City finds that "many people freeze when they start writing, and although they know the subject thoroughly, they spend an inordinate amount of time in getting started." He overcomes this inertia by talking into a tape recorder in a free-association sort of way. During the playback he jots down key sentences or phrases and he finds that his paper begins to arrange itself.

ADVANTAGES OF EACH METHOD

Each person in the survey found advantages for him in the method he was using. We have already examined the reasons why some prefer no outline. Now we will look at the advantages claimed by the other methods.

Topic Outline (60 per cent)

As shown in Table I, the topic outline was the most used with 60 per cent using a topic outline, consisting of words or phrases. The topic outline will be divided into two types: single words, and phrases alone or in combination with single words.

Word Outline (8.75 per cent). The simplest of all outlines is the one using single words for headings and sub-headings. Ease and flexibility of operation are the main reasons given for its use.

Richard M. Davis, Associate Professor, Air Force Institute of Technology, Wright-Patterson Air Force Base, Ohio, recommends the word outline as "easy to make, easy to check, and easy to alter."

Phrases and Words (51.25 per cent). Over half of the 80 surveyed used a topic outline in which they used phrases or a combination of single words and phrases. Those that used the combination estimated that they used an average of 40 per cent individual words and 60 per cent phrases.

Efficiency and speed were the reasons given for using this type outline. As Louis Percia, Publications General Supervisor, The Mitre Corporation, Bedford, Mass., puts it, this type outline "saves time, does not require much effort, but organizes thought."

Combination Outline (30.00 per cent)

Twenty-six use an informal outline that is a combination of methods. If a sentence comes to mind that might be used in the final paper, it is written down in the outline. Don Kelly, Hydrologist, U. S. Geological Survey, Woodbridge, Virginia, finds that "by using a combination I remain flexible, avoid following the same pattern as a cop-out, and can set inspirational sentences down at once."

Those that used a combination outline were asked to estimate the percentage of each that they use. The percentages of each are as follows:

Phrases		41.4 per cent
Sentences		35.5 per cent
Words		20.9 per cent
Others	(Tables, diagrams, Illustrations, and PERT charts)	2.2 per cent

Sentence Outline (5 per cent)

Four, or 5 per cent, find that the sentence outline works best for them.

John R.Golaszewski, Jr., Technical Writer, RF Surface Group, Radiation, Inc., Melbourne, Florida, gives as the advantages of the sentence outline: "speed, efficiency, clarity, aptness of thought and complete content." The length of his sentence outline is about 10 per cent of the finished report. He finds the sentence outline perfect for short-notice assignments.

Although Charles Strong, Instructor, University of Missouri, does not use the sentence outline for writing papers, he points out that it is useful in classification: "Whereas I believe that the formal outline is a valuable exercise in classification and a succinct means of displaying one, I believe that it is a hindrance in writing." He wonders, however, if people might not have been so "brainwashed to the effect that the formal outline is an indispensable prelude to good writing that they metaphorically put on cap and gown when questioned and will champion the formal outline." Such effect in this survey, if any, however, appears to be minimal.

SUMMARY

Despite its championship by English teachers, the sentence outline is used by only a small percentage of professional report writers. There are many other effective ways of organizing the material for a report–some using outlines and some not.

David W. Hills, Head, Editorial Branch, Naval Missile Center, Point Mugu, California, sums it up: "Each of us works differently; what is good for one may not work with someone else. I feel that outlines are required more by some people than by others. The clearer the thinker, the briefer the outline."

5 EXPERIMENTS IN TECHNICAL WRITING

The authors decided to test these five "rules" of good writing. Some are true; others are not.

JOSEPH H BLOOM and SETH A GOLDSTEIN
Rensselaer Polytechnic Institute

In these experiments, the general procedure was the same: two versions of a piece of technical writing—one the original version, the other the special variation (abstracted, summarized, repetitive, "easier," or pictorial)—were presented to the same or comparable audiences. Tests measured comprehension, reading speed, or retention.

Are Abstracts Valuable?

Should you always write an abstract for your report? Should journals insist on an abstract preceding the article?

We tried to answer these questions by exposing the same audience to two versions of a description of aircraft detection systems. The abstracted version, though shorter, took comparatively longer to read (4 minutes for 572 words in the abstracted version, compared to 14.5 minutes for 3100 words in the original).

The key to this apparent contradiction lies in high "information density"—an academic term defining content per given number of words. We found that the rapid reader suffers most from high information density. When we tested him on how much information he had absorbed, his score was lower than that of the slow reader. It would appear that unless he is told that he is reading an abstract, the fast reader often skims right past pertinent information.

These results question the value of abstracts except for the purposes of information retrieval. It is possible that a lot of abstracting wastes the writer's time and many people object to reading such highly condensed material.

Summaries: Useful or Not?

In technical writing the abstract condenses everything from the beginning to the end, the summary picks out only the highlights or orients the reader to important facts. Reports, for example, usually have summaries at beginning and end. Therefore, the same question raised in connection with abstracts can be asked here: Is too much emphasis being placed on the summary?

We gave two versions of a piece of technical writing—one with a summary and one without—to the same audience and found that reader comprehension and retention improved noticeably when summaries preceded the main article. Apparently they soften that first rough contact with a technical subject.

Do Repetition and Emphasis Help?

All writers agree that a certain amount of repetition and emphasis of important facts is absolutely necessary. But too often, the technical writer confines this to the opening and closing statements of his article. He ends it there, hoping the reader can take care of himself throughout the rest. Well, the reader may—if he knows as much about the subject as the writer does. More often than not, he loses track of the salient facts in the mass of detail.

We found that reader retention increases with two practices: (1) constant repetition of important facts and ideas, and (2) highlighting these same facts and ideas with attention-getters—underlining, capitalization, charts, color. To prove this, we presented three comparable audiences with three versions of a technical advertisement about textile dyestuffs as taken from a commercial magazine. One was the original as designed by the company; a second repeated important information over and over again; a third reduced the information to tables and listings.

The result? The second and third groups, those receiving the repetitive and highlighted versions, retained considerably more information than the first group.

Do Shorter Words Increase Reading Speed?

Not according to the results of these experiments. Two groups of college students were presented with a paper on a highly academic subject. It was full of abstract words, long sentences, and abstruse ideas. One group received the article in its original form—written for a college-level audience according to a popular reading formula. The other group read the same article rewritten for a sixth-grade level. Results showed that the reading speed and comprehension were approximately the same in each case and, if anything, the use of shorter words hampered the process.

It might be concluded from this research that the best way to communicate is to use the level of your readers as a guide: don't oversimplify a piece of copy that is to be presented to a high level audience, and, conversely, don't go "highbrow" on what is to reach a lower level of readers.

Is a Picture Worth 1000 Words?

Pictures or sketches may work both for or against the technical writer. Their effect was pointed up in the last experiment of our series.

A building was described in words and one group of readers was asked to draw the floorplan from this verbal description. Another group was shown exterior views of the building and asked to draw the floorplan. There was one correct solution. Both groups came out with about the same number of correct plans. But the group shown the picture added, in many cases, things that were not asked for, indicating that their imaginations were more stimulated than those in the first group.

The technical writer can apply the findings of this experiment as he wishes. Sometimes it may be that this visual stimulation complicates rather than clarifies the reading material; at other times it undoubtedly enhances the total effect.

Where These Experiments Are Heading

Where will these experiments lead us? Admittedly the Rensselaer tests do need further application, and each academic year will add more to our understanding of the communication process. Some topics slated for the future will deal with the use of rigid outlines, technical reporting techniques, further study of readability formulas, and the human factors in writing.

Part V
Following Through

MICHAELSON, in the second article in this section, states: "the discriminating reader...examines a paper not only for what it contains but for what it lacks." The preceding sections of this book examined what the report should "contain"; this one focuses on the final phases of analysis and on the last minute details that must be completed to ensure that there are no pieces "lacking." Following through, as the articles in this section attest, is what makes incomplete reports complete and complete reports better.

Gould begins by identifying ten problems that are common to many technical reports. Of these, several are treated in depth in the preceding parts of this book; others—those considered as additional to rather than part of the report-writing process—are considered in the articles that comprise this part. Michaelson, using an analogy to solid-state theory, then explains the need to identify and eliminate omissions in information (gaps) and ambiguities in writing (traps) that affect the communication process.

The next seven articles discuss various "gaps" and "traps" that result from a failure to pay close attention to follow through. Kennedy explains how to write titles that are informative for readers and "machineable" by indexing systems. Sekey uses Shakespeare's *Othello* as a text to illustrate the distinctions among and purposes of abstracts, conclusions, and summaries. Making use of the concept of "context" introduced earlier by Miller, Broer argues for the integration of text and figures. These components of the report, he says, are interdependent and, therefore, authors should follow the premise that "figures + captions + text: *one* story." However, because "tables lack the contextual clues that running text provides," Buehler describes how to design effective, self-contained tables. Her guidelines offer the writer a means for turning an "amorphous mass of numbers into a crisply organized presentation of data." In the next article, an analysis of several advertisements shows how we can use subheads to produce reports that are interesting, informative, and easy to read.

Although the Goodrich and Roland article would have fit nicely into the part on research results, we've included it here because checking the accuracy of reference citations should be considered a necessary step in the follow through. The "extremely high" percentage of errors these authors found in the citations they checked can and should be eliminated. Careful proofreading, using the techniques suggested by Turner, will help an author remove these and other errors that are often made as a result of carelessness.

In the next-to-last article, Clauser re-emphasizes a point made explicitly or implicitly by every article in this book: *Keep Your Reader in Mind*! As he says: "it is the reader's reception of our writing that measures the extent to which we successfully communicate our ideas." This attention to our reader remains a crucial consideration—irrespective of whether we are in the planning, writing, or follow through stage. Especially important, however, is that the paper never be considered as finished until you have reread it with your reader in mind.

The final article in the book details a method for successfully completing what might be considered as the ultimate step in following through: getting your article published. White provides a "journal analysis matrix" designed to increase the chances of having your article published. In addition, at the same time that this matrix serves as a step in following through, it also serves to remind us of the matrix Stratton discussed in the first article of this book. As such, it is a reminder that our ultimate success as authors depends upon our conscientious attention to each and every step in the writing process, and that to be successful as authors requires that we understand the task of writing, master the skills, and practice the art.

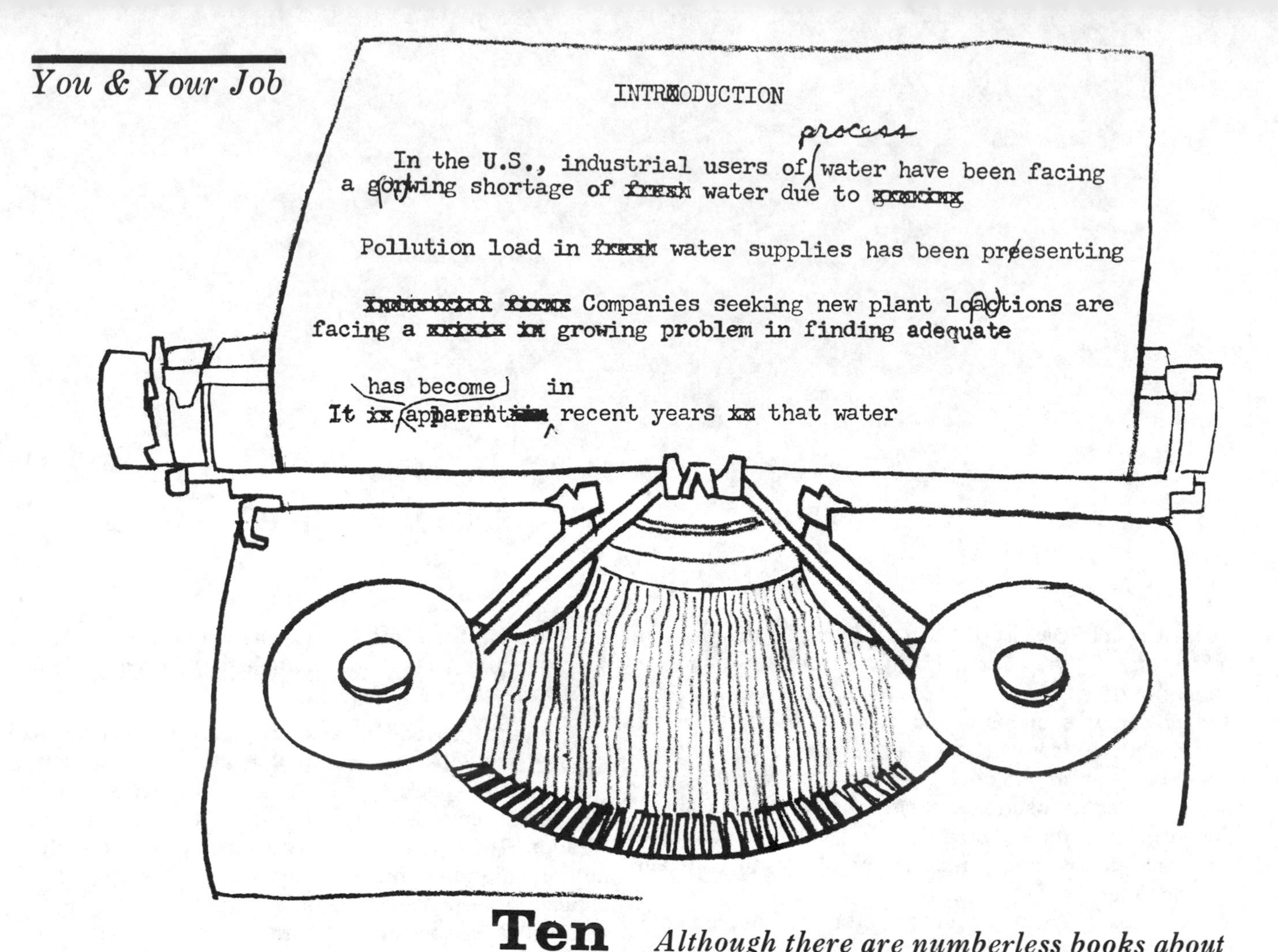

Ten Common Weaknesses In Engineering Reports

Although there are numberless books about technical writing, many engineers still commit basic writing errors. Here's an expert's examination of the most common mistakes.

JAY R. GOULD, *Rensselaer Polytechnic Institute*

One company that I recently worked with had become concerned over the quality of the reports being turned out by its engineers and scientists. After examining more than 100 of these "problem" reports, a pattern of common errors began to emerge. These ten common mistakes are outlined below.

If you feel that your report writing could stand improvement, these ten points can be used as a handy checklist. The next time you write a report, look over your first draft to see if you can improve on any of these specifics:

1. Abstracts

The abstract is a vital part of the report; it will probably be read by more people than read the report itself. Abstracts are also valuable for indexing and filing.

The fault with most abstracts is that they are skeletal outlines; they only define the subject matter. Here is a skeletal abstract:

This report describes and compares some of the less orthodox processes used in the manufacture of salt. It contains a historical review of older processes together with their advantages and disadvantages.

This abstract is so short as to be noncommital. It contains very little that is specific; the author has staked out some territory and that's about all.

To correct: Go through your data, pick out the salient facts, and especially conclusions and recommendations, if any. Make the abstract fit your piece of writing. Here is an *informative* abstract for the same report:

This report describes and compares some of the less orthodox processes used in the manufacture of salt: (1) the Alberger Process in which a fresh brine feed and recycle brine is heated in a number of shell and tube exchangers; (2) the Morton Process, similar to the Alberger Process, except that only one heater and one flasher are used; (3) the International Process in which all heating is by direct injection of steam into the brine; and (4) the Richards Process in which hot brine is flashed under vacuum in a closed flasher instead of being cooled in contact with an open pan. Advantages and disadvantages of the four processes are discussed.

2. Introductions

If you find that the beginnings of your reports are disjointed, forget the word "introduction" and think of what you should be doing for the reader. Remember, he may think in almost opposite terms from you.

To correct: This check list should help:

1. Relate the introduction to the needs of the reader. Try to imagine what he will want to know first. It may be the cost of the project, the competition involved, or how the project came about.
2. Show the purpose and scope.
3. Identify the subject matter.
4. Relate the subject matter to other projects.
5. Show the basic method and procedure followed in carrying out the project.

This excerpt from an introduction combines many of these features:

Steady business improvement during last year's final months enabled most companies in the chemical process industries to make a good showing in the year-end financial reports. Sales and profits hit close to the high figures reported for the previous year.

This report gives the sales and earnings for five process industry groups: chemical and allied products; paper and allied products; petroleum refining; rubber products; and stone, clay, and glass products.

3. Headings and Titles

Uninterrupted pages of text make much technical writing difficult to read, and equally difficult to retain the information.

To correct: Don't hesitate to break your report into headings. Headings serve at least three purposes:

- They organize the material into logical units and force you to stick to your subject matter.
- They make reading easier by letting in some air and white space.
- They make it easier for the reader to refer back to particular sections.

4. Paragraphing

Faulty paragraphing stems from haphazard arrangement of items within the paragraph as well as an apparent lack of knowledge of what a well-constructed paragraph can do. Faulty paragraphing is the cause for at least one-third of reports being turned back for revision. Keep these two things in mind:

- Each paragraph is a unit. If the main idea in a paragraph can be expressed in a single, highly compressed sentence, you have done a good job.
- Paragraphs can be constructed according to function. For example, perhaps the most useful device for paragraphing is the topic sentence. This is a preliminary over-all statement that can be followed by details inherent in the topic statement.

If your writing lacks good structure, look to your paragraphing. Pay more attention to your topic sentences and the supporting details, as in this example:

Some accidents are caused by a combination of laziness and ignorance. An operator may lack the energy and personal drive to find out the governing circumstances and not have the technical knowledge to appreciate some special requirement.

5. Listings and Tabulations

The example below shows how a set of instructions might look if written in a straight narrative form:

To operate the digitally controlled tube tester, you should first turn on the main switch and wait for the green light. Then you insert the tubes from left to right. After that, you move the calibrate switch to run position, the socket switch to manual position, and the test switch to automatic position. Then you push the start button.

To correct: Take your rough copy and see how you can clarify it by adding listings and tabulations. Don't ask the reader to dig out the information. Help him by numbering the various steps, by underlining, italics, capitalization, spacing, and other devices.

Here is how the author of the instructions on the tube tester could reorganize his material:

THE DIGITALLY CONTROLLED TUBE TESTER

OPERATION

1. Turn on MAIN switch and wait for green light.
2. Insert tubes from left to right.
3. Move switches:
   a. The CALIBRATE switch to RUN position.
   b. The SOCKET switch to MANUAL position.
   c. The TEST switch to AUTOMATIC position.
4. Push START button.

6. Point of View

Examine this piece of writing:

We are glad to provide you with information about the coloring of negative slides. A copy of your completed report would be greatly appreciated when you are through.

You should do the coloring process in three steps: Apply a water-insoluble lacquer; after the lacquer has dried, we always paint the enclosed area with a water solution; after the dye has dried, the lacquer should be removed.

When you examine this section closely, you find that the author has used various approaches. He starts out by assuming a "you" attitude then he editorializes with his use of the pronoun "we." He runs through a gamut of styles: active, passive, personal, impersonal, and conditional. He has been guilty of a shift in *point of view*.

To correct: Adhere to a single point of view, as in this corrected example:

We are glad to provide you with information about the coloring of negative slides.

We find that the coloring process is done best in these three steps:

1. Apply a water-insoluble lacquer immediately around the lines or letters to be colored.
2. After the lacquer has dried, paint the enclosed area with a water solution of an aniline dye.
3. After the dye has dried thoroughly, remove the lacquer with a suitable solvent.

7. Transitional Material

Industrial writing should be sparse and to the point. But it's possible to be so sparse and so lacking in guides and signs that you are practically unintelligible. For example:

The construction of a technical paper is essentially the same as the construction of a report. The paper usually consists of a main presentation plus a short abstract. The technical paper does not contain the separate elements of the formal report. It should be broken up into logical sections. Many technical papers follow the outline of abstract, purpose, conclusions, and procedure.

It is difficult to follow through the author's intentions in such a condensed piece of work.

To correct: Supply phrases and guide words that lead the reader from one thought to another—and provide headings, listings, and tabulations when needed. Here is the same paragraph with the addition of a few such devices.

The construction of a technical paper is essentially the same as the construction of a report. But the paper usually consists of one main presentation plus a short abstract. Although the technical paper does not contain all the elements of the formal report, it can usually benefit by being broken up into logical sections. As a result, many technical papers follow this outline:

(1) Abstract
(2) Purpose
(3) Conclusions
(4) Procedure

8. Figures and Tables

Improper positioning of tables and figures in the text can create confusion in the mind of your reader.

To correct: Determine whether the tables or figures should be placed in the text itself or should be reserved for the appendix.

An *internal* table is one that must be used closely with textual material. In such a case, place it in the report proper as close as possible to the reference in the text.

A *record* table contains supplementary material, helpful but not necessary to the discussion at hand. Place it with all other supplementary material in the appendix or other terminal sections so that it doesn't distract from the main presentation in the central body of your report.

9. References and Bibliographies

Many writers are inconsistent when recording references and bibliography.

These citations came from an engineering report I read recently.

Kunin, R., and Myers, R. J., Ion Exchange Resins. New York: John Wiley and Sons, Inc., 1950.

Smith, H. R., and P. D. Jones, Elastic Creep of Automobile Tires, Holt, Rinehart, and Winston, 1958.

Two items, and two methods of recording.

To correct: Pick a system and remain faithful to it in any given report.

10. Keep the Reader in Mind

This check is really an over-all test of the effectiveness of your reports. After you have completed the first draft, see if you have aimed at the reader in these three particulars:

(1) Have you devised a method for getting the reader to start reading? This points up the importance of a good introduction.
(2) Have you devised methods to keep the reader reading? That is, is your organization and structure as clear and logical as possible?
(3) Have you decided what you want the reader to retain? This will be a guide to how you should emphasize material in abstracts and summaries, appendixes, tables and graphs.

Key Concepts for This Article

For indexing details, see Chem. Eng., Jan. 7, 1963, p. 73 (Reprint No. 222). Words in bold are role indicators; numbers correspond to AIChE system.

Active (8)	**Passive (9)**
Writing	Engineering
Improving	Reports

Information faults in manuscripts for professional journals are caused by two kinds of structural defects: the information gap, **in which essential information is omitted, and the** information trap, **in which important points are not emphasized and are obscured in the paper.**

INFORMATION GAPS AND TRAPS IN ENGINEERING PAPERS†

By HERBERT B. MICHAELSON*

Abstract: An analogy to the energy band structure in semiconductors is given. An information gap corresponds to the "forbidden gap" in semiconductor theory and an information gap corresponds to the "trapping states" of impurity atoms. Illustrative examples of gaps in manuscripts are poor reader orientation, lack of proper information in illustrations, and inadequate conclusions. Examples of traps are semantic confusions, superfluous illustrations which obscure the issues, and critical points of the paper which are "buried" in the manuscript.

Introduction

After an author has written a manuscript for an engineering or scientific journal, he invariably reviews his draft copy for information content. If the author tackles this job in a spirit of self-criticism, he usually discovers gaps in his exposition and finds himself filling in the missing bits of information. If, on the other hand, he reads his work with self-satisfaction and with pride in a task well done, he immediately falls victim to what psychologists and physicians call *euphoria.*

*Assoc. Editor, *IBM Journal of Research and Development,* IBM Corp., New York, N. Y.

†Presented at 1959 Dual National Symposia, IRE, PGEWS, Los Angeles Sessions.

The dictionary meaning of euphoria is "a sense of well being and bouyancy." The euphoric author tends to view his manuscript through rose-colored glasses. He takes pleasure in finding the virtues and strong points of the paper. The tinted lenses protect the author's eyes from glaring omissions and ambiguities in his writing, and he subconsciously fills the gaps from his mental store of background information. Euphoria begets indiscrimination.

A manuscript which suffers from clogged information channels might well obscure a sound piece of engineering work. We might think of two kinds of communication defects in such a paper. The first is the *information gap,* a piece of information which was left out either unintentionally or because it had been considered unnecessary. The second is the *information trap,* an ambiguity which ensnares the unsuspecting reader.

The purpose of this paper is to show the nature of these information faults in manuscripts and to illustrate the kinds of defects commonly encountered. The "in-

Reprinted from *IRE Trans. Engrg. Writing and Speech,* vol. EWS-2, pp. 89-92, Dec. 1959.

formation problem" will be interpreted briefly in terms of the energy band theory for semiconductors, where we find gaps and traps in abundance!

A Band Model of Written Communication

From a well-balanced paper, the information flows freely to the reader. This movement of ideas is analogous to the concept of current flow in solid-state theory. Physicists think of electron flow in terms of the available energy levels; we can consider the flow of ideas in terms of available information levels in a manuscript.

In Fig. 1 we see the conventional energy band picture for semiconductors. When electrons in the valence band acquire enough energy, they may jump the energy gap into the conduction band. Impurities in the semiconductor provide trapping levels. A valence electron may be trapped in the lower impurity level before it can reach the conduction band.

In Fig. 2, we have our analogous band picture for written communication. In this intrinsic case, we have no impurities or trapping levels. Instead of the energy levels in the solid, we have *information levels* in the manuscript. At the bottom, corresponding to the valence band, is what we have called the *valid data band,* which contains the essential information of a manuscript. By means of careful and complete exposition, an author adds information levels as he writes his paper, permitting ideas to move up into the *communication band* at the top. When important information is lacking, there is an *information gap.* The wider the gap, the fewer ideas communicated to the reader. So far, we have a close analogy to electrons jumping into a conduction band and to the resulting flow of current!

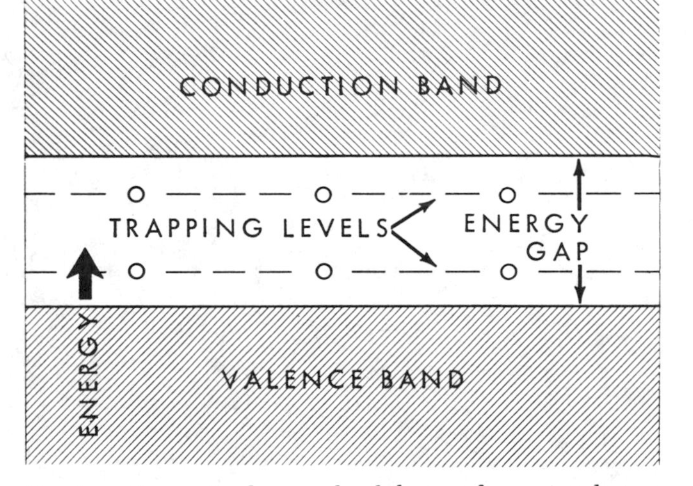

Fig. 1 — Conventional energy band diagram for semiconductors.

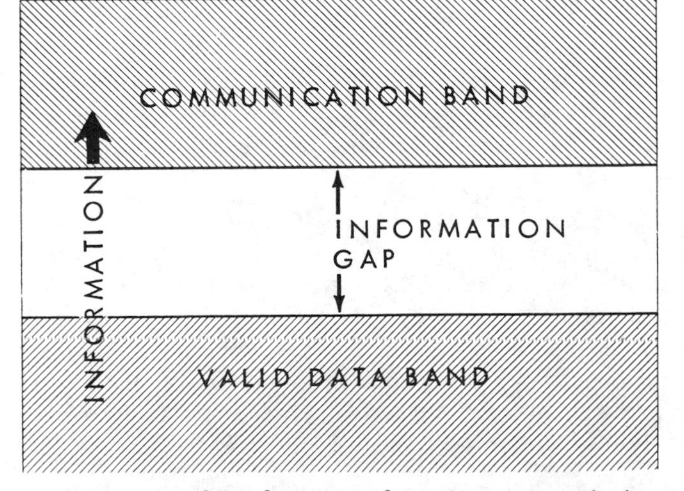

Fig. 2 — Proposed band structure for written communication. Missing information leaves a gap in the manuscript, preventing ideas from reaching the reader via the communication band.

In semiconductor theory, the addition of trapping states increases the flow of electrons and holes because the impurities provide smaller energy gaps. In the information-band structure in Fig. 3, however, "impurities" (manuscript defects) widen the gap. According to our model for information flow, the traps do not aid communication; rather, they are a hindrance and a source of frustration, confusion and general discomfort to the reader. The fact that our analogy to energy band structure seems to fall down at this point is hardly surprising, however, since information traps by their very nature play a disrupting role!

Regardless of the contrary nature of our information traps, solid-state theory could contribute a great deal more to our understanding of writing problems by analogous concepts of recombination (of previously published data), of lifetimes (spent in rewriting), of holes (in the author's arguments), of imperfections (in exposition), and other possible analogies. We can profitably confine our attention, however, to the nature of gaps and traps and their practical relation to sound manuscripts.

Information Gaps

Before we leave our analogy to solid-state theory, we might observe that an area of omitted information in a manuscript is like a "hole" in a semiconductor. Items not included in a manuscript actually have properties! These "information vacancies" can be recognized, with a

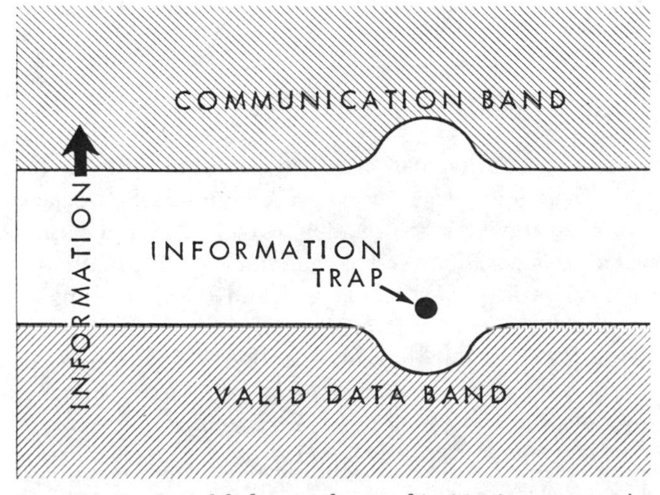

Fig. 3 — Band model showing how ambiguities in a manuscript will trap information before it can reach the communication band.

little practice, and are looked for by editors and by the most critical class of readers, the manuscript referees of the various professional societies. The discriminating reader thus examines a paper not only for what it contains but also for what it lacks.

How can an author strengthen his manuscript against these probes for weakness? There certainly are no cookbook rules for including or omitting certain classes of information. Obviously, much depends on the kind of paper being written, on the nature of the intended journal, and on its readership. And, because writers are not automatons, a great deal depends on the writer himself—on his literary style and his sense of proportion.

If the critical reader sees in the manuscript an unsatisfying picture of the results, he will look for certain general kinds of information which are commonly omitted. These will be discussed here. Our selection of information gaps, unfortunately, is neither a catch-all nor a cure-all, and every paper must be judged on its own merits. Indeed, we are happy to give the reader the opportunity (and the pleasure) to find the information gaps in *this* manuscript!

Let us now examine the various categories of information gaps. The first is a favorite target of the editor's blue pencil: *poor reader orientation.* The target, however, seems to vary in size and importance, depending on who is aiming at it! There are at least two distinct schools of thought on the matter of reader orientation. One is that a paper written for specialists should dispense with painfully obvious and space-consuming background material. According to this way of thinking, a paper may make an immediate frontal attack on the problem at hand, without prior pleasantries or wordy explanations. Many such papers are published by reputable journals like the *Physical Review* or the *Transactions of the Mathematical Society*. Some of the IRE TRANSACTIONS accept highly specialized papers written in this fashion.

A second school of thought, exemplified by journals like the PROCEEDINGS OF THE IRE, holds that authors should give ample background material for the benefit of those readers who are nonspecialists. Early in the paper, the author should outline the extent of the prior literature, the nature of the problem, and the route by which he will approach it. If the author does not clearly show the motivation of his work, most readers (including the knowledgeable specialists) will be confronted with the kind of information gap shown in Fig. 2. Jumping across this gap into the communication band can be a forbidding leap!

A second class of gaps concerns the identification of the actual points of novelty in a paper. Neglecting to distinguish what is new from what has already been published is the biggest (and most unkind) gap of all! The writer who carelessly omits important literature references ignores proper acknowledgment and unwittingly claims originality for ideas not his own.

An especially difficult area of information gaps, often underestimated by authors, lies in the wording of labels and captions for illustrations. Journals have widely different standards for caption writing. Some will permit a full paragraph of explanatory material. Others are content to identify illustrations by figure number alone. Proponents of the latter policy blandly assume, of course, that every reader will pursue every word of the paper.

Ideally, a self-explanatory caption is reasonably short but describes the figure without requiring reference to the text. The extra effort put forth in writing careful captions will be well spent because the figures should illustrate the important points in a paper, and titles for the important points should not be lacking in information.

One important decision an author must make is the amount of detail he should include in his manuscript. The decision is never a simple one. There are no rules for solving this universal problem, and the author must make a rather intuitive choice as to the amount of data his readers will need. In reviewing the rough draft, he can look for four kinds of information gaps that plague all authors. Specifically, the paper should give enough theoretical or experimental information for the following purposes:

1) to support the basic concept;
2) to explain the methods sufficiently so that others can use them;
3) to show the limitations of the results and, if necessary, to show the agreement with known results;
4) to show the reliability of the data (or the validity of the concept).

These supporting details seem obviously necessary for any well-balanced technical paper, particularly those dealing with research or development work. Yet bad gaps of information commonly found in manuscripts frequently belong in one of these four categories.

Whatever the sins of omission, we may think of them in terms of the missing information levels in the middle empty band of Fig. 2. The fewer the empty levels in this band, the more communication of ideas to the reader.

Information Traps

A writer is fortunate if he has been able to include all the essential information in his paper, leaving no unpleasant gaps. He will be doubly fortunate if all his important data, unimpeded and undiverted, get through to the reader. The traps in the manuscript are usually camouflaged by verbiage and tend to escape an author's notice.

According to our information-band structure illustrated in Fig. 3, an ambiguity in a manuscript will trap a piece of specific information, preventing it from rising to the communication levels in the upper band. We might think of the ideas in a paper as originating in the lower band and occasionally becoming trapped before reaching the reader by way of the upper communication band. For readers, the traps are an occupational hazard; for writers, a "reputational" hazard. Every writer should seek out the traps while his manuscript is in draft form. Before an author can find verbal traps, he must learn to identify them.

There are two kinds of information defects. One type is the semantic confusion, which concerns word meanings or the reader's reactions to them. The other is the structural fault, in which information is misleading because it is organized poorly.

Of the several kinds of semantic confusions, questions of terminology are among the easiest to recognize. To illustrate these, we may cite a few random examples.[1] The holes which carry current in semiconductors are not the same as holes (lattice vacancies) in semiconductor crystal lattices. To talk glibly of "holes" can lead to confusion. As another example, the distinction between constituents of alloys and impurities in metals has always been poorly defined. As the percentage composition decreases, at what point does an alloying element become an "impurity?" Or, to illustrate another question of terminology, what frequency characteristic for audio amplifiers deserves to be called "high fidelity?"

A more common trap for information is the "high-order abstraction" of the semanticist. Expressions like "improved performance" or "fair agreement with other published data" are vague generalities, unsatisfying to the reader who looks for meaning and significant results in a paper.

Another language difficulty is awkwardness of literary style. The reader who has to hack his way through a tedious verbal undergrowth becomes too weary to find the author's contribution to the literature. Before such information can reach the communication band in Fig. 3, it becomes caught in a trap of verbiage.

By "structural faults" in a manuscript, we mean the kinds of over-all construction which tend to "bury" the critical points. Two well-known ways of emphasizing the important things in a paper are:

1) to avoid placing them in central portions of sentences, paragraphs, or sections of a paper;
2) to use contrasting length to attract the eye. The key points in a paper thus might be placed in an unusually short paragraph or an unusually long one.

Some authors consider it unnecessary to summarize their results at the end of the paper. Item 1) is a strong argument for writing a "Conclusion" section, because placing summarizing material at the end of a paper gives a stronger argument. If the essential findings are scattered throughout the paper, the literary constructions lack emphasis and the paper is weak. Information is thus diverted—trapped by poor organization. Of course, a section labeled "Conclusion" is not necessarily informative. In order to have any value, a concluding section should summarize the nature and validity of the author's findings. Remarks on the significance and important applications of the work will usually strengthen the Conclusion.

The character and number of illustrations for a technical paper may actually hinder the flow of information. One example is the chart or photograph which does not clearly show the real points of interest. The graph chart with poorly proportioned coordinate scales, for example, will be ineffective. Another case is a photograph of a machine in which the novel features are hidden beneath its cover or casing.

Too many illustrations sometimes defeat their purpose. The fewer the illustrations, the more the eye is attracted to them and the more they emphasize their message. When a reader is confronted with an excessive number of visual aids, he tends to lose sight of the important data. When those data become trapped in this way, the reader loses perspective.

Conclusion

One of the joys of writing a formal paper is to fill in the information gaps which were actually missing from the *work.* An author will sometimes experience a sudden insight when writing his manuscript after months of laboratory work. Inevitably, seeing the results expressed formally on paper becomes an opportunity for further analysis. Writing is thus a creative part of engineering and scientific work.

Finding the traps can be rewarding, too, in unexpected ways. Information traps are never far removed from defects in reasoning. Self-editing, then, should not be regarded as a superficial polishing operation, nor should the work of a professional editor be so regarded.

Our comparison to energy band theory for semiconductors suggests a "flow" of information from manuscript to reader. If we consider one final analogy, the overlapping bands which represent conduction in metals, we find in Fig. 4 the ideal model for information flow in a manuscript; no gaps, no traps.

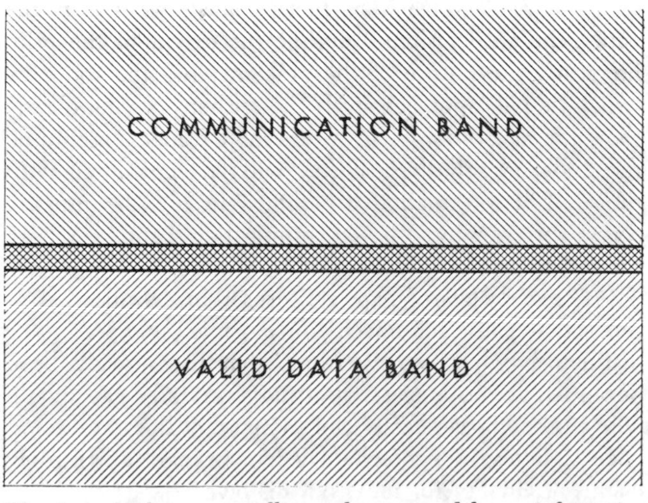

Fig. 4 — Analogy to metallic conduction model: no information gaps or traps.

[1]Numerous other examples were shown in earlier articles by the author: "Semantics and syntax in technical reports," *Chem. and Engrg. News,* vol. 28, pp. 2416-2417, July, 1950; and "Semantics in report writing," *TWE J.,* vol. 1, pp. 1-3, Summer issue, 1957.

Writing Informative Titles for Technical Literature — An Aid to Efficient Information Retrieval

The advent of indexing schemes (many of them automated) that rely upon the words appearing in the title of a technical document has made the meticulous writing of titles more important than ever. Much of the effectiveness of such indexing—and thus the probability of efficient retrieval by readers later on—depends upon the care and the accuracy with which the author originally wrote his title. Described here are approaches to titling technical papers suggested by a technical library.

ROBERT A. KENNEDY
Bell Telephone Laboratories, Inc.
Murray Hill, N. J.

Received October 22, 1963
(Based on a presentation given at the American Documentation Institute 26th Annual Meeting, Chicago, Ill., Oct. 6-11, 1963)

Keyword-in-context[1] or permutation indexing is used at the Bell Telephone Laboratories for a variety of information announcement, record, and search purposes[2]. Much of the material processed — technical memoranda and other internal reports, published papers, talks, computer programs, specifications—originates within the company. Titles, usually acceptably substantive but sometimes editorially supplemented, form the basis of the material permuted. Since the machine indexing program was begun in late 1960, regular exposure to and use of the series of permuted awareness bulletins, cumulated catalogs, and special-purpose indexes has contributed to a detectable, not insignificant improvement in both the informativeness and machinability of the titles assigned by authors. To encourage further improvement, the Technical Information Libraries recently prepared and distributed to over 5000 of the Laboratories' scientific and engineering personnel a few suggestions on choosing titles for their papers. The substance of this brief guide is presented herein.

CONTENT OF TITLE

Consider the title as a one-sentence abstract. Without attempting to summarize the content of your paper, make the title reflect the subject as definitively and concisely as possible. Include some reference to the important topics under which you, as a user as well as a producer of information in this field, would expect, or reasonably hope, to find your paper indexed.

CHOICE OF TERMS

Choose title terms which are as highly specific as the content and emphasis of the paper permit. Thus, when appropriate:

USE	*RATHER THAN*
a nonreciprocal ferrite coupler	a ferrite device
a vanadium-iron alloy	a magnetic alloy
a 7090 FAP subroutine	a computer program

TITLE STRUCTURE

Provide sufficient context, but only enough, to clarify the relationships between the selected technical keywords. Remove words which tell the reader little or nothing. Use short connectives like *of, for, the, on* . . . , rather freely. Use conventional phrases like *Introduction to* or *Analysis of* or *Status of* only where they are important in indicating the nature and level of the paper. Avoid such generally unhelpful phrases as:

A report on . . .
Some problems associated with . . .
A study of the factors affecting . . .
Some thoughts on . . .

SUBTITLES

When several papers are written under one over-all title, identify the successive parts as Part-1, Part-2, etc. Assign a suitably specific subtitle for each part. Where reasonable, avoid repeating in the subtitle topic words already given in the main title.

LENGTH OF TITLE

Balance brevity against descriptive accuracy and completeness. Consider whether a two- or three-word title is not possibly too cryptic. About four important words is often a good choice. If a prospective title exceeds 14 to 15 words (or about 100 characters), see whether it might not be shortened to this length without serious information loss.

Reprinted from *IEEE Trans. Engrg. Writing and Speech,* vol. EWS-7, pp. 4-5, 1964.

SYMBOLS, EQUATIONS, FORMULAS

Where convenient, use words or other machinable equivalents for expressions containing superscripts, subscripts, lower case characters or other special notations which cannot be directly or clearly duplicated on standard keypunching and computer equipments. For examples:

USE	*RATHER THAN*
divalent samarium	Sm^{2+}
yttrium iron garnet	$Y_3Fe_2(FeO_4)_3$
diopside	$CaMgSi_2O_6$
trimethylammonium ion	$(CH_3)_3NH^+$
chi-square	χ^2
per cubic cm	cm^{-3}
critical temperature	T_c

FILING SUBJECTS IN RELATION TO TITLES

Assignment of one or more filing subjects to each Technical Memorandum (the major internal report series) has long been practiced by Laboratories' authors. Although the machine permuted indexes used depend primarily on titles, filing subjects continue to be very useful. One of their important functions is to indicate the general field or level of investigation and application of a document. To this end filing subjects should be chosen which do not merely repeat title words but put into broader perspective what is more closely specified in the title. In the machine indexing operation such filing subjects can be coupled to the title to provide additional subject access points useful to both the specialist and the nonspecialist inquirer.

A guide to the choice of filing subjects is planned.

INTERNAL HEADINGS

Principles similar to those suggested above for constructing titles are often helpful in choosing informative section headings, chapter headings, and the like.

SOME GOOD REPORT TITLES

Sample Preparation by Potassium Pyrosulfate Fusion for X-ray Spectrochemical Analysis.

Gallium Phosphide Crystal Growth by Vapor Phase Iodide Transport.

Interaction of Phonons and Spin Waves in Yttrium Iron Garnet.

Effect of a Nonideal Regenerator Slicing Characteristic on the Error Rate of a Binary PCM System.

Properties of Thin Tantalum Films Sputtered in a Partial Methane Atmosphere.

Thermal Resistance of the 2N1675 Diffused Silicon Power Transistor.

Transmission Tests on a 50-Pair, 19-Gauge Expanded Polypropylene Insulated Cable.

A Tunnel Diode Oscillator for Use in an Audio Frequency FM Parametric Amplifier.

Comparison of Serial and Parallel Transmission of Digital Data Over Telephone Channels.

Procedure for Welding, Cleaning, Inspection and Sealing of Flexible Corrosion-Resistant Steel Hose Assemblies.

REFERENCES

1. H. P. Luhn, *Keyword-in-Context Index for Technical Literature (KWIC)*—IBM Advanced Systems Development Division report RC-127, August, 1959.
2. R. A. Kennedy, "Library Applications of Permutation Indexing," *J. Chem. Doc.*, 2, 181-185, July, 1962.

Abstract, Conclusions and Summaries

ANDREW A. SEKEY

Abstract–The functions of the Abstract, Conclusions and Summary sections in engineering and scientific reports and papers are compared and contrasted. Each is found to serve a specific purpose and not to be interchangeable with the others. As an illustration mock specimens of these sections are developed and compared for Shakespeare's Othello.

INTRODUCTION

MOST scientific and engineering papers are preceded by an Abstract, and end with a Conclusions or Summary or both. The functions of these parts of a document are vaguely understood by most readers and even writers. Yet only too often they are treated as if they were cloaks, worn alternately by different bodies on different occasions. For example, a recent text [1] on technical writing treats Abstract and Summary as practically interchangeable. Again, Summary and Conclusions frequently appear as a single unit at the end of articles. The purpose of this note is to illustrate that provided the paper is extensive enough to justify inclusion of all three sections, they *can* all have different identities.

In order not to restrict the subject matter to any one professional discipline, I shall take as an illustration Shakespeare's Othello, pretending that everything that happens on the stage is an "experiment," whose circumstances and implications can be analyzed objectively.

THE ABSTRACT

The main function of the Abstract is to help the reader decide if he wants to read the document. (I shall use the term "document" as a general label for paper, article, report, thesis, etc.) This has become all the more important since the advent of abstracting journals, which give readers access to a multitude of abstracts while they do not have the articles themselves also within easy reach. Thus the decision, based on an abstract, to read the entire document may involve the reader in effort, time delay and expenditure that is far more than merely turning the pages of the journal in his hand.

A good abstract must thus tell the reader what he will find in the document and, if possible, also what he might reasonably expect to find but will not. (For example, any illusion that the title of this note may have given that it is a manual for writing should have been dispelled by the Abstract.) Thus not only should the Abstract present the main result(s), but also the method by which they were achieved–e.g. analysis, experiment, computer simulation–and their significance. Yet this difficult task must be accomplished within a strict limitation of the number of words, usually between 50 and 200.

Abstract of Othello

OTHELLO

"Othello, a Moorish commander of the Venetian Army, marries after a passionate courtship Desdemona against the wish of her father, a Senator. During their subsequent stay in Cyprus Othello's jealousy is fiendishly aroused by Iago, his shrewd but vengeful lieutenant. Iago's scanty–and false–evidence eventually convinces Othello of his wife's unfaithfulness and he strangles her, then commits suicide. Thus is blind jealousy shown to possess the power to destroy even a just and brave man." (76 words)

Notice the several "key words" [2] appearing in this version: Moorish, passionate, Cyprus, jealousy, vengeful, strangles, suicide. They are names, events, features etc. closely associated in our minds with Othello. Likewise, a well worded technical abstract will invoke numerous associations which, like so many coordinates in a multidimensional subjective space, will "localize" in the reader's mind the place where the article might fit in.

THE CONCLUSIONS

Most technical documents report an investigation or observation, and if well written will contain separate sections for facts and opinion. (Scientific papers prefer: Materials and Methods, Results, Discussion, Conclusions; while an engineering report might have: Design Considerations, Manufacturing Process, Field Trials, Analysis of Data, Conclusions.) In any event, in the section called Conclusions the author reduces the data and other information presented to a few well founded statements of general validity, without reiterating the supporting facts.

It must sometimes be tempting to draw striking conclusions that will evoke astonishment and admiration, but such temptation must be tempered. The conclusion is no more valid than the evidence on which it is based, and by overextending himself the author will, eventually, lose credibility. While the scope of one's own work should not be underestimated, speculative generalizations should be made modestly and sparingly. In the terminology of the courtroom, if the Discussion is the debate between the prosecutor and the defence attorney, the Conclusions is the word of the judge, with his feet firmly planted on the evidence presented, but his eyes scanning future horizons.

Manuscript received February 14, 1973.
The author is with Tel-Aviv University, Tel-Aviv, Israel.

Reprinted from *IEEE Trans. Prof. Commun.*, vol. PC-16, pp. 25-26, June 1973.

Conclusions for Othello

OTHELLO

(A study in the psychology of jealousy)

"The play shows how the emotional stability of a seemingly indestructable strong man can be shattered once a point of insecurity is found. Othello's initially latent preoccupation with the color of his skin—foreshadowing like feelings in today's multiracial societies—is carefully nurtured by Iago to the point where Othello believes it to be the cause of Desdemona's desertion. That Othello allows himself to be driven by Iago to the murder of his loving wife without ever even suspecting his motives for incriminating Desdemona, is the tragic outcome of Othello's honesty, naivete and thus vulnerability."

While in this sample—which is more like a "moral of the story" than a Conclusion—there is no previous belief to confirm or to shatter, in scientific documents the Conclusions may well answer to questions posed in the Introduction. Though such questions might have been rhetorical in that they had occurred to no one but the author, the degree to which they have been solved is a measure of the success of the work reported, while the remainder should act as a stimulant for the readers to pursue the topic further.

THE SUMMARY

If the Abstract is like the restaurant's menu card inviting the reader to try and taste, the Summary may be likened to the package of colour slides brought back from the summer vacation. By projecting them one may relive the adventures, select the important ones and discard the rest, and place the entire trip into perspective.* Thus while the Abstract is written for the uncommitted inquirer, the Summary is strictly for insiders who now know as much about the case as the author does.

But what, one may ask, is its purpose then? Can it add anything to the Conclusions that the reader could not find for himself by re-reading the Abstract? I shall try to show that it frequently can.

Scientific papers are usually written as a linear progression from hypotheses through experiments or analysis to results. (This convenient tradition takes much excitement out of reading, though, for the true process of discovery is anything but of this type. [3, 4]) Yet one of the most effective ways of anchoring a new notion in the reader's mind is by illuminating it from a different angle. I am not thinking here of interpreting the observations reported in the context of various theories—for which the Discussion is the appropriate place—but of making the Conclusion "plausible" in an altogether different framework. This may be done in but a few sentences, giving a final twist to the document that sends the reader home thinking further about what he thought he already fully understood.

*No pun intended!

In more conventional writing, the summary merely fulfills the last of the three tasks of the proverbial Army instructor: "Tell 'em what you're goin' to tell 'em; tell 'em; tell 'em what you told 'em." The Summary is thus a recapitulation, a synthesis, as opposed to the Abstract, which is an exposition.

Lastly, if the document is one whose purpose is to initiate action, the Summary may be the most appropriate section in which the author may make his recommendations. (Some writers prefer to lump them with the Conclusions, but I feel that since the latter are drawn from the evidence reported—e.g. that one type of engine is noisier but cheaper than another—while the recommendations are often based on "environmental constraints"—e.g. that the competition is building a quiet engine—, they are better presented in different sections.) And now, the example.

Summary of Othello

OTHELLO

"Othello was victorious when fighting the Turks or pacifying the Doge, but not in laying his self-doubts to rest. He knew his worth as a commander and statesman, yet could be led to believe that Desdemona prefers the insignificant Cassio to himself. Nonetheless, Iago could not have succeeded with the monstrous plot built on this weakness of Othello without the one crucial piece of "evidence": the handkerchief.

Yet why did Iago engineer this disaster, which ultimately cost him his own life? Shakespeare suggests that he was settling an old account with Othello* while Verdi, whose opera based on this plot provides much insight to the story and its characters, has *his* Iago declare this Credo: "I believe in a cruel God, who has created me in his image. . .I am evil because I am a man;" So, ultimately, we still do not know where the root of Othello's tragedy resides: in his jealousy, his bad luck, Iago's evil mind? Or did we witness not a tragedy but Divine Justice, for if Othello did indeed once seduce Iago's wife as alleged, isn't he but punished for that deed in a terrible but fitting and just way? Yet the answer, if there is one, must lie in the play itself."

ACKNOWLEDGEMENT

I am indebted to Mrs. M. Karni for stylistic improvements on the original version of this paper, and to Prof. Mark Beran for helpful comments.

REFERENCES

[1] H. M. Weisman, *Basic Technical Writing*. Merrill, 1968.
[2] J. E. Morris, *Principles of Scientific and Technical Writing*. McGraw-Hill, 1966.
[3] A. Koestler, *The Act of Creation*. Hutchinson, 1964.
[4] J. Watson, *The Double Helix*. New York: Atheneum, 1968.

*"For that I do suspect the lusty Moor
Hath leap'd into my seat, the thought whereof
Doth, like a poisonous mineral, gnaw my inwards;"

Integrating Text and Figures

JAN W. BROER
Philips Research Laboratories

'Integration' of figures (illustrations) and text is discussed; balance, efficiency, and pleasure are pragmatic attributes. Technical writers and editors usually consider figures a special difficulty. To remedy this, a conformity between processing figures and processing sentences is assumed; it suggests a metaphoric transfer of knowledge from perception to technical writing. A case of similarity grouping and a contrast illusion illustrate that perception research may help technical writing. The jungle of normative rules gets perspective, ideas for new rules arise.

MEET THE MEMBERS

In our world of technology and science the 'paperboys' are continuously dressing up and regrouping the huge data collectives. As an established school of information transformers we technical writers and editors play our game of filtering and trimming, and we know the rules. But do we? Just consider how many rules there are. In a hundred-page textbook on technical writing you may easily find two hundred dos and don'ts. Two per page. Who is so familiar with them that he bears them all in mind everywhere, all the time? The main set of rules and tools is our writer's competence to thread together: words to sentences, sentences to paragraphs, paragraphs to sections, and sections to reports or papers. All that is done in agreement with the

ABC + D standard,

whereby writers try in a Dynamic way to be Clear, Brief and Accurate.

In our tool-kit we carry another instrument with great potential. Handling it, however, is another matter; when we try, we tend to make fools of ourselves, or so we think. Usually we play safe and believe our dignity restored by remarking that, properly speaking, it is other people's competence to handle that tool. It is the tool to arrange lines to figures, figures to movies, movies to TV-visions, TV-visions to what? Maybe video-disk productions; 'they' should know, of course.'They' and their products are not in the family. An unfortunate situation.

Look at the textbooks on technical writing again. And find out how much, or rather how little, the books tell you about figures. You will realize how distant the figure/sentence relationship appears to be. One figure is worth a thousand words, people say. For the thousand words there is the whole gamut of dictionaries, thesauri, grammar, semantics, and pragmatics. For that figure, however, there is next to nothing; scattered pragmatics seems to be all.

How can we overcome our uneasiness about figures? We technical writers suspect the answer: we advocate 'integration', the integration of figures and text. Do we know what that means? I think we do, vaguely. We want figures and sentences to operate in an efficient and pleasant sort of 'togetherness'. Figures and sentences should cooperate in a 'well-balanced' way for the transfer of information. The word 'well-balanced' is a thorn in our flesh. Silent movies, without any text, are one extreme of imbalance; bibles, without any illustra-

FIGURE 1. LACKING UNITY OF WORD AND PICTURE

tions, are another. Well-founded criteria for balance appear not to exist; one might perhaps say that almost anything continuous merits a curve, instead of words, and that any five or more parts of a construction involving different functions are worth an illustration. The odd misunderstandings that affect people in a condition of imbalance are neatly brought out in Fig. 1. Following the English psychologist Cecil A. Mace in his book "The Psychology of Study" (Penguin, 1968) we could define the criteria of efficiency in terms of the degree to which a paper enables readers to attain their, or the author's, goals and in terms of the economy of effort (including speed) with which the readers do so. The aspect of pleasure in the togetherness is closely related to the motivating power the paper manages to build up, which encourages the readers to persevere.

'UPSTAIRS/DOWNSTAIRS': NO

The conservatives among us see the visual piece, in all its alluring forms, as nothing but the dependent auxiliary of the verbal element, as the jaunty parlourmaid among the reliable house staff.

Some of us, who obviously accept the togetherness, also hold that figures + captions should form one readable unit themselves, completely independent of the text of the paper. Such subtlety makes the writer offer two stories where his readers would, I think, be better served by one.

The proponents of this 'upstairs/downstairs' character of a paper defend themselves with two arguments. The first is the notion that prospective readers are apt to use the pictorial part of the paper to quickly decide whether or not to read it all. Certainly, I would reply, but the very presence of all those picture versions could turn a prospective reader away from any solid reading activity. He might become the perpetual browser, with no time left to study things in the classic way. The way of pondering words and sentences to analyze statements and to weigh reasoning and assumptions, slowly and seriously. It would be sad to lose that.

The second argument is the notion that writing the paper is easier when you do it from the figures, a matter of heuristics. Looking at the skeleton story of figures + (elaborate) captions helps much in writing. That is true, I agree, but it is also true that one should by no means include all the heuristic matter in the final version of one's paper. Brevity forbids it. Hence, no compromise about integration; figures + captions + text: one story.

INTEGRATION: YES

To begin with, there is a good and pragmatic reason for integration. A figure and in particular its elements, the lines, require integration with text, because as visuals on their own they are not unique. A figure without an explanatory sentence is usually poly-interpretable. The example in Fig. 2 shows a vase or, perhaps,

FIGURE 2. POLY-INTERPRETABLE (FROM E. RUBIN, DANISH PSYCHOLOGIST)

the badge of the League of African Twins. You cannot know it, unless I give you the sentence that settles the matter.

The pragmatic reason for integration does not stand alone. There is more to it. The Gestalt psychologists of the 1920's realised that already. The structures underlying figure processing and sentence processing might be simply identical. Considering the eyes as two swellings of the brain that seek the sun light, one might expect an analogy between, say, eye optics and mental processes, between light rays and words, between optical images and sentences. In his book "Pattern Recognition, Learning, and Thought" (Prentice-Hall, 1973), Leonard Uhr of the University of Wisconsin says, "I don't think there is any clear-cut distinction between perception and cognition. Perception talks about the first few transformational stages, cognition about subsequent stages. They almost certainly use the same mechanism".

The English author A.C. Parkinson in his "Pictorial Drawing for Engineers" (Pitman, 1962) says,"... in the rational order of things every object is 'visual' before 'functional'." The most interesting word in the second quotation is 'rational'. If the transition from 'visual' to 'functional' is dominated by our rationality, the process must be a smooth one, without any serious discontinuity.

SO: MEET THE FAMILY

In some deep-lying structure, the processing of figures and the processing of sentences would appear to be the same thing. All in the family, then; a fortunate situation. It suggests a close relationship between the dos and don'ts for figures and sentences. In that perspective, the thing to do could be to apply the results of perception research to technical writing, in one way or other. Metaphoric transfer of knowledge or 'seeing the analogy' - that's the way. The least that can be expected is some ordering in the mass of technical writing rules, the most a complete duality between the two fields. At any rate, assuming the family ties are real, we should be able to find situations in both fields that are strongly analogous. Let me give you two examples of such duality, and show you what can be done with it.

Example 1

"Similarity grouping" is a well-known effect in perception. In Fig. 3 we see a multitude of black dots on a white background.(Both Fig. 3 and Fig. 4 are taken from a paper by Glass and Switkes (1976) on pattern recognition.) Our brain makes us 'see' more: it fuses the impressions of the dots into an impression of a pattern of circles. The individual dots are grouped into a larger unit. How does this come about? A close look shows that each dot and a near neighbour have positions that lie on those circles. As a matter of fact, the picture was made by superimposing, one upon the other, two random patterns, each containing half the total number of black dots. The patterns are completely identical, but one is displaced slightly with respect to the other by rotating it through an angle of a few degrees about the centre. This small rotation is responsible for the impression of circles.

If we now replace one of the superimposed patterns by its negative, that is, white dots on a half-tone grey background, the correlation of the pairs in space is not affected - they still lie on the circles, but we no longer see any circles (Fig. 4), just a chaotic assembly of pairs consisting of one black and one white dot. In technical writing we talk about 'details' instead of 'dots' and about 'the main line of thought', instead of 'the circles' in the material.

Papers, and in particular reports, are written that contain numerous pairs of details whose closely spaced members weaken each other in some way. This weakening sometimes results from an excessive desire for objectivity. You know the type: every statement is closely followed by a qualifying one of the "yes, but" kind; even the hair-raising "yes 'yes, but', but" addition is frequently seen. Perfectionists love it. The poor reader then finds it impossible to identify the main line of thought in the matter. How come? The reason might be similar to the one that made the viewer miss the circles when the dots were divided into black dots and white ones. Remedy: restore the harmony at the level of the details, that is eliminate, or at least soften, the contrast between the members of the various pairs. The main line of thought will re-appear.

FIGURE 3. SIMILARITY GROUPING

FIGURE 4. SIMILARITY GROUPING DISTURBED

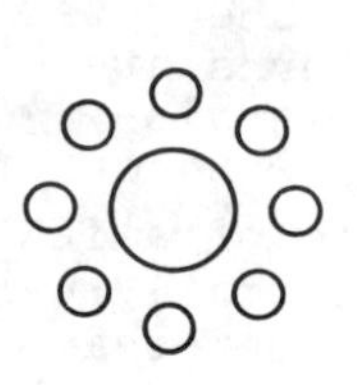

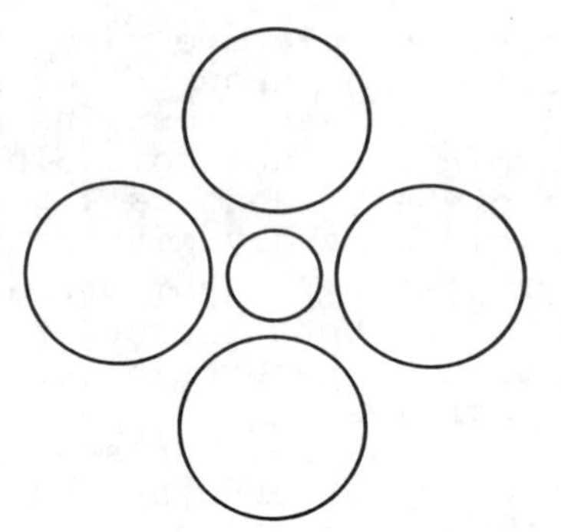

FIGURE 5. SIZE CONTRAST ILLUSION

Example 2

Fig. 5 shows a well-known contrast illusion involving the size of circles (Humphrey, 1971). The circle surrounded by the large circles looks smaller than the circle surrounded by the small circles. Measuring their diameter, however, you will find that the two central circles are the same size. Depending on the order in which you look at the two configurations you will say that the central circle has become either bigger or smaller. The changing contrast in sizes between the central circle and the surrounding ones causes the effect. In estimating what size a central circle is, the observer uses two data: the stimulus value s and the expected value E. The stimulus value s is the sensory datum corresponding to the size of the central circle. The expected value E is an 'average size' datum that follows from the sensory information corresponding to the surrounding circles, the sensory context. What happened when you looked at the two configurations can now be reformulated: the size is underestimated when the deviation s-E is negative and overestimated when s-E is positive; in estimating size, therefore, the observer amplifies the deviation. In simpler words: observers exaggerate - and so, I dare say, do readers.

Let's now apply the situation of Fig. 5 to technical writing. The results obtained tell their own tale if we look at Figs. 6 and 7. Taken in that order Figs. 6 and 7 represent the classic induction/deduction type of paper in science. The paper begins with a host of small parcels of information, all treated at the same level. Their description typifies them as details of a 'peripheral nature'. At their centre they induce a major idea, the core C_1, a sort of hypothesis. In the second part the paper attributes to that same core, now labeled C_2, a few huge collectives of data at close range, that is, information deduced directly from C_2.

Now let us reverse the order and take Fig. 7 before Fig. 6, a kind of revision of the paper. The revised paper begins with a few huge collectives of data that are supposed to have a unifying idea C_1^* at their centre. In the second part the paper now provides a host of little parcels of information, precious little gems, of course, peripheral in nature, and in agreement with the existence of the same unifying idea, now labeled C_2^*.

If you accept the structural identity of perception and technical writing, you may say that the contrast illusion in perception now proves that the revised paper is more stimulating reading than the original one. The reason is simple: $C_1^* < C_2^*$, whereas $C_1 > C_2$. The second version leads the reader to conclude that the unifying idea is really important, after all. All is well that ends well. The first version, on the contrary, makes him conclude that the hypothesis core is not as important as he originally thought it was. Great boast, small roast. Of course, the result obtained applies only if the author's main objective is indeed to stress the core idea.

Another application of the contrast illusion to technical writing is shown

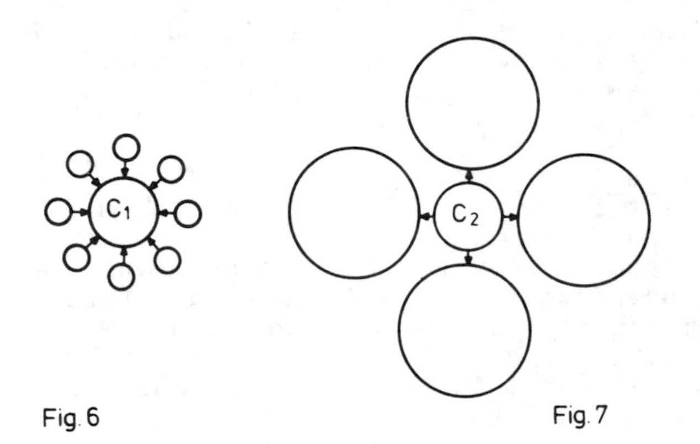

INDUCTION/DEDUCTION TYPE OF PAPER

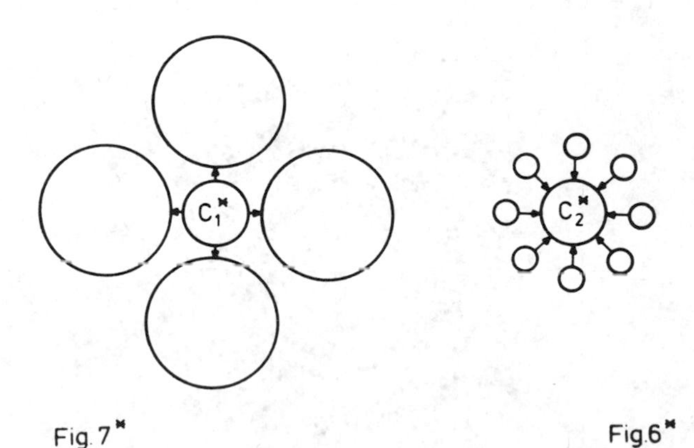

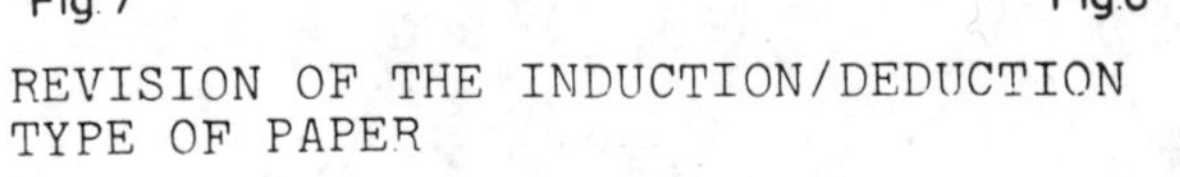

REVISION OF THE INDUCTION/DEDUCTION TYPE OF PAPER

in Fig. 8. If the writer wants his readers to focus their attention upon some statement e, he can do so by expressing e in the form of a short sentence, s(e), surrounded by a context comprising a few long ones, L_i. His readers then have to process some paragraph like $L_1 - L_2 - L_3 - s(e) - L_4$. The writer could have adopted the equivalent trick of wrapping up e in a long sentence, L(e), and surrounding it by a context of short ones; the paragraph then would have read: $s_1 - s_2 - s_3 - L(e) - s_4$.

Suppose now that the writer considers the statement of such importance to his readers that he falls back on another technical writing tool of respectable age: he repeats statement + context, in a different form. The problem then arises: which order is better? The contrast illusion supplies the answer:

$$L_1 - L_2 - L_3 - s(e) - L(4) \text{ first}$$

and $$s_1 - s_2 - s_3 - L(e) - s_4 \text{ second,}$$

is the better order. It reinforces e, because the importance attached to e increases (s(e) → L(e)) during reading. An example of the better order is given in the appendix.

THE RETURN

Our 'all-in-the-family' look at figures and sentences has resulted in some pairs of strongly coupled situations, one in the field of perception and the other in the field of technical writing. Or should I say the 'jungle' of technical writing? If we plod through the multitude of textbooks in our own particular field, I'm afraid we indeed frequently get the impression of being in a jungle. Analytical depth is in short supply. Instead, we find a lot of normative material: a jungle of rules of the 'you-must-because-I-say-so' type. The trouble with a jungle is that nobody seems to know how to go through it rationally. To read one book (Vernon, 1970) in the obviously better cultivated field of perception can be an eye-opener. One easily spots the many similarity pairs in the picture of perception and that of technical writing. Figs. 3 and 4 again have a message for us. If we writers and editors bring perception well within our horizon, we shall manage to see the integration of figures and text as one part of our field, now well-structured on a basis that is more than pragmatic. We'll have progressed from seeing dots to perceiving lines.

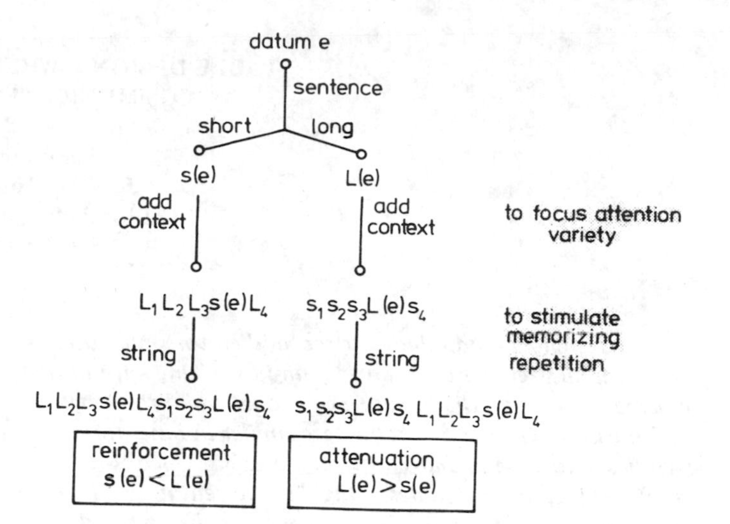

FIGURE 8. PERCEPTION (FIGURE 5) HELPS TECHNICAL WRITING

APPENDIX

An example of
L_1-L_2-L_3-s(e)-L_4 followed by
s_1-s_2-s_3-L(e)-s_4

Human beings are aroused by variety; the little flasher at the side of your car attracts more attention than the powerful, but nonstop shining headlight. This is a truth, however, that applies to more sectors of life than traffic only, which many scientists of the present generation have come to appreciate. Of course, presupposed are accuracy, brevity, and clarity, the characteristics that cannot be missed before any variety may be brought in. Variety is the spice. The writers of the best textbooks have made use of it, be it subconsciously, since Old Testament times at least, one could say.

To score my point I want to repeat myself and say this:

It's like a good meal. Ingredients of quality make it. The benefits are widely felt. In the long run, however, the real and remaining attraction comes from the variety, the dynamics that ensures continued reinforcement of the partaker. Indeed, good textbooks are a case in point.

REFERENCES

Glass, Leon, and Switkes, Eugene, (1976), Pattern recognition in humans: correlations which cannot be perceived. Perception 5, 67-72.

Humphrey, N.K., (1971), Contrast Illusions in Perspective. Nature 232, (July 9), 91-93.

Vernon, M.D., (1970), Perception Through Experience, Methuen & Co., London.

TABLE DESIGN – WHEN THE WRITER/EDITOR COMMUNICATES GRAPHICALLY

Mary Fran Buehler
Jet Propulsion Laboratory
California Institute of Technology
Pasadena, California

In designing data tables, writers and editors must use more graphics-related techniques than in working with other text materials: e.g., the handling of spatial and directional relationships. Graphs and tables are similar in basic form; each has a horizontal and a vertical scale. Writer/editors can use this similarity in designing data presentations. Patterns are discussed for one-, two-, and three-dimensional tables. Some guidelines are given for tabular legibility in general (i.e., in printed books) and also for legibility in reports published in microfiche form.

DESIGNING WITH DATA

Most writers and editors naturally enough – seem to be more at home with words than with the details of graphic design. And yet there is one text element that demands not only expertise in the language but also expertise in handling such visual aspects as spatial and directional relationships, the proper management of repetition, the use of white space, and the effects of photographic reduction. All these abilities may be required in the design of effective tables.

But that is not all. Table design also requires the ability to analyze data relationships correctly, to understand the needs of the potential reader, and (probably most important) to present tabular data in a form that will be readable and usable under a variety of publication conditions.

Small wonder that so many problems occur in the design of tables.

Adding to the problem is the lack of a standardized body of knowledge on table design in technical communication. A sampling of publications in the technical communication field (various references and style guides) reveals that there is not even a universally adopted nomenclature for the parts of the table.

The definitive reference on table design remains the *Bureau of the Census Manual of Tabular Presentation* (1), which seems to be the model for many other, less complete treatments. Although *Tabular Presentation* was published in 1949, it is still the only single, comprehensive source of authoritative advice on table design. I have adopted its nomenclature both in the material presented here and in the chapter on tables in the handbook *Report Construction* (2), from which some of the graphic elements in this paper (e.g., Figures 1, 3, and 5), have been derived.

The desirability of adopting a common set of principles and a common nomenclature for table design seems self-evident. And the role of the Bureau of the Census – with its experience in the design of data tables – would seem to make the Bureau a self-evidently valid guide to follow. Of course, each publications organization should adapt the basic principles to suit its own needs.

Armed with a set of basic table design principles, the writer/editor can tackle the more difficult – because more varied – aspects of table design. What makes a table easy to read? How can comparisons and relationships be brought out most clearly? How can table design accommodate the constraints of publications processes, especially micropublishing? ("Micropublishing" refers to the widespread practice of photoreducing publications and distributing them in microform, especially in microfiche.)

These questions could be approached from many points of view. Probably the most basic requirements for a table are (1) the table must be rational (that is, it must make sense) and (2) the table must be legible under all conditions of use. These requirements obviously intermesh: a rational table design makes a table easier to read, and a legible design is necessary for the table to make any sense at all. But for ease of discussion, these elements are discussed separately below.

RATIONAL DESIGN

Tables must show relationships clearly, and one of the most powerful means of showing relationships is to emphasize similarities and differences. This is done by grouping similar items together (or treating them in the same way) and separating dissimilar items (or treating them differently); for a fuller discussion, see "Similarities and Differences: A Key to Clarity" (2, pp. 15-17). This principle of grouping items to bring out similarities and differences should underlie all considerations of table design, and should be understood to be an integral part of the discussion below.

Tables are concise: that is part of their usefulness. But, at the same time, tables lack the contextual clues that running text provides. Because they lack context, tables must be totally clear and self-contained; the reader should never have to search about in the text to find what the table means. (A table that is not self-contained is unfortunate enough in a printed book; in a microform – for instance, in microfiche – an inadequate table may force the hapless reader to search through several microfiche frames to find the information needed.)

The first step in designing a table is to visualize the table's skeleton and other vital parts. In the most basic way, tables resemble graphs. Each form – the table and the graph – has a horizontal and a vertical scale; these scales form the skeletal structure of both tables and graphs. Where the control of the two scales comes together, an exact value is indicated. The horizontal and vertical scales of the graph are the abscissa and the ordinate, respectively; the horizontal and vertical scales of the table are the boxhead and the stub. These scales and their relationship to the data are shown in Figure 1.

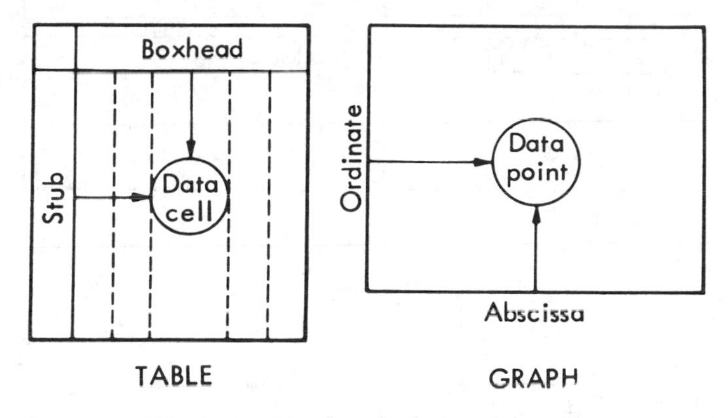

Figure 1. The Pattern of Tables and Graphs

The similarities between tables and graphs emphasize some important elements of table design:

1. The control of the vertical scale (the stub) should extend straight across the table.

2. The control of the horizontal scale (the boxhead) should extend straight down to the bottom of the table.

3. The information in the boxhead itself should not read across; it must read down, in order to identify the data that it controls.

The pattern shown in Figure 1 is that of a simple, two-dimensional table that presents the relationships between two variables (Figure 2). But often tables are not this simple: they must deal with three variables or more.

Some advice on multidimensional tables (primarily oriented toward social science research) can be found in the chapter "Three and More Dimensional Tables" in Hans Zeisel's *Say It With Figures* (3), not to be confused with Darrell Huff's *How To Lie With Statistics* (4), which contains much sound advice on drawing inferences from data but is more concerned with graphs than with tables. Two- and three-variable tables are also discussed in the appendix "Basic Principles of Table Reading" in Morris Rosenberg's *The Logic of Survey Analysis* (5).

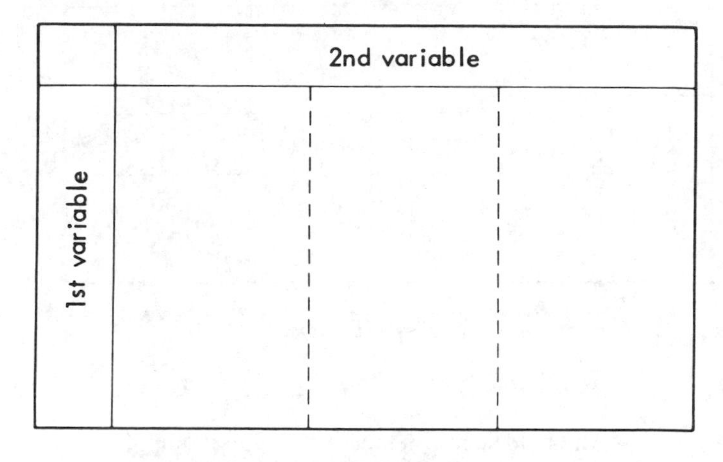

Figure 2. A Two–Dimensional Table

Recognizing that there are different ways of presenting multidimensional data, let us consider one simple and widely used pattern. First, we need to identify the basic parts of the table (Figure 3). Along with the stub and the boxhead, we now see the data field, which contains the actual numbers that we want to present. And in the data field, in the form of field spanners, we have a provision for identifying our third variable. A simplified pattern is shown in Figure 4.

Figures 3 and 4 illustrate several important aspects of rational table design:

1. The table's skeleton – the stub and the boxhead should remain intact and identifiable, regardless of the number of field spanners in the data field.

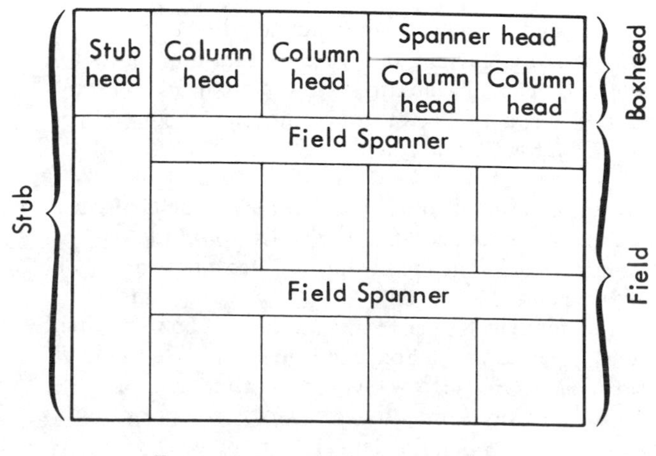

Figure 3. The Parts of a Table

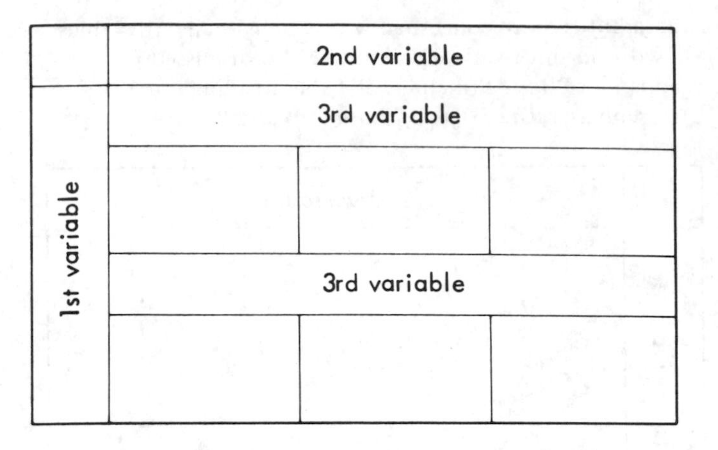

Figure 4. A Three-Dimensional Table

2. The field spanners must be used in groups of two or more, and the topmost field spanner must lie directly under the boxhead (otherwise there will be a slice of data in the data field that will not be identified with the third variable).

3. Great care must be taken to differentiate between field spanners and the boxhead. The top field spanner – lying directly under the boxhead – may be inadvertently mixed into the boxhead and may, in fact, wind up as the top layer of the boxhead. This is one of the most obvious and easily corrected errors in table design, but unfortunately it is all too common.

4. The field spanners should not cut through the stub. Running the field spanner through the stub is less serious than putting a field spanner in the boxhead. But the boxhead and the stub – as emphasized above – should remain intact. Field spanners should remain in the field: they help to define the field visually and thus help the reader to see the difference between the data and its identifiers. (An example of the importance of pointing up similarities and differences.)

Right and wrong ways to handle field spanners are shown in Figure 5. Note, in Figure 5a, that the top field spanner has been incorporated into the boxhead and the bottom field spanner runs through the stub. Figure 5b shows the correct arrangement.

Figure 5 also illustrates another requirement of good table design: in each column, the variable *and* the unit of measurement are identified. Units of measurement should never be included in the data field, where they simply make the numbers harder to read. (Also, of course, when the unit is placed in the boxhead it need be stated only once for each variable; this saves space, time, and money in the form of keyboard strokes.) A rare exception to this rule is a table in which the data are varying widely in magnitude; in that case it may be simpler to put the units in the field than to show the variation in the size of the numbers. But this solution should be handled with extreme care.

Along with two- and three-dimensional tables, there is also a class of tables that do not conform to the principles outlined above. These tables, which usually consist of two or more listings, do not have a stub and do not read across. They should, however, have an adequate boxhead. These tables are, in essence, one-dimensional. (An example: a table listing the advantages and disadvantages of a method or product.) One-dimensional tables are perfectly correct and can be very useful. Field spanners in

(a) WRONG

Table 2. Test results for Models A and B

Test number	Model A		
	Var (unit)	Var (unit)	Var (unit)
1 2 3 4			
Model B			
1 2 3 4			

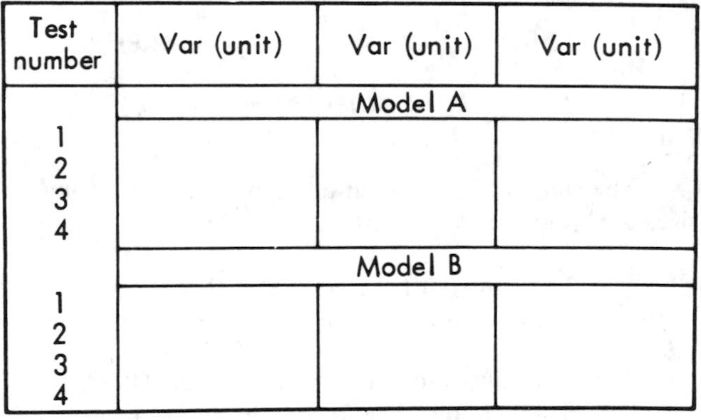

(b) RIGHT

Table 2. Test results for Models A and B

Test number	Var (unit)	Var (unit)	Var (unit)
	Model A		
1 2 3 4			
	Model B		
1 2 3 4			

Figure 5. Right and Wrong Ways to Handle Field Spanners

the one-dimensional table should be treated differently from those in more complex tables: if the field spanner is extended completely across the data field in a one-dimensional table, it will cut the table in two. A suggested solution is shown in Figure 6.

Some other points to remember:

1. Every table should have a table number and title (except for short listings of three or four lines).

2. Every table must be cited in the text and should appear as soon after the citation as feasible.

(But do not break up the text with many pages of tables; group such tables at the end of a report or, more likely, in the Appendix, and put a summary table in the text proper.)

3. Table titles should give the reader a clear indication of the table's contents; no two table titles should be identical.

4. It is helpful to indicate the top and bottom of the table with horizontal rules and to set off the boxhead and the field spanners with rules. Vertical rules between columns may be omitted unless required for clarity (also see the discussion on legibility below). Boxing the perimeter of a table with vertical rules may be aesthetically pleasing, but is not required for clarity or good form. Horizontal rules in the data field should ordinarily be avoided, except to define field spanners or, if absolutely necessary, to indicate groupings of data.

5. Table footnotes should be indicated unambiguously (superscript lowercase letters are good indicators on numerical data) and the footnotes

Table 3. Comparison of Models A and B

Advantages	Disadvantages
Model A	
________	________
________	________
________	________
Model B	
________	________
________	________
________	________

Figure 6. A One–Dimensional Table With Field Spanners

should be placed at the bottom of the table, never integrated into the general footnote system of the overall publication.

LEGIBLE DESIGN

The legibility of tables can be considered from two points of view: (1) the requirements of tabular presentation in general and (2) the special requirements for micropublishing. With the increased reliance on micropublishing today, the latter requirements assume increased importance.

General Legibility

Much of the information in this section rests on research results presented by Miles A. Tinker in *Legibility of Print* (6), especially on his chapter "Formulas and Mathematical Tables" (6, pp. 196-209). Professor Tinker reviews and analyzes the results of 238 legibility research studies, several of which are specifically applicable to table design.

Tinker points out that tables are more difficult to read than running text precisely because the context is missing, and numerals must be read as individual units rather than parts of a familiar whole (i.e., a word). Research results include the following:

1. Arabic numerals are so much easier to read than Roman (because Roman are more complex and less familiar) that Roman numerals should never be used for table numbers. (In fact, Tinker recommends Arabic numbers for virtually all publications uses, including chapter headings, volume indicators, etc., and even for such non-publications uses as historical dates on public monuments) (6, p. 41).

2. All-capitals type is harder to read than lowercase (6, p. 57). The reason appears to be that people recognize words by their shape, and the shape of a lowercase word is more distinctive. For example,

   table design

   as compared with

   TABLE DESIGN

3. A 1-pica space between columns seems to promote legibility as well as a 1-pica space plus a rule, which would seem to indicate that vertical rules are not necessary for good legibility if at least 1 pica of space is left between columns.

4. A space after every 5 or 10 items down a column aids legibility. The 5-line spacing is slightly more effective than the 10-line spacing.

On the question of type size, Tinker found that 8-point type was more legible than 6-point type. But because both these type sizes fall below the requirements for micropublishing legibility discussed below, 8-point type should not be considered optimum for any publication that will be distributed in microfiche form.

Micropublishing Requirements

Micropublishing provides a fast, economical, spacesaving means of disseminating information. But microforms have their own constraints for legibility: in microfiche, for example, the reduction ratio for a page of text is 24:1, so that a 105-mm by 148-mm microfiche can carry 98 report pages 8½ by 11 inches. (For additional information on microfiche, see the *National Standard, Microfiche of Documents* (7). Microfiche are ordinarily used by projecting the fiche, frame by frame, on the screen of a reader. ("Hard" or printed copies can also be made by enlarging the fiche frames to page size.)

The legibility recommendations below are drawn from the "Guidelines for Copy Preparation for Microfiche," in NASA's *Technical Publications Program, A Working Guide* (8). Writers and editors working within other organizations may have their own micropublishing guidelines; for those who do not, the NASA guidelines may be useful. The *Guide* points out that about 80 percent of NASA's scientific and technical documents are available only on microfiche. The need for legibility in this form is obvious. Some points to remember:

1. Avoid broken, dirty, or uneven typing impressions. Table rules, if used, should be strong and unbroken; faint lines may be lost.

2. Do not use type smaller than 10 points.

3. Do not use paper with show-through from the reverse side.

4. Do not use a page size larger than 8½ by 11 inches.

5. Leave a margin of at least 1 inch around the image area.

6. So far as possible, do not rotate figures or tables 90 degrees; use a "portrait" format rather than a "landscape" format. Reason: some microfiche-reading equipment will not permit convenient rotation of frames.

7. Foldout tables present obvious problems for microfiche reproduction. If possible, sectionalize an outsize table into successive 8½- by 11-inch pages. (This will probably require repeating the stub and the boxhead on each page of the table.)

In short, "Image clarity for viewing in a reader as well as making blowback copy depends largely on the thoughtfulness and tidiness with which the original copy was prepared" (8, p. 15).

SUMMING UP

My personal bias toward table design has perhaps become obvious in this discussion. To me, designing tables is one of the most challenging, sometimes maddening, but ultimately satisfying elements in technical communication. Turning an amorphous mass of numbers into a crisply organized presentation of data can indeed be a rewarding experience. Perhaps this is so because, in table design, the writer/editor is allowed – and, in fact, required – to use some of the graphic elements and techniques that he or she cannot ordinarily use in working with words and sentences.

But if tables are different, in this sense, from other forms of technical communication, they are also similar, in a more basic sense. As in any other kind of technical communication, the writer/editor must think, above all, about the reader – the person at the other end of the communication link. Perhaps, because tables are so concise and self-contained, tables require even more attention to the reader's needs. And perhaps this is why the creation of an effective table is so satisfying.

REFERENCES

1. Jenkinson, Bruce L., *Bureau of the Census Manual of Tabular Presentation,* United States Government Printing Office, Washington, D.C., 1949.

2. Buehler, Mary Fran, *Report Construction,* reprinted by the Institute of Electrical and Electronics Engineers, 1976, and available from the IEEE Professional Communication Group. The chapter "Tables," pp. 30-33, was also reprinted in the *IEEE Transactions on Professional Communication,* Vol. PC-20, No. 1, pp. 29-32, June 1977.

3. Zeisel, Hans, *Say It With Figures,* New York: Harper & Row, 1957.

4. Huff, Darrell, *How To Lie With Statistics,* New York: W. W. Norton & Co., Inc., 1954.

5. Rosenberg, Morris, *The Logic of Survey Analysis,* New York: Basic Books, Inc., 1968.

6. Tinker, Miles A., *Legibility of Print,* Ames, Iowa: Iowa State University Press, 1963.

7. *National Standard, Microfiche of Documents,* ANSI PH5.9-1975, National Micrographics Association, 8728 Colesville Road, Silver Spring, Maryland, 1975.

8. *Technical Publications Program, A Working Guide,* National Aeronautics and Space Administration, Washington, D.C., 1979.

IM

SUBHEADS BOOST AD INFORMATION, INTEREST, READABILITY LEVELS

We read a lot of trade books—or, rather, skim through them—looking for examples of particularly good advertising and particularly bad advertising, our purpose being both to inspire you to do better work and to help you avoid bad practices that might waste your advertising investment.

We have just examined a batch of tear sheets that have been accumulating over several months and have made what we consider to be an amazing discovery. An ad with subheads is almost always an ad that is informative, interesting and easy to read.

We can only guess why this is so; apparently it is a consequence of a writer's sensitivity to the reader's informational needs and his reading behavior—an understanding that transcends the advertiser's compulsion to "say nice things" about himself and his product and the advertising man's compulsion to demonstrate his creative talent.

A subhead "carries the reader along"—assuming that the main headline and/or the illustration have gotten the reader started. Usually, the copy is long (although we will see later some ads with medium-length copy in which there is a subhead for all of or most of the paragraphs) and the presence of the subheads makes the very length of the type less visually formidable.

Also, the writer appreciates the fact that the reader has many things to engage his attention, and what the writer is telling him may not be of major or immediate importance to him; hence, the desirability of inserting a stimulating subhead every so often as a sort of "booster pump" on the transmission line.

And a third point—if well-written, any one of the subheads acts as a flag or hook that may stop the skimming reader even if the main head or the picture has failed to touch one of his hot buttons. So a subhead can serve to get more ads read as well as get more parts of an ad read.

• A striking example of the use of subheads to keep the reader interested is the Hyster ad, "Our most important moving part." The part referred to is the driver, shown mid-ad. Here are the subheads:

1. Human engineering, and why we think it's so important.

2. The doctor who helped JFK's back helped design seats for Hyster.

3. Everything your driver touches is designed to save him time.

4. Does it really make a difference?

The copy is beautifully written: "Remember the problems President Kennedy had with his bad back? To help correct those problems, he worked with one of the world's best orthopedic specialists.

"That same specialist was retained by Dreyfuss Associates to help design seats for Hyster lift trucks.

"Obviously, Hyster pays more than casual attention to the job keeping your driver comfortable, alert, and productive."

Again (answering the subhead "Does it really make a difference?"):

"Any lift truck that's the right capacity might get the job done. Eventually. But most drivers get more done in less time on a Hyster lift truck than on anything comparable" copy continues.

MACHINERY INSTALLATION HYSTER

"No matter how good a lift truck may look on a spec sheet, you never really know what it can do until somebody gets behind the wheel and puts it to work. That's when your investment in a Hyster truck starts paying off."

Dave Newman, Cole & Weber Inc., Portland, wrote this copy and he got excellent cooperation from Bill Bowman, who designed the layout and spec'ed the type.

• "What every machine tool buyer, distributor and builder should know about machine tool installation" is good long headline on a Machinery Installation Systems ad. The copy is broken into several sections with their own subheads:

"To perform properly machinery must be properly installed." "Improper installation causes chronic problems." "Everyone gets hurt in the process." "Problems are so easy to avoid." "There must be a catch."

• A Hewlett-Packard spread has a main subhead to go under the two-word main head. "Distributed confusion":

"There are almost as many approaches to distributed processing as there are computer companies offering them. By letting your needs dictate the right solution, Hewlett-Packard can help you clear up the confusion." Then, scattered through the text are the following:

"Putting an entire network of computers at your fingertips." "How a small computer handles big computer jobs." "Turning raw numbers into usable information." "It takes more than a good product to make it a safe buy." "The HP 3000: The system that makes it all seem simple."

• The least interesting part of the Uniroyal spread is its headline: "Discovered by Uniroyal." The subhead under the illustration is a good one, though: "A fungicide that helps wheat crops without harming the environment." And the smaller subheads scattered through the copy are successful, we think:

"The costly search." "Built-into-the-plant protection." "Bigger roots for tougher crops." "Over 1,400 Uniroyal discoveries."

• The subheads in the Borg-Warner ad take their cue from the main headline: "When Burlington Northern needed 85 locomotives, the tracks led to BWAC":

"When Niagara Mohawk needed a new fleet of trucks, we got them on the road." "When Buttes Gas and Oil needed an offshore drilling platform, BWAC waded right in." "When it's time for you to explore leasing, we'd like to show you the way."

• Subheads punch up the selling points in 3M's "You're looking close-up at a better way to prepare surfaces":

"Better cleaning." "More corrosion resistance." "245% better adhesion." "Adapts to semi-automatic or fully automated operations." "Let us

HEWLETT-PACKARD

A fungicide that helps wheat crops without harming the environment

The costly search

Built-into-the-plant protection

Bigger roots for tougher crops

Over 1,400 Uniroyal discoveries

UNIROYAL

UNIROYAL

When Burlington Northern needed 85 locomotives, the tracks led to BWAC.

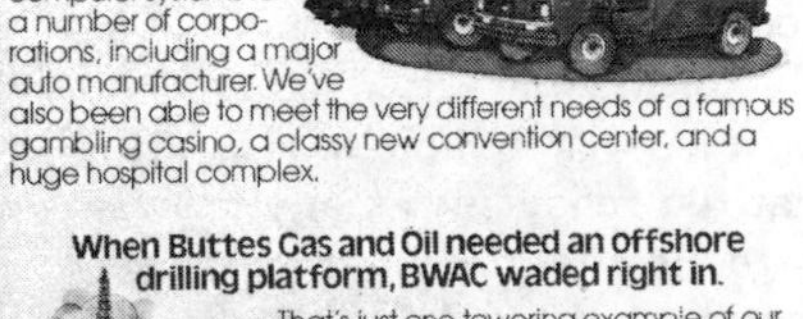

computer equipment as well as complete computer systems for a number of corporations, including a major auto manufacturer. We've also been able to meet the very different needs of a famous gambling casino, a classy new convention center, and a huge hospital complex.

Many big-ticket businesses have known about Borg-Warner Leasing, division of Borg-Warner Acceptance Corporation, for a long time. Now, we'd like you to know us, too. We're a major equity source. We can lease you just about anything you might need for your business. We're big, diverse and creative. And you don't have to take our word alone. You can take it from a lot of other great companies!

When Niagara Mohawk needed a new fleet of trucks, we got them on the road!

If it flies, floats or drives, we've probably leased it already. Our extensive vehicle leasing and leveraged leasing programs have provided huge fleets of trucks and cars, locomotives, boxcars, cabooses—just about everything from pleasure boats to shrimp boats, school buses to 727's! Other versatile leasing programs have supplied

When Buttes Gas and Oil needed an offshore drilling platform, BWAC waded right in.

That's just one towering example of our specialized equipment leasing capabilities. You'll find others in almost any industry you can think of, from mining to manufacturing to farming. (We've actually leased an entire pig farm, from the first silo down to the last little piglet.)

When it's time for you to explore leasing, we'd like to show you the way.

In addition to the programs above, Borg-Warner Leasing has a wide variety of other tailored financial services, including installment sales financing, vendor programs, sales and leaseback, and more. Now that you know we're here, and you know what we can do, you should also know where to write for information: Borg-Warner Leasing, division of Borg-Warner Acceptance Corporation, One IBM Plaza, Chicago, Illinois 60611.

BORG-WARNER

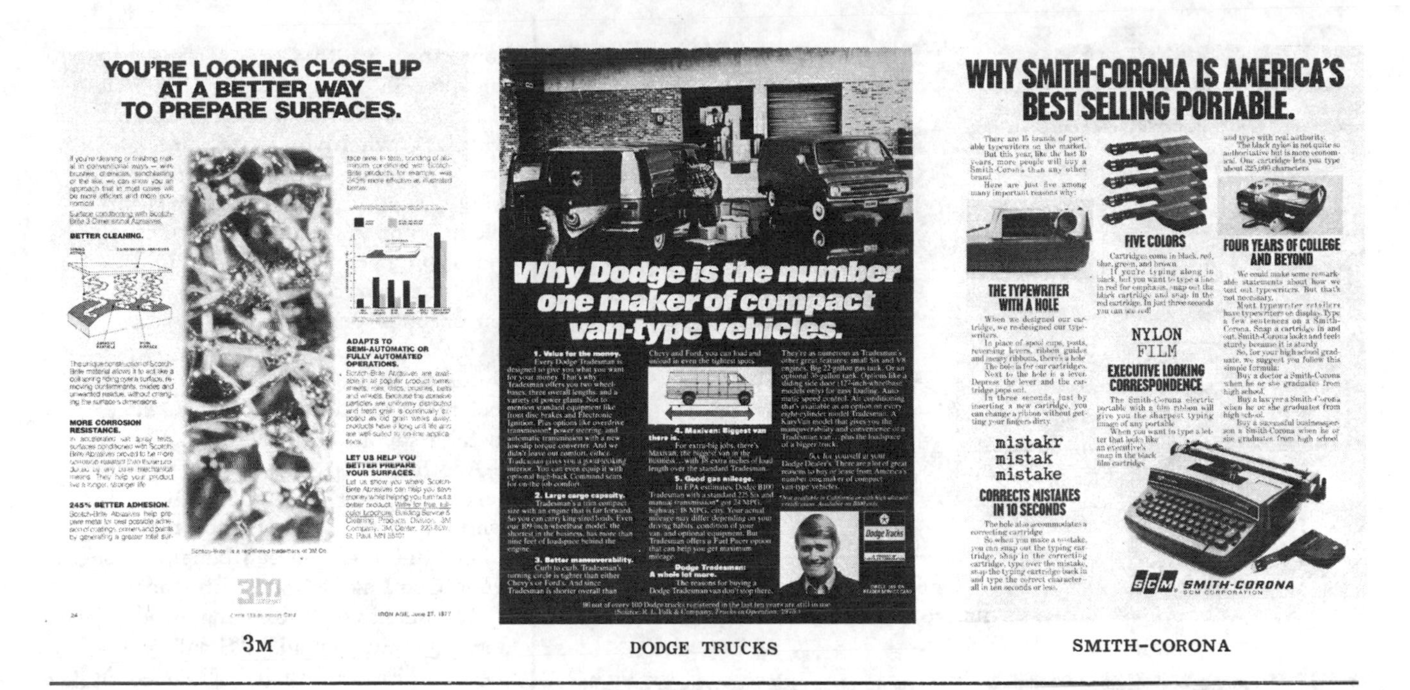

3M DODGE TRUCKS SMITH-CORONA

help you better prepare your surfaces."

• As you know, we don't care for text set in reverse, but that's not the fault of the Dodge Truck copywriter who answered "Why Dodge is the number one maker of compact van-type vehicles" with these subheads:

1. Value for the money.
2. Large cargo capacity.
3. Better maneuverability.
4. Maxivan: Biggest van there is.
5. Good gas mileage.
6. Dodge tradesman: A whole lot more.

• Combining subheads with small illustrations works out well.

Here's how "The truth about rebuilt fuel injectors for your Detroit Diesel" begins:

"An injector is just an injector, right? If it fits it works, right? Some people would have you believe just that. And why not? They're making a pretty good living. selling rebuilt injectors that fit into your Detroit Diesel engine. Injectors that look just like reliabilts—the injectors designed for your Detroit Diesel. You might even be running a set of these look-alikes right now—without knowing the harm they can do.

"Well, enough is enough. We're here to set the record straight."

Which the ad proceeds to do in a series of picture-copy units under these subheads:

"How to spot the real thing." "Ours vs. theirs." "Why we write the mm. output on every box by hand." "Save $3.00 today, spend $300.00 tomorrow." "Think of it as a heart transplant."

The ad is an insert, and the message continues on the back, but the above will give you a good idea of what a splendid job was done on this ad, handled in-house.

• Another fine ad of this *genre* is SCM's explanation of "Why Smith-Corona is America's best selling portable." We wish more copywriters would write the way Irwin Warren, R. K. Manoff Agency, New York, does when he tells about "The typewriter with a hole":

"When we designed our cartridge, we re-designed our typewriters. In place of spool cups, posts, reversing levers, ribbon guides and messy ribbons, there's a hole. The hole is for our cartridges. Next to the hole is a lever. Depress the lever and the cartridge pops out.

"In three seconds, just by inserting a new cartridge, you can change a ribbon without getting your fingers dirty."

And here is how he described

DETROIT DIESEL ALLISON

OWENS-ILLINOIS

WESTERN UNION

Four beautiful reasons to drive an AMC Fleet.

Five practical reasons to drive an AMC Fleet.

The $350 fleet allowance

Practical cars that aren't less car.

Excellent resale value.

Guaranteed Value Plan.

The Buyer Protection Plan® II.

AMC fleet & leasing representatives.

AMC

how the Smith-Corona "corrects mistakes in 10 seconds": "The hole also accommodates a correcting cartridge. So when you make a mistake, you can snap out the typing cartridge, snap in the correcting cartridge, type over the mistake, snap the typing cartridge back in and type the correct character—all in ten seconds or less."

Oh, the beauty of short, common words to make things clear. Credit, too, to Lars Anderson for an easy-to-take layout.

• Another good one is Owens-Illinois' "ComPak. If you're about to send a new product to market, you're the one who can benefit most from ComPak." Good subheads:

"If you have an existing product, don't go away mad." "Small potatoes?" "Faster than a speeding bullet?" "The machine evolved from man." "The price is right." "Why are we working so hard to sell you less of our product?"

• The Western Union ad, "Keys to the world's executive suites," is more picture than words, but there are still bold subheads:

"Multiple message machine." "Electronic Mail." "Many new services."

• Half of the AMC spread is illustration of "Four beautiful reasons to drive an AMC fleet." The other half is copy, headed "Five practical reasons to drive an AMC fleet," and it has these subheads:

"The $350 fleet allowance." "Practical cars that aren't less car." "Excellent resale value." "Guaranteed Value Plan." "The Buyer Protection Plan II."

• And the Signode ad, "Handle them one at a time. Or all at one time"—which is mostly illustration—has these subheads to brighten the text area:

"Signode helps eliminate overhandling." "Secure product in compact, stackable units." "Unitizing works for all kinds of products." "Signode mades unitizing work for you."

• "An oil that can help save fuel in diesels" is the big headline in Shell's ad. Then a subhead leading off the copy: "Shell multigrade oil recognized as a timely contributor to fuel economy by leading truck fleet

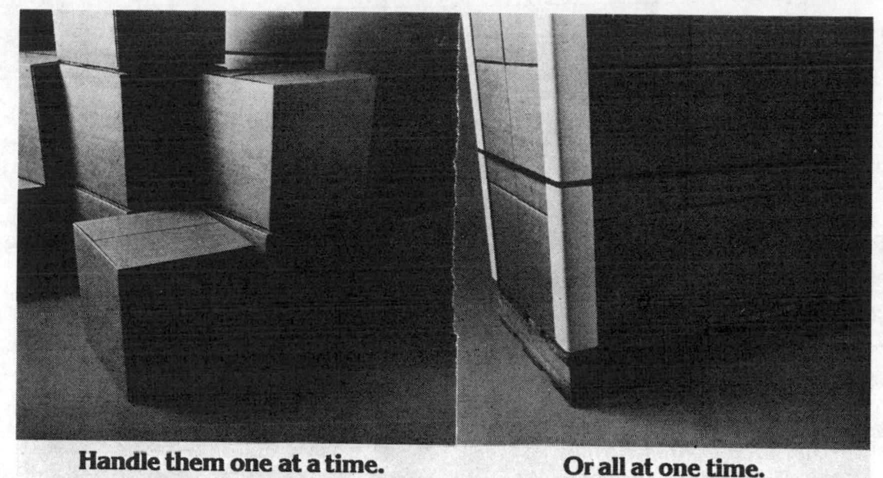

SIGNODE

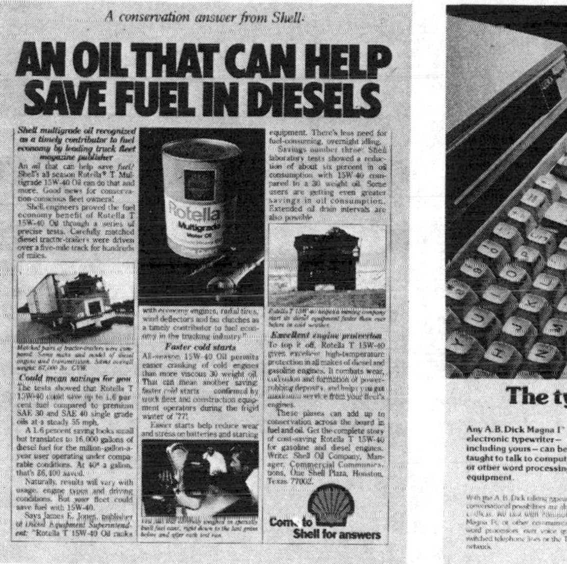

SHELL

The typewriter that talks.

Any A.B. Dick Magna I® electronic typewriter—including yours—can be taught to talk to computers or other word processing equipment.

Use your own phone.

Responds automatically, day or night.

Adapts in 15 minutes.

Want to start a conversation rolling?

ABDICK

A.B. DICK

magazine publisher." Down in the copy, these subheads:

"Could mean savings for you." "Faster cold starts." "Excellent engine protection."

Just because the ad has a big picture is no reason not to consider using subheads.

• "The typewriter that talks" is the fine headline on an A. B. Dick ad. The opening paragraph is set in bold-face, making a sort of subhead: "Any A. B. Dick Magna I electronic typewriter—including yours—can be taught to talk to computers or other word processing equipment." And each paragraph has its heading:

"Use your own phone." "Responds automatically, day or night." "Adapts in 15 minutes." "Want to start a conversation rolling?"

This ad is just about perfect—the copy thanks to Milt Lynnes, Marsteller Inc., Chicago; the layout and typography, thanks to Jack Lundstrom. ■

Copy Chasers

ACCURACY OF PUBLISHED MEDICAL REFERENCE CITATIONS

JUNE E. GOODRICH, B.S.
Mayo Clinic and Mayo Foundation
CHARLES G. ROLAND, M.D.
Jason A. Hannah Professor of the History of Medicine
McMaster University, Hamilton, Ontario, Canada

ABSTRACT

Among 2,195 reference citations, published during 1975 in ten major US medical journals, 634 (29%) were found to be erroneous on direct checking of the original source. The percentage of error within individual journals ranged from 14 to 50 per cent. Such a high error rate would seem to seriously diminish the usefulness of published reference lists and, possibly, raise questions about the accuracy of other portions of the literature also.

A largely unquestioned assumption on the part of readers of scientific and technical articles is that the references cited therein are cited accurately. That is, one assumes that one can take the information in the reference list and use it to find the needed citations.

We have questioned this assumption, at least in its application to the medical literature. Our null hypothesis may be stated thus: that reference citations published in major North American medical journals provide accurate information for readers.

Material and Methods

We selected ten journals (see Table 1), including three of the major large-circulation general medical journals and seven others representing some of the chief specialty disciplines in medicine.

The first article containing references from each of ten consecutive issues of each journal published in 1975 was used. All references were verified by one of the authors (J.E.G.) directly, by examining the cited journal or book and comparing the published citation with the original publication. All errors found were classified by type (Table 1).

TYPES OF ERRORS

Incorrect name—Misspelling included wrong initials and omissions of hyphens, umlauts, and other accent marks. Judgments about the addition of names not actually authors and the omission of names of actual authors were assessed in the light of style for each individual journal. Thus "et al," for example, was not considered an omission of a name or names if the particular journal regularly used "et al" in reference citations.

Table 1. Journals Studied and Errors Found in the Lead Article of Ten Consecutive Issues

		Incorrect name				*Name of journal*		*Wrong*			
Journal	*No. of Refs.*	*Missp.*	*Omit*	*Add*	*Incorrect Title*	*In-correct*	*Wrong Journal*	*Vol*	*Pp*	*Yr*	*Others*
Am J Psychiatry	327	10	3		38			3	9	1	
Am J Publ Health[a]	157	2		1	9			2	3		No vol. or pp--9 times
Anesthesiology	261	29	4	1	64			4	21	2	Ref. repeated
Ann Intern Med	387	43	1		70			7	17	1	Ref. repeated
JAMA	117	15			37			3	4		
J Bone Joint Surg	311	7			20			3	8	2	
J Med Educ	84	3			10			1	3		
N Engl J Med	194	4			16			2	5		
Pediatrics	236	27	2		42			2	6	1	
Surg Gynecol Obstet	121	11			37	1		2	2	1	

[a] Many journals and books not available in Mayo Clinic Library for checking.

Incorrect title—Any deviation from the title as given in the original article was considered an error (for example, adding or omitting an "s," a hyphen, a mark of punctuation, an article, or a subtitle).

Name of journal—Deviations from *Index Medicus* style were not considered errors, but rather journal style.

Wrong entry: volume, year, or pages—If the article was not found where given in the reference, an attempt was made to track it down so that the proper error could be checked (for example, if the volume and year did not agree, which was wrong?). If beginning and ending pages were included in the reference, an error was checked if one or the other was wrong.

Other errors—The errors listed above covered most of the problems encountered. However, some oddities appeared. One journal completely omitted the volume number three times and the page numbers six times. Two other journals repeated a reference in the list.

Results

Table 1 lists the journals studied, with the total number of references and the errors found in the various categories. Table 2 gives the total references and errors, and the percentage of error for each journal. Some reference citations had errors in more than one category.

Table 2. Percentage of Error for Each Journal

Journal	*No. of refs.*	*Total errors*	*Per cent error*[a]
JAMA	117	60	50
Anesthesiology	261	126	48
Surg Gynecol Obstet	121	54	45
Ann Intern Med	387	140	36
Pediatrics	236	80	34
Am J Psychiatry	327	64	20
J Med Educ	84	17	20
Am J Publ Health	157	26	17
N Engl J Med	194	27	14
J Bone Joint Surg	311	40	13

[a] This is not to say that this percentage of the references checked had errors, because some references had two (or even more) errors.

Here are two examples of time-consuming errors typifying the problems that can result.

Reference as published:

Bahnson, C. B., and Wardwell, W. I. Personality Factors Predisposing to Myocardial Infarction. In Psychosomatic Medicine, Proceedings of the First International Conference on the Academy of Psychosomatic Medicine. Excerpta Medica Foundation, 249-257, 1966

Corrected:

Bahnson, C. B., and Wardwell, W. I. Personality Factors Predisposing to Myocardial Infarction. Excerpta Medica International Congress Series No. 134, 1966, pp 249-256

There were thirty-nine volumes of this journal dated 1966. The title of the book is not on the outside, just the series number. So it was necessary to check *each* volume until discovering the correct one, a very time-consuming and discouraging process.

Reference as published:

Roberts, Brooke; Rosato, F. E.; and Rosato, E. F.: Heparin—A Cause of Arterial Emboli? Surgery 53:803-808, 1964

Corrected:

Roberts, B.; Rosato, F. E.; and Rosato, E. F.: Heparin—A Cause of Arterial Emboli? Surgery 55:803-808, 1964

Here the volume and year didn't agree, so which was correct? One must resort to the index of the volume, or try the volume corresponding to the year given, until the correct combination is found.

Discussion

Obviously, the percentage of errors found was extremely high. We have not attempted to make value judgments by indicating which errors are "serious" and which are "trivial." Most of the errors found were sufficient to require the expenditure of extra time to resolve the discrepancy. On the basis of wasted time alone, multiplied by some finite but unknown number of readers who might seek any given reference, the existence of so many errors is disturbing.

Errors can be so substantial as to raise the likelihood that some readers, particularly those lacking a first-rate medical library, might not find the erroneous citation at all. The effects of this occurrence are incalculable but probably not insignificant.

Quite aside from these disturbing specific consequences of erroneous reference citations, there is the general question of credibility. What reliance can readers have in authors and in journals which, despite their combined efforts, permit the publication of such slipshod work? The reference list is part of the article. The article is the literary embodiment of the research. If the reference list is significantly inaccurate, what about the data in the Results section?

The question of the relative responsibility of author and of journal is moot. Both parties should have vested interests in ensuring optimal accuracy. Yet our results show that the efforts are insufficient, at least in the journals studied during this period.

36 Aids to Successful Proofreading

RUTH H. TURNER

Abstract—Nearly all proofreading is worthwhile, much is essential. Efficient proofreading techniques are grouped for different types of material. A style sheet simplifies work on long documents. A dictionary should always be available—the vagaries of English spelling are a major source of errors.

HOW do you handle your proofreading chores? Are you one of those "Oh-I-didn't-notice-that-mistake" people? If you find proofreading a distasteful part of your job, then here are three dozen assists to make the task, if not fun, at least more bearable.

The proofreading technique you use will be determined by the kind of document. Is it a legal description of a plot of land or a memo to a department head down the hall? Is it to be reproduced 5,000 times for distribution at a conference or a set of notes for a speech tonight? Maybe it is the annual departmental budget with its column upon column of million-dollar figures. Now, you face the titanic task of being sure that every number is right.

Which of these documents warrants the attention of a pair of proofreaders? Which one can be given a quick scan before you say "This is ready for my signature?"

Another concern is the amount of time available for proofreading. How essential is accuracy? What is your own feeling about mistakes? If you are anxious to improve your work but time is of the essence, try out some of these ideas, adapt them to your own circumstances and tastes, and feel happier in your job.

The aids are classified as follows: short narrative copy (letters, memos, etc.), long narratives (consisting of ten or more pages), technical material and statistical tables. A concluding section warns the unwary of a list of "boners" that can so easily trip up a typist, whether a rookie or an executive secretary. Some of the suggestions can be used for proofreading more than one kind of copy, as well as for other material that is not included in these classifications.

Here they are—36 aids to successful proofreading:

Short Narrative Copy (Letters, Memos, etc.)

1. Pay attention to dates. Do not assume they are correct. Check spelling of months, even the correctness of the year, especially if other than the current year. During January, the typist's fingers and brain must undergo a retraining period; keep a watch on the results until the new habit has taken hold.

Adapted and reprinted with permission from *The Secretary*, vol. 38, no. 2, p. 12, Feb. 1978; copyright 1978 by The National Secretaries Association (International), Kansas City, MO 64108.

The author, who died last year, was a freelance writer living in Grants Pass, OR.

2. Do not overlook the name, address, subject or reference line, signature line, even the list of recipients of copies. Errors can sneak into these items as well as into the body of the letter.

3. When you read the body of the letter, examine the ending and beginning of lines to be sure a little word like "that" or "and" has not been repeated.

4. Concentrate as you slowly read for typing errors. This is no place for speed-reading.

5. If you have copied from typed material and have ended each line at the same place, check your copy by laying it over the original and holding them up to a strong light. This procedure will reveal errors without the need for reading.

6. Personalized letters or memos which are identical except for names, addresses or short insertions can be proofread by carefully examining the personalized portions, and then applying suggestion 5 above.

7. Possibly you and another person can arrange to proofread each other's material. However, this procedure must not be allowed to bottleneck your schedule. Also, this won't work if either of you is sensitive about having another person find your mistakes.

8. In order to concentrate specifically on typographical errors, read backwards. Of course, this method will not reveal omissions or duplication of copy, and so must be followed by a second reading for content.

Long Narratives (Manuals or Reports Consisting of Ten or More Pages)

9. Make a style sheet. Some professional proofreaders use this technique to assure consistency throughout a manuscript relative to spelling of specialized terminology or jargon, unusual capitalization and punctuation, treatment of headings, format for outlines or tables, use of names and titles. Any items which are unique to your project can be jotted down on this style sheet.

Keep this sheet for use in later revisions of the manual or for production of additional manuals in a series.

10. When you are ready to proofread, do it in steps. For example, check all the headings and titles first. Have you consistently followed the style sheet?

11. Scan page numbers. Have any been missed or duplicated?

12. Read the body of the manual against the original, sentence by sentence. This is a tedious, slow task and can be speeded up if you know the material well enough to catch duplications, omissions and typing errors with a continuous, concentrated reading.

13. Make sure any references to other parts of the manual are correct. For example, "See page 6 for a similar list." Possibly, in the final typing, that list moved to page 7.

Reprinted from *IEEE Trans. Prof. Commun.*, vol. PC-22, pp. 19-20, Mar. 1979. (Reprinted by permission, Copyright © 1978, The Secretary, official publication of Professional Secretaries International, Kansas City, MO.)

14. You may also be responsible for verifying title and page numbers of other reference material. If so, look up each item in the book referred to if at all possible.

15. When copying from a draft that has handwritten inserts, be sure none is overlooked.

16. If the draft from which you are working has been written on scraps of paper (and how often this does happen), count the number of paragraphs in the draft and compare it with the number in the typed copy, to be sure that you have all of them.

Technical Material (Such as Codes, Scientific Data, etc.)

17. Two people, one reading aloud, the other checking copy, may be needed to verify this kind of data.

18. If such an arrangement is not available, do your own proofreading, sentence by sentence.

19. If the draft is in typwritten form, lay a ruler under each line. This will help you keep your place on both the original and the copy. This technique also helps when the inevitable interruptions occur.

20. Outlines can best be proofread by breaking the task into components: check headings, then the mechanics (numbers, letters, indentations), and finally a continuous reading of each item against the original. Also use suggestion 17 above.

Statistical Tables, Budgets, Itemized Data

21. Suggestion 17 above is the surest technique for verifying this kind of copy.

22. If you must do your own proofreading, the following method is helpful: If the columns have been typed by tabulating *across* the page, then proofread your copy *down* the columns. Do this by folding the original from top to bottom along the column and laying it next to the corresponding column of the copy. Then read item by item.

23. If columns were typed down the page, then proofread across the page, folding the original beneath each line and laying it under the corresponding line of typed entries, so that you can easily read the original against your copy.

24. Count the number of entries in each column, and compare with the number in your copy. If there is a difference, hunt down the culprit.

Watch for These Boners

25. Make the dictionary your best friend. Is it *privilege, privelege* or *priviledge; gauge* or *guage; discresion* or *discretion; chief* or *cheif?* Never be embarrassed to look up a word, lest you be more embarrassed at an error in work completed.

26. Watch closely for omission of "ed" or terminal "s."

27. Study related words until you understand differences in their meaning and use: *affect/effect, imply/infer, immigrate/emigrate, advice/advise, counsel/council/consul.*

28. Mentally repeat syllables of long words containing many vowels: *evacuation, responsibilities, continuously, individual.* It is easy to omit one.

29. Be careful of those double letters. *Supersede* is the only word in the English language ending in "sede," and *succeed, proceed* and *exceed* are the only ones ending in "ceed." All the others (*precede,* etc.) end in "cede."

Double letters can so easily *embarrass* you, especially if the *occasion* to use *accommodation* or *occurrence* becomes *necessary.*

30. Silent letters also complicate a typist's life: *rhythm, rendezvous, malign.*

31. If punctuation causes you difficulties, check these marks after you have completed all other proofreading. Be careful not to omit a closing parenthesis or quotation mark.

32. Be consistent in use of commas. If you have to read a sentence more than once to get its meaning, a comma may correct the problem. For want of a comma, the meaning may be lost.

Are apostrophes going out of style? So often they are omitted or used incorrectly. Use an apostrophe to show the omission of a letter in a contraction (*isn't, won't*) and to show possession (*Helen's* pencil, *Mr. Brown's* letter).

Refer to the dictionary to determine whether to use a compound word or a hyphenated word. If the word does not appear as either one in your dictionary, use a hyphen.

35. Be consistent in use of capital letters. Follow an established practice for the type of document or its subject matter if one has been set up. If not, make a style sheet of your own.

36. Interchanged words are easy to overlook: *their/there, were/where.*

There is one last suggestion—proofread tomorrow what you worked on today.

Writing is for Readers

H. R. Clauser

Reading is an indispensable part of written communication. To communicate, our words, phrases, and sentences must be read and understood. Writing that does not meet this requirement is noise.

It follows, then, that it is the reader's reception of our writing that measures the extent to which we successfully communicate our ideas; thus the reader is a proper and important person to study when we set out to improve our writing. So it is he, the reader, rather than you, the writer, who will be the principal subject of this article.

In recent years reading courses have become almost as popular as courses on how to win friends. And rightly so, for they can improve reading speed and comprehension. But in one way, reading courses are based on a partially false assumption—that is, that all writing is good writing and that writers are aware of the rules of good reading. We all know that this is far from being the case.

So, to help our readers and make it possible for them to practice what they have learned from reading courses, we should write in accordance with the rules of good reading. If we follow this practice, we will quickly discover that the rules also help us, the writers, for as it turns out, the rules of good reading are identical with those of good writing. Let's study the rules for effective reading.

READING AND WRITING WITH A PURPOSE

The following statement introduces one of the chapters in a course on reading: "When an author writes an article or a chapter, he has a purpose in mind for conveying his thoughts to others . . . When you are about to read a selection, you should decide what that purpose is, and adjust your reading to it." This is sound advice, and I'm sure many readers, whether or not they have taken a reading course, try to apply it.

But, unfortunately, readers are frequently double-crossed by writers whose writings seem to have no discernable purpose. Reading such writing is like retrieving a ball of yarn that is aimlessly rolling and unwinding on a patch of uneven ground. The reader does not know what to expect. He cannot determine the level or pattern of the thoughts or information the writer wants to convey. Consequently the reader may become discouraged and give up; or, having retrieved the unraveled ball of yarn, he may wonder why he did.

So, for effective communication there must be a definite purpose behind our writing to give it form and a sense of direction. To determine what this purpose is, we must ask these two questions before beginning to collect or select our material: What is it for? and Whom is it for?

Answering these questions will go a long way toward giving your writing a sense of unity and focus. There are often several answers to the questions. You may have several aims in mind. The solution, then, is not to ignore the questions and try to serve all possible purposes. Difficult as it may be, you must still arrive at answers that will define the purpose of your work for the reader.

There is another important reason for answering the foregoing questions. Writing with a definite purpose not only benefits the reader, but also helps you, the writer. As you will see later, it serves as the benchmark, the basic guide for your writing task. It sets the tone, the pace, and the technical level; it helps determine the amount of detail to be included; and it suggests how the writing should be composed.

READING AND WRITING IDEAS

One of the first things the students in reading courses learn is to read ideas and not just words. They are taught to master the art of searching for and quickly grasping the author's ideas.

The key to helping the reader in his search for our ideas is the paragraph. The principles of good reading teach him that each paragraph viewed as a whole represents one and only one unit of thought or idea. So the writer should approach each paragraph as if it were the only writing before him at that moment, and it should be written with these two questions in mind:

1. What is the basic *subject* or thing I want to cover in this paragraph?

2. What is the important *idea* or thing about the subject I want to convey?

Paragraphs constructed using these questions as a guide will make it easy for the reader to spot the basic thought and to separate it from supporting, but often subordinate, detail.

In every piece of expository writing, there is supposed to be a hierarchy of ideas and subjects, and the student taking a reading course is taught to uncover that hierarchy and quickly pick out the main ideas. Thus, to help him in this task, our writing should clearly distinguish between the importance of various ideas we are trying to convey to him. There are a number of ways to do this. A well chosen title for an article, a report, or a memo can often give the reader at least a good preview of the principal idea or ideas covered. More commonly, the subheads can function as statements of either the main ideas or subjects and, at the same time, show the structure of the written piece. For example, in this article the subheads represent the major subjects covered. Other familiar methods are abstracts, summaries, and lists of major points covered. The accompanying box, for example, highlights the main ideas of this article.

HERE ARE THE HIGHLIGHTS

- In written communication, no one plays a more important part than the reader.
- The rules of good reading are identical with those of good writing. In terms of writing they are:

  1. Write with a purpose.
  2. Write ideas, not just words.
  3. Write paragraphs as units of thought.
  4. Highlight the main ideas.
  5. Include only relevant detail.
  6. Select amount and kind of detail to suit the audience.
  7. Relate detail to the main ideas.
  8. Permit conscious inaccuracy when it leads to better understanding.

A method not so common but which could be used more often to advantage is leading off each major section of the paper with a statement of the main idea. **The idea can be emphasized by underlining it. In printed matter the main ideas can be set in boldface** type, as is shown in Figure 1.

Here is a comprehensive survey of

Permanent Magnet Materials

- *Magnet steels* • *Alnicos*
- *Other magnetic alloys*
- *Ceramics* • *Fine particle magnets*

By [illegible] J. Fabian, *Associate Editor, Materials in Design Engineering*

• Permanent magnet materials can be used in a wide range of applications where a self-contained source of potential energy is needed. Their use ranges from simple door latches having a high holding force, on up to complex motor stators having a strong, inherent magnetic field. A great number of permanent magnet materials have been developed to meet the needs of such varied applications. The purpose of this article is to describe the various metal and ceramic materials that are available today.

In the past, the magnet steels (described first below) were the only permanent magnet materials available to the designer. Although they are still used to a limited extent today, they have been replaced in many applications by the newer permanent magnet alloys and ceramics which possess superior magnetic properties.

Magnet steels

are generally low in cost and are available in a wide variety of shapes and sizes. Although suitable for some applications, they are not used where high magnetic strength, small size or light weight is required.

At one time the magnet steels were the only permanent magnet materials available to the designer. Although they are still used to a limited extent today, they have been replaced in many applications by the newer permanent magnet alloys and ceramics which possess superior magnetic properties.

Figure 1—Highlighting the Main Ideas

READING AND WRITING DETAILS

Perhaps the most difficult part of reading and, likewise, of writing, is dealing with detail. Often it is relatively easy to set down the main ideas of a piece of writing. But when that has been done, writer's (or reader's) cramp frequently sets in. We either feel that there is nothing more to say, or we feel hopeless and indecisive in facing the mountain of detailed material before us.

In dealing with detail, our first concern is to avoid burdening the reader with unnecessary detail. Out of the mass of facts on hand, we must sift out that which is relevant. "What has this to do with it?" This is the question we must constantly ask because irrelevant details seldom do any good and may do some harm.

Indeed, one of the unkindest tricks the writer can play on the reader, particularly one who has taken a rapid reading course, is to suddenly present him with an isolated, irrelevant fact. In the paragraph below, the last sentence is such an isolated fact, with no relation to the rest of the paragraph. Examination of the entire article showed that it also had no connection with anything else in the article. This example illustrates a particularly dangerous type of irrelevancy. Because "hardenability," the irrelevancy, in some instances could be related to the subject of "carbide content," the reader is doubly confused.

> A further control of carbide content can be obtained by composition variations. The carbide content is increased by the addition of chromium or by lowering the carbon equivalent, and it is decreased by the addition of copper or nickel. *Hardenability is significantly increased by the addition of molybdenum or chromium and is moderately increased by the addition of copper or nickel.*

Sprinklings of such irrelevant facts through technical papers and reports is common in technical writing. Perhaps the most common cause is that most of us do some thinking as we write and sometimes even write in order to think. Therefore it is not unusal that, from the many associations generated by our thinking, some unwanted ones will drop unnoticed into our paragraphs. The cure is to seek them out when you revise and never to succumb to the temptation to write the way you think or talk.

HOW MUCH, WHAT KIND OF DETAIL?

Besides having to decide what is relevant to the subject about which we are writing, there are other problems to face. One of these is deciding the amount and kind of detail needed.

The amount and kind of detail needed to amplify our main ideas is closely related to the audience. Every field, every subject, consists of several stages of sophistication. And the stage that your audience occupies will largely determine the kind and amount of detail you can safely include. For example, in the general field of science, there are five stages of sophistication (see Figure 2), according to Bachelard in his book, *La Philosophie du Non*. The double lines separating the stages represent the real barriers that exist between stages, and when writing in the general science field, we must respect them.

Although it is relatively easy for a person in one stage to move to a lower stage and back again, it requires a good deal of effort to hurdle the barrier into the next higher stage. Thus, if the audience is largely in Stage 3, the detail you present must be largely of the Stage 3 variety or less. Or if you are addressing a Stage 4 subject to a Stage 3 audience, only a small amount of detail can be successfully communicated. And, according to W. M. Thistle,[1] to try to transmit any detail at all back through two or more of the stages is almost an impossible task.

What should be done when different amounts of detail are desired by various members of a mixed audience? Should the paper, report, or article contain practically everything—the laboratory procedure, the test set-up, the reasoning steps, the rejected alternatives, and an-

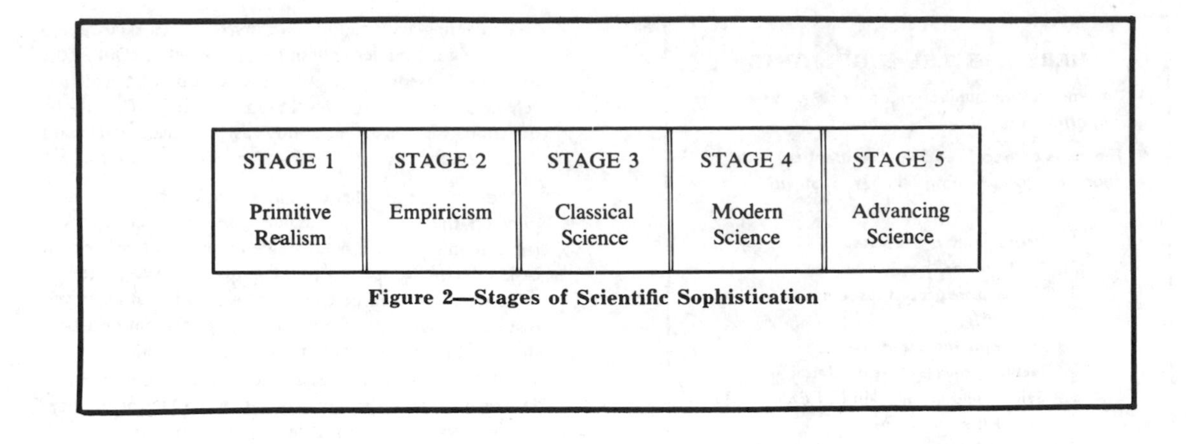

Figure 2—Stages of Scientific Sophistication

swers to all questions or criticism that might be raised? If we include all such details in the conventional manner, we reduce the impact of our major findings and run the risk of losing important readers. On the other hand, if we omit some of the detail, we can be criticized for being incomplete or for not fully substantiating our findings.

It is a difficult decision to make. But rather than playing it safe and always including "everything," the problem should be faced each time we have a writing job to do. In many cases the problem can be solved by making use of appendixes. Another way is to treat some of the less important detail as optional reading. This is easily done in printed material by enclosing the optional reading in a box, as shown in Figure 3. When other methods of reproduction are used, the technique of optional reading is more difficult to apply. Nevertheless, it is still often feasible to insert boxes even in typewritten pages.

HOW ACCURATE THE DETAIL

Closely related to the amount and kind of detail is the question of how accurate detail must be when conveying ideas in one stage to readers who are in a lower stage of sophistication. Many scientists and technical men insist on maintaining the same precision in their writings regardless of the audience they are addressing. They would rather remain speechless than be caught describing high polymer thermoplastic fibers as resembling tiny corkscrews hooked together and arranged in parallel, intertwining rows, or referring to an edge dislocation in a crystal as being similar to a scar in the flesh.

But a certain degree of conscious or deliberate inaccuracy is necessary and permissible in order to make our detail palatable and understandable to other than our wise and knowing colleagues. In an editorial in *Science* some months ago, Warren Weaver[2] proposed the idea of "communicative accuracy" to gauge the degree of inaccuracy permitted in scientific writing. He suggested: "A statement may be said to have communicative accuracy, relative to a given audience of readers or hearers, if it fulfills two conditions. First, taking into account what the audience does and does not already know, it must take the audience closer to a correct understanding. . . . Second, its inaccuracies (as judged at a more sophisticated level) must not mislead, must not be of a sort which will block subsequent and further progress toward the truth. Both of these criteria, moreover, are to be applied from the point of view of the audience, not from the more informed and properly more critical point of view of an expert."

One of the important techniques they are taught is to relate detailed facts to the main ideas to better understand and retain the detail. Nila Smith, a leader in the field of reading instruction, puts it this way: "As you read, think of the main idea as a magnet drawing the particles toward it—the 'particles' being the smaller and detailed ideas. Then visualize the main idea together with its cluster of subideas as a unit in itself. In a nutshell, these are the processes which will enable you to grasp and hold in mind a series of minor factual details."

Translating this reading rule into a rule for writing, we can say that each detail must be directly and intimately related and connected to its main subject or idea. This means every detail in its proper place. And its proper place is with the idea it is supporting.

HOW TO CONVEY DETAIL

Deciding the kind, amount, and accuracy of the detail is only half the job. We must still help our readers to read, comprehend, and remember the details we have decided to offer them. To tell us how we can do this, we again can refer to what is being taught to our readers in their better reading classes.

To emphasize the importance of this rule and also to summarize many of the points covered in this article, there is reproduced in Figure 4 a small section of a long article in which the author has failed the reader.

In reading this piece of writing, even the brightest readers would have difficulty, first, in sorting out the main subjects and ideas, and second, in associating the details with the main ideas. But those willing to take time, after some study would see that there are three major subjects and ideas covered:

A. The thermal problem encountered in satellites. The thermal problem involves protecting the delicate electronic equipment from three sources of radiation heating—the sun, the earth, and the atmosphere.

B. The properties of the materials and surfaces involved in these problems. The satellite skin must have high reflectivity and low absorptivity to protect the payload from radiation heating.

C. The materials and surfaces used to solve the thermal problems. A lightweight material whose surface is treated to provide high reflectivity and low absorptivity, and to resist abrasion from meteoric dust, seems to offer the best solution.

In Figure 4 the details on each of the three main subjects (labeled A, B, and C) are scattered throughout the section. Although this piece of writing could profit by considerable revision, its readibility would be greatly improved by merely relating the details to the main ideas.

manganese steels, are water hardening. In general, tungsten steel magnets are forged, then machined to shape.

Chromium steels

Chromium magnet steels owe their development to the shortage of tungsten during World War I. As shown in the table (see next page), the maximum energy product values of the chromium steels are not as high as those of the tungsten steels. However, magnetic differences between the two materials are not appreciable and the chromium steels have the advantage of being lower in cost.

Compared to carbon steels, the tungsten and chromium steels are more stable after temperature changes and mechanical shock but, nevertheless, they still show some magnetic variations upon aging. Because of their high remanence the tungsten and chromium steels are useful for magnets requiring a small cross-sectional area.

Cobalt steels

The cobalt steels were largely developed in Japan where investigators found that additions of about 35% cobalt greatly increase the coercive force of steels containing 5 to 9% chromium and small amounts of tungsten. In general, an increase in cobalt content is accompanied by an approximately linear increase in coercive force and an increase in residual induction and maximum energy product. Higher coercive forces can be obtained with the cobalt steels than with the carbon, tungsten or chromium steels.

A wide range of magnetic properties can be obtained with the cobalt alloys within suitable composition limits. As shown in the table, manufacturers have established a number of standard alloys, the principal ones containing 3, 9, 17 and 36% cobalt. All of these alloys are martensitic and contain iron-carbon precipitates throughout the crystal lattices as a result of quenching from about 1475 to 1750 F.

How Permanent Magnets Behave

In order to design a permanent magnet that will perform a given function, the designer must have a basic understanding of how ferromagnetic materials behave. The following description, supplied by Indiana Steel Products Co., explains how the magnetic properties of materials are obtained, and explains their significance.

Understanding the hysteresis loop

Fig 1 shows how the induction in a magnetic material changes as the magnetizing force is varied. When demagnetized material is subjected to a gradually increasing magnetizing force up to H_{max}, the induction in the material increases from zero to B_{max}. If the magnetizing force is then gradually reduced to zero, the induction decreases from B_{max} to B_r on the vertical axis. This B_r value is known as the residual induction.

If the magnetizing force is reversed in direction and increased in value, the induction in the material is further reduced, and it becomes zero when the demagnetizing force reaches a value of H_c, known as the coercive force. A further increase of this negative force causes the induction to reverse direction, becoming $-B_{max}$ at $-H_{max}$. If the magnetizing force is reversed from this point to H_{max}, the change in induction is along curve $-B_{max}$, $-B_r$, B_{max}. This cycle gives the complete hysteresis loop.

Such a curve applies to all magnetic materials, the difference in materials being largely a matter of the values. Materials having a low coercive force are low-energy materials, and those having a high coercive force are high-energy materials. These are commonly known as soft and hard materials, respectively, but the terms low-energy and high-energy are more representative of the characteristics of magnetic materials.

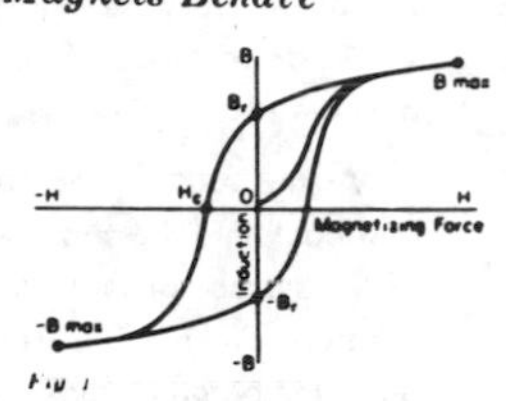

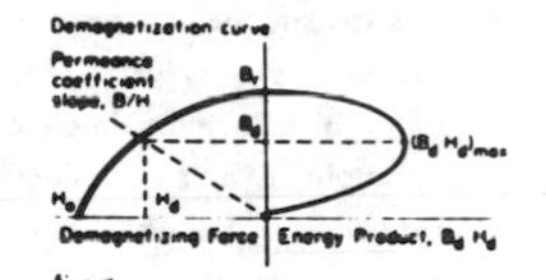

Obtaining the maximum energy product

The section of the hysteresis loop from B_r to H_c is of major interest to designers of permanent magnets. This is known as the demagnetization curve and is shown in Fig 2. At the right of this curve is the conventional energy product curve, which is the product of B and H as taken from the demagnetization curve and plotted against B. The product of B_d and H_d at any point on the demagnetization curve indicates the useful energy produced per unit of volume. In the cgs system where B_d is in gauss and H_d is in oersteds, $B_dH_d/8\pi$ is equal to energy product in ergs per cu cm. The external energy is zero at both B_r and H_c, and reaches a peak value at a point known as the peak energy product, $(B_dH_d)_{max}$. This point represents the maximum external energy that can be produced by a unit volume of a given material. It is a criterion for comparing different permanent magnet materials. Furthermore, in the case of fixed air gap applications the design which causes the magnet to work at the maximum energy point will require the least volume of magnet material.

Before the Alnico alloys were introduced, the 36% cobalt alloy was the best permanent magnet material known, the maximum energy product being 1.0 x 10^6 gauss-oersted. The alloy is still used today in a limited number of applications; however, it is relatively expensive for the amount of energy that it produces.

Figure 3—Detail as Optional Reading

A

B

A

Because of the essential lack of atmosphere at orbiting altitudes, the heat transfer problems of satellite bodies are considerably different from those described for vehicles within the atmosphere. Free molecules which exist at altitudes of 100 miles or higher do not develop convective heating. Thus, the only significant heat transfer process involves radiation heating of the skin due to the solar flux. The thermal problem in this case does not primarily involve the structure of the satellite but the delicate electronic equipment which is the payload. For this reason the concern is with temperature changes from normal involving rather low values.

B

C

The primary external sources of heat to the satellite include the radiation from the sun and the reflected and emitted radiation from the earth and atmosphere. By far the most important parameters of the structure are the emissivity and absorptivity of the skin. High polish may be used to give a high value of reflectivity and thereby a low value of absorptivity. Another approach involves the plating of a light metal shell with a material of high infrared reflectivity covered by a transparent layer having a hard surface which would be resistant to meteoric dust. A low value of absorptivity generally signifies a low value of emissivity. This condition presents a problem if internal heating is developed by electronic equipment because of the difficulty of radiating the internal heat. An alternate procedure would be to rely on high emissivity and adopt spinning techniques to continually rotate the position of radiation heating to a position of radiation cooling.

B

As the satellite revolves about the earth, it passes from the earth's shadow into sunlight and back into the shadow again in periodic fashion. If the reflectivity and emissivity values of the surface are not favorable, a low heat sink body may develop quite respectable temperature cycles. The peak temperature may exceed 1000°F on the sun face side and the base temperature may drop below -200°F on this same face. The earth face undergoes less drastic swings because it never sees the sun in full incidence of its rays and it always sees a relatively warm earth even while in the shadow region. A spinning body, or a body of high reflectivity, may develop very mild temperature cycles. The Explorer satellite, for example, is a low deflectivity, spinning type, while the I. G. Y. satellite depends on high reflectivity. Possible abrasion by meteoric dust would be expected to be critical for a high reflectivity type, leading potentially to its destruction by melting for the case of a light metal skin. The spinning type should not be as sensitive to such abrasion.

C

From a materials viewpoint it is obvious that a lightweight metal may be expected to serve as satellite material. However, depending on the design, surface treatments are required to ensure either high reflectivity or absorptivity values which are an absolute necessity for the protection of electronic components as well as the body itself.

Figure 4—Confusing Organization

CONCLUSION

Writing, as we have been told many times, is one of our most important tools. Much of what we do has little importance until we can communicate it to others.

As this entire article has tried to show, the reader is the most important consideration whenever we write. We cannot expect him to organize our thoughts and put them in the right order; to sort out the relevant details; to relate the detail to the main ideas; and to translate our vague language into correct ideas. As we have seen, even with the best writing, the reader must work to get our ideas from the printed page. So let's ease his job as much as we can by practicing the rules of good reading when we write. Ω

REFERENCES

[1] M. W. Thistle, Popularizing Science, *Science*, April 25, 1958.
[2] W. Weaver, Communicative Accuracy, *Science*, March 7, 1958.

INCREASING ARTICLE ACCEPTANCE

LEIGH CREE WHITE
Editorial Consultant
Voorheesville, New York

ABSTRACT

The journal analysis matrix helps a writer learn about an editor's preferences and as a result the writer can more carefully craft his work to approximate journal style. This crafting raises the possibility that the editor will accept the article. The material explains how to set up a matrix, and includes some possible categories for analysis.

One step to success in writing for publication which many a writer ignores or treats lightly is a thorough analysis of the articles in the journal in which he or she wishes to publish. Chances of manuscript acceptance rise with an accurate analysis because the author gives the editor more nearly what he needs.

The "journal analysis matrix" which I recently introduced to students at Rensselaer Polytechnic Institute in a graduate course, Writing for Publication, provides an effective systematic approach for minutely examining content and form of journal articles. By detailing the information and then incorporating it into the article structure, the writer demonstrates professional attention to the style, format, and content that gives a particular journal its special emphasis.

Before explaining the matrix in detail, let me mention two traditional techniques for analyzing journals or magazines. One consists of a review of the *Writer's Market* or *The Writer*. These reference books contain guides to understanding not only professional and technical journals but also popular outlets for scientifically or technically based articles. Information includes journal title, address, editor; journal policies on copyright, payment, query letters; scope of content and desired length of manuscripts.

The other traditional technique relies on information from the journal pages, details on audience description, guidelines for writers, and pertinent submission information. Some journals publish this information prominently on the inside of the front and back covers. A quick glance through these pages will verify for an author whether an abstract is required, what system of documentation is acceptable, and what format conventions are used. Some questions a writer might ask in this "quick-glance analysis" are: What seems to be the journal's policy on use of subheads, graphs, charts, and other illustrations? Are titles scholarly, all encompassing, or entire sentences? Knowing these policies points the writer in the direction of the format a journal desires. But the serious writer will go one step further and develop a journal analysis matrix along the lines explained below.

TO SET UP THE ANALYSIS

Rule one sheet of plain paper 8½ X 11 inches in 1-inch squares. Turn the paper horizontally so the one set of larger spaces is on the left. Divide the sheet in half with a double line. There are four tiers of squares above the line and four tiers below. Across the top in the first line of blocks label the categories to analyze, and on the row under the double line, list the second set of categories. Develop your own categories shaped around the particular style or content troubles you have with writing. The following categories are possibilities (see Figure 1).

1. Journal name, editor, address, telephone number. This keeps handy the references needed for addressing manuscripts. Use all three squares directly under the heading.
2. Contents. Again, using all three squares under the heading, summarize in two or three lines the major thrust of the journal contents. Note its general audience.
3. Article title and author. Review three articles per analysis sheet. This is usually a sufficient number. One article title and its author go in each square under the heading.

From This Point On, All the Information in Any Horizontal Line of Squares Refers to One Article

4. Theme of the article. List the general theme of the article to get a feeling for the subject matter preferred by the journal.

JOURNAL NAME, ETC.	CONTENTS	ARTICLE TITLE	THEME	OTHER COMMENTS	VIEWPOINT	PERSON	VOICE	ANECDOTES	EXPOSITION & OPINION

Figure 1. First tier of journal analysis matrix.

FURTHER CATEGORIES

In the lower tier, the headings help you take a further look at journal policy through the article analysis. Suggestions for additional categories, shown in Figure 2, are detailed in the following material.

1. Article. Repeat the titles of the articles being explored to make this part of the chart easier to follow.
2. Length. Some journals use articles of varying length; others stay within certain limits. If this fact is not stated in the guidelines, figure out the average number of words per line times the lines in the article for a total word count.
3. Illustrations. For some journals and magazines, the inclusion of photographs or drawings may be the deciding factor in the editor's acceptance of the article. What kinds of illustrations, if any, are used in the articles being examined? Do graphs or charts accompany the article?
4. Line of reasoning. Does the reasoning in the article go from Some journals choose a majority of how-to articles, others like exposition, still others want original research or a combination of several types of content.
5. Other comments. In this square, refine the theme further. Perhaps the author did an exceptional job, or contrarily, he missed the development of the points.
6. Viewpoint. How does the author attack his theme? What major conclusions does he draw? Write these views here.
7. Person. While many professional journals prefer articles written in the third person, some use second, while others permit first. Knowing this helps to focus the article as the editor wishes. And the editor is the one the writer must satisfy.
8. Voice. Is there a majority of active or passive voice in sentence structure? In either case, go thou and do likewise.
9. Anecdotes. This device is not used much in technical

ARTICLE	LENGTH	ILLUSTRA-TIONS	LINE OF REASONING	BY AUTHOR-ITY	TYPE OF ARTICLE	REFER-ENCES	HUMOR	LEAD	QUOTES

Figure 2. Second tier of journal analysis matrix.

writing, but anecdotes (a fiction technique adapted to non-fiction) can help sell a scientifically based article to a popular magazine. If the article contains anecdotes, assess them in this square.

10. Exposition and opinion. Must the article contain strictly verifiable facts or can opinion enter the discussion? This varies within professions and is a good point to check. the specific to the general or the general to the specific? Some editors demonstrate a preference that the writer can pick up with a careful check of several articles.
5. By authority. Will the journal accept an article that interviews a professional or academician or must the writer be the originator of the research, survey, or theory? This is particularly important for student writers to learn. They may have to coauthor the paper with a "professional" in order to get published.
6. Type of article. Does the article explain a process in which the reader can participate easily by following the directions? Does the article report on research? Does it develop a theory?
7. References. Are references for the article required by the journal and do they follow a standard style?
8. Humor. Not many professional journals use outright humor, though sly wit can be found occasionally. Popular magazines delight in an article with some light humor, even though the subject is a serious one. If wit comes easily, use it; if it is an effort, forget it.
9. Lead. What techniques does the author use in the first paragraph (called a lead) to gain the readers' attention? Along with that, does the journal require an abstract?
10. Quotes. Perhaps the journal being analyzed desires numerous quotes obtained in interviews with authorities.

Some journals insist that the author receive a signed release from the originator of these quotes. Usually this information will be in the writer's guidelines which many professional journals print in each issue. Popular magazines do not often include extensive guidelines in their pages, but authors can write for such information.

ANALYZING A PROFESSIONAL JOURNAL

When a writer finishes analyzing three articles from a journal in this manner, he knows the style the journal prefers. He may learn, for instance, that abstracts are a must, but not all the articles will contain references; that the journal uses photographs, graphs, and charts; that the writing flows smoothly with plenty of sentences in the active voice; that points in an article are frequently enumerated in lists; that strong leads abound, but article endings may be weak.

I used the journal analysis matrix recently to analyze a professional journal (Figure 3) so that acceptance chances for an article I was writing would be higher. In Figure 3, I note some of the journal's characteristics. (The name of the journal and titles of the articles have been changed to avoid influencing another writer.)

The writing style in this journal tends to be formal, yet not stilted. The articles frequently foster a definite point of view, though not often with the use of the first person. Some articles involve the reader directly with the imperative, how-to approach, while others report formally on research. Another type of article presents exposition of theory and opinion.

JOURNAL NAME, ETC.	CONTENTS	ARTICLE TITLE	THEME	OTHER COMMENTS	VIEWPOINT	PERSON	VOICE	ANECDOTES	EXPOSITION & OPINION
Journal of Mythical Futures	research, teaching techniques, theory	New Land Use Law Provisions Jones	revisions in law, implications for builders	not complete interpretation	foresees problems ahead	third	both passive & active	no	omniscient
P. T. Barnham editor Dept. Arch. CPD	directed to profs, students, architects	Style of the Future, Workshops Smith	in-service education --why it has failed	develops another model	programs, need to be more realistic	third first	both but more active	a few minor ones	yes
vol. 7(2) 1978		An Exper. Forecasting Housing Starts McMahan	results of an experiment	wanted to see if working with builders helped	concern for motivation improvement, tested one method	third	mostly passive	no	gave suggestions for further research

Figure 3. Journal analysis matrix of the Journal of Mythical Futures.

Interpretation of the leads of articles in the journal showed that the majority began with a forceful statement to catch the reader's attention. A few leads dryly presented facts in such a way that only someone vitally interested in the subject would continue reading. I concluded that a finely developed writing style was not essential for article acceptance.

Although the editorial statement on the inside front cover states that beginning authors are welcomed, all the authors in the issue I analyzed carried titles after their names. The statement, though, leads one to believe that students might have an opportunity, an important insight for both professors and students. Use of a journal analysis matrix points out some direction for the writer that could be missed using only the two common analysis techniques mentioned earlier.

THE VALUE OF AN ANALYSIS

With the journal analysis matrix as a reference, a writer can mold his own approach to a journal's established patterns. Since it is more satisfying to receive an acceptance letter than a rejection form, it makes sense to analyze a journal before writing an article for it. After all, it increases the chances of acceptance.

Graduate students at RPI in the course Writing for Publication are encouraged to do a thorough analysis of the journals they wish to write for. The journal analysis matrix adds one more tool to their box of analysis techniques. Student and professional writers alike will find the matrix helpful in getting articles into print. Does it work? The publishing of this article is one example of its success!

REFERENCES

1. H. A. Estrin, Writing for Publication, *Journal of Technical Writing and Communication, 5*:2, 1975.
2. D. Newcomb, *A Complete Guide to Marketing Magazine Articles*, Writer's Digest, Cincinnati, 1975.
3. H. B. Jacobs, *Writing and Selling Non-fiction*, Writer's Digest, Cincinnati, 1967.

Other Articles in Communication by This Author

4-H Safety Goes Specific, *Farm Safety Review, 10*:5.

Jobs Via Convention Roads, *Journal of Home Economics*, Spring 1954.

The View Is So Different, with J. F. Keim, *Extension Service Review*, May 1954.

I'm Leaving Extension, *American Association of Agricultural College Editors, 39*:2.

A Wide-Angle View, *Extension Service Review*, July 1957.

Author Index

Subject Index

Editors' Biographies

Craig Harkins is site communication manager at IBM's San Jose, California, development and manufacturing facility. He has been with IBM for more than 20 years in a variety of communication, education, and management assignments. Before joining IBM, he was a computer operator on a UNIVAC I for Pacific Mutual Insurance Co., Los Angeles; a reporter-photographer with the *St. Petersburg* (Florida) *Evening Independent*; a news broadcaster-writer for WDAE, Tampa; and an information specialist in the U.S. Marine Corps.

He has A.B. and M.A. degrees in English from Colby College and New York University. He has a professional diploma in communication arts from Columbia University, and a Ph.D. in communication from Rensselaer Polytechnic Institute.

Dr. Harkins has served as national secretary for the IEEE Professional Communication Society, and as a member of the board of directors for the Society for Technical Communication. He is a member of the International Communication Association. He has served on the editorial board for *IEEE Spectrum.* His articles have appeared in many publications, including: *Audio-Visual Communications, IEEE Transactions on Professional Communication, Proceedings of the IEEE, Technical Communication,* and *Technical Photography.*

Daniel L. Plung attended the Bronx High School of Science and the City College of New York, where he received a B.A. in English. He received his M.A. and Doctor of Arts Degrees from Idaho State University.

Prior to entering graduate school, he was a supervisor at a market research firm; his principal responsibility was report preparation and review. After completing his doctorate, he taught as an Assistant Professor of English and as a Lecturer in Speech at Idaho State University. Currently Dr. Plung is Technical Editor for Exxon Nuclear Idaho Co., Inc., operators of the Idaho Chemical Processing Plant at the Department of Energy's Idaho National Engineering Laboratory.

He has presented papers at several conferences and his articles have appeared in many publications, including: *Journal of Technical Writing and Communication, Teaching English in the Two-Year College, Journal of Business Communication, Rocky Mountain Review of Language and Literature, Readings in Business Communications: Strategies and Skills, Critque,* and *IEEE Transactions on Professional Communication.*